PRAISE FOR *DECARBONIZING LOGISTICS*

"*Decarbonizing Logistics* is one of those books that you always believed existed already – until you find out that it didn't. It leads you to question almost everything you read before on the same subject. Alan's book is both comprehensive and holistic. It describes the issue in all its complex aspects, presenting the best literature and patiently dissecting the most dubious claims that are numerous in this field. Beyond the rigorous analysis of the facts, Alan leaves ample scope for optimism on the possibility of achieving the ambitious objective, if one is ready to face reality. This is a problem of unusual complexity and several of the shiny solutions suggested so far have failed to produce sustainable results."
Sergio Barbarino, Chair, the European Technology Platform for Logistics (ALICE)

"Professor McKinnon has made another groundbreaking contribution. The greenhouse gas footprint of logistics is large. Reducing it relies on several mechanisms because logistics involves many activities and participants. This book disentangles this complexity and proposes a clear framework for reduction. It identifies interventions by shippers, service providers or the public sector. The book lays the foundations of initiatives to come. It should appeal to a wide range of policymakers, academics and practitioners."
Jean-François Arvis, Lead Economist at the World Bank and co-author of its *Logistics Performance Index*

"The UK's rising transport emissions are an indicator of the importance and timeliness of this book. The clear analytical approach, using the latest models – whilst avoiding complex language and mathematics – provides practical, evidence-based advice for a wide range of users, including logistics companies, regulators, politicians, policy makers and researchers."
Baroness Brown of Cambridge DBE FREng FRS, Deputy Chair of the UK Committee on Climate Change

"Continued advances in logistics have enabled ever-more globalized production of goods and services, bringing higher incomes, new jobs and more consumer choice in developed and developing countries alike. This progress has come at a price: the CO_2 emissions associated with moving the raw materials, inputs and consumer goods are causing climate change with detrimental effects for the same populations that benefit from the growing trade. Alan McKinnon's book could not come at a timelier moment. We need to decarbonize logistics if we want to ensure that in the long term its negative impacts don't outweigh its contribution to global wellbeing. The book's analysis, combined with concrete policy recommendations to reduce the carbon intensity of logistics, provide invaluable tools for national policy makers and the international community."
Jan Hoffman, Chief, Trade Logistics Branch Division on Technology and Logistics, UNCTAD

"This book is a comprehensive analysis that goes far beyond explaining why we need to decarbonize logistics – a sector that is expected to triple in volume by 2050. It synthesizes a huge and diverse literature and shows that there is no shortage of strategies and carbon-reducing initiatives. Policymakers and business leaders committed to bringing emissions down to levels consistent with the COP21 Paris Climate Change Agreement will find a wealth of technical information and practical examples to help them update regulation and design programmes and action plans."
Wolfgang Lehmacher, Head of Supply Chain and Transport Industries, World Economic Forum

"Not only has McKinnon articulated the urgency of aacting on logistics and climate, he has provided concrete strategies and examples to mobilize action. Without question, *Decarbonizing Logistics* will influence business, government and other players in transforming the global logistics sector into a more efficient and environmentally sustainable one."
Sophie Punte, Founder and Executive Director, Smart Freight Centre

"The timing of this book on the market is impeccable – with many governments currently grappling with how best to reduce their transport-related greenhouse gas emissions based on their nationally determined contributions committed under the Paris Climate Agreement, and the International Maritime Organization recently agreeing to reduce global shipping-related greenhouse gas emissions by 50% by 2050. The book is extremely well compiled, building on many years of practically orientated research experience by the author. It is truly

international in its presentation. I would highly recommend the book for transport policymakers, logistics firms, shipping companies, local government representatives, vehicle manufacturers, internet retail businesses and IT specialists working in this arena."
Professor Ralph Sims, co-ordinating lead author of the transport chapter in the 5th Assessment Report of the Intergovernmental Panel on Climate Change (2014) and Massey University, New Zealand

"This book finally provides us with the missing piece of the puzzle. We now have analysis and recommendations for reducing emissions across freight transport specifically, rather than overall transport. *Decarbonizing Logistics* will become the reference for logistics professionals within business, think tank and policymaker circles. I congratulate and thank Alan McKinnon for capturing his vast knowledge and experience within the bounds of this new book."
Rasmus Valanko, Director of Climate and Energy, World Business Council for Sustainable Development

Decarbonizing Logistics

Distributing goods in a low-carbon world

Alan McKinnon

First published in Great Britain and the United States in 2018 by Kogan Page Limited

2nd Floor, 45 Gee Street
London
EC1V 3RS
United Kingdom

c/o Martin P Hill Consulting
122 W 27th Street
New York, NY 10001
USA

4737/23 Ansari Road
Daryaganj
New Delhi 110002
India

Hardback ISBN 978 0 7494 8047 9
Paperback ISBN 978 0 7494 8380 7
E-ISBN 978 0 7494 8048 6

British Library Cataloguing-in-Publication Data

A CIP record for this book is available from the British Library.

Library of Congress Control Number

2018015409

Typeset by Integra Software Services Pvt Ltd, Pondicherry
Print production managed by Jellyfish
Printed and bound in Great Britain by CPI Group (UK) Ltd, Croydon CR0 4YY

CONTENTS

LIST OF FIGURES

LIST OF TABLES

PREFACE

We live in an age when many academics believe that only journal papers matter and that the book is an outdated medium. So when, as a senior professor, you set out to write a book, you must have a good reason. In my case, the reason seemed obvious. Someone needed to review and synthesize the vast literature that has accumulated on the decarbonization of logistical activities. The formal role of assessing the global state of knowledge on climate change-related topics rests with the Intergovernmental Panel on Climate Change (IPCC). Between 2012 and 2014 I was one of the lead authors of the transport chapter of the IPCC's fifth Assessment Report. Within the available time and word limits, it was not possible for that chapter to explore in sufficient depth the growth of carbon emissions from freight transport and the potential for reversing it. Hence the need for a whole book devoted to the subject.

The subject merits a publication of this length because logistics is so fundamental to our way of life and considered to be one of the hardest sectors to decarbonize. With freight volumes expected to triple by 2050 and freight vehicles among the heaviest users of oil, many analysts, policy makers and managers struggle to see how logistics emissions can be brought down to levels consistent with the COP21 Paris Climate Agreement over the next few decades. To appreciate the scale of the challenge and the prospects of meeting it, one must assemble evidence from a multitude of sources, as I have tried to do in this book. This has not been easy as the relevant literature is copious, diverse and expanding by the day. It also reveals major differences of opinion on the best technological and managerial pathways to low-carbon logistics. My aim has been to provide a comprehensive and balanced overview written in a style accessible to everyone with an interest in the field. On one thing I am sure we can all agree on – there is no shortage of tried and tested, carbon-reducing initiatives that companies and governments can take in the short to medium term to drive down logistics emissions. What we need is their commitment to implement these initiatives as soon as possible.

It has been 25 years since I published my first journal paper on CO_2 emissions from distribution operations, at a time when global warming was scarcely mentioned in the logistics literature. Since then, climate change has evolved from being just another environmental problem to being potentially the greatest threat facing our civilization. As the climate science becomes more alarming, the need for logistics emissions to drop sharply becomes more pressing. In this book I have tried to convey a sense of the urgency with which we will have to transition to a low-carbon economy with all that this entails for the logistics sector.

Over the years that I have been researching this topic I have had the pleasure of working and liaising with specialists in many different fields and organizations. They have very helpfully collaborated with me on projects, shared their data, answered my questions and generally shaped my thinking on many aspects of the subject. They are too numerous to mention by name, but I fully acknowledge all the valuable contributions that they have made along the way. The one person whom I will single out for extra special thanks is my wife Sabine, who never wants to hear me utter the words 'I have a book to write' ever again.

Climate change 01

The nature and scale of the challenge

A rapidly warming planet

It is important to start any book about decarbonization with a review of the climate science to explain why it is so vital that we cut carbon emissions from human activity and do so rapidly. Following a thorough assessment of the available scientific evidence, the Intergovernmental Panel on Climate Change (IPCC, 2014a) concluded that 'warming of the climate system is unequivocal' and that:

> Anthropogenic greenhouse gas emissions have increased since the pre-industrial era, driven by economic and population growth. This led to concentrations of carbon dioxide, methane and nitrous oxide that are unprecedented in at least 800,000 years. Their effects are extremely likely to have been the cause of the warming since the mid-20th century.

The nub of the problem is that since the mid-19th century we have extracted and burned vast amounts of fossilized carbon in the form of coal, oil and gas that were deposited over a period of 300 million years. We have also deforested much of the planet, releasing carbon dioxide (CO_2) from the current stock of vegetation. All this has dramatically increased the concentration of CO_2 in the atmosphere over what, in planetary terms, is a remarkably short period. It has risen from roughly 250 parts per million (ppm) in 1850 to just over 406 ppm at the start of 2018 (Scripps Observatory, 2018). CO_2 is very effective at trapping heat in the atmosphere even at these very low concentrations. This greenhouse gas effect was demonstrated in laboratory experiments by the Irish physicist John Tyndall in the 1860s. The discovery of a causal connection between CO_2 concentrations in an air mass and its average temperature therefore long preceded the observation of a statistical correlation between these variables at a global scale in the second half of the 20th century (Walker and King, 2008).

As the IPCC quote explains, it is not only CO_2 emissions that are warming the planet; they account for roughly 76 per cent of GHG emissions by gas (EPA, 2017). Other GHG gases, including methane, nitrous oxide (N_2O) and various hydrofluorocarbons (HFCs), are reinforcing the effects of CO_2 to varying degrees. The 'global warming potential' (GWP) of these gases depends on the amounts of energy they absorb and the length of time they spend in the atmosphere. Methane, nitrous oxide and HFC123 have GWPs respectively 21, 310 and 11,700 times greater than CO_2 over 100 years (UNFCCC, 2017) but are emitted in much smaller quantities.[1] A major concern, however, is that as the planet warms, much of the frozen methane currently trapped in the tundra and under the sea bed (in so-called 'clathrates') will vapourize and vent into the atmosphere. The release of methane has in the past played a critical role in natural cycles of climatic change. There is a real danger that by raising methane levels, human activity will accidentally trigger another of these cycles. We might then cross a climatic 'tipping point' the damaging effects of which would be irreversible. This is one many climate-related 'tipping points' identified by Schellnhuber (2009) as posing a major environmental threat to mankind.

Since pre-industrial times, average global temperature has risen by 1.1°C (WMO, 2017a), a rate of increase that is unprecedented and around 50 times faster than the Earth would typically recover from an ice age (Clark et al, 2016). At a global level, 2016 was the warmest year on record and the 40th year in succession in which the average temperature exceeded the mean value for the 20th century. This prompted the director for climate change research at the World Meteorological Office (WMO, 2017a) to observe that we are seeing 'remarkable changes across the planet that are challenging the limits of our understanding of the climate system... We are now in truly uncharted territory.' While the average global temperature figure is a widely used indicator of the degree of warming, it gives a misleading impression of the magnitude of the climatic impacts. Even small changes in this average temperature can have a major destabilizing effect on weather systems. Already we have seen a significant increase in the frequency, intensity and duration of extreme weather events, often with devastating consequences. Longer-term trends are also becoming more pronounced, such as the 20 cm increase in average sea level since the start of the 20th century (WMO, 2017b). It is in the Arctic, which is widely regarded as the barometer of climate change, that some of the clearest evidence of climate change has emerged. The summer-minimum ice cover in the Arctic has been shrinking at an average rate of 13.4 per cent per decade over the past 40 years (NASA, 2017). This is highly significant because when the reflective albedo of the ice is replaced by dark ocean more heat is absorbed, creating a positive feedback loop that accelerates the contraction of the ice sheet.

The International Energy Agency (2016: 29) argues that 'in the absence of efforts to stabilize the atmospheric concentration of GHGs, the average global temperature rise above pre-industrial times is projected to reach almost 5.5°C in the longer term and almost 4°C by the end of this century.' Stern (2015: 9) notes that 'global mean temperatures regularly exceeding 4°C above pre-industrial have likely not been seen for at least 10 million years, perhaps much more.' This, he argues, would be likely to transform the 'physical and human geography of the planet'. For example, a 4°C rise would raise sea level by between 6.8 and 10.9 metres and 'submerge land currently home to 470–760 million people globally' (Strauss, Kulp and Levermann, 2015: 6). A temperature increase of this magnitude would be catastrophic in its impact on human and natural systems. Lynas (2007b) describes just how intolerable life would be in a world 4, 5 or 6°C warmer.

The governmental response

National governments and international organizations are committed to averting such a climatic disaster and to keeping future global temperature increases with a 'safe limit'. Two degrees celsius has been widely adopted as that safe limit, partly because at higher figures 'the probabilities of nonlinearities and tipping points are believed to increase greatly' (Stern, 2015: 14). Adapting to a temperature increase of this magnitude will still pose major economic, environmental and political problems, particularly in those parts of the world most exposed to the rigours of climate change. For low-lying islands, even 2°C of global warming poses an existential threat from rising sea levels. For example, much of the island nation of Kiribati in the South Pacific, which has a population of 110,000, is at serious risk of inundation by 2100. In recognition of the severity of these local impacts, the UN COP21 Paris Accord in December 2015 committed participating countries to keeping the temperature increase 'well below 2°C', with 1.5°C now being widely advocated as a new limit.

The challenge for climate modellers is to relate these temperature limits to atmospheric concentrations of GHGs. They are now able to quantify with reasonable accuracy the relationship between the quantity of GHGs in the atmosphere (or 'carbon budget') and the probability of it causing average temperature to rise by certain amounts (Meinshausen et al, 2009). On this basis it is possible to estimate how much more GHG can be emitted by human activity in future decades for us to have a given chance of keeping the temperature increase within 1.5 or 2.0°C by 2100. IPCC (2014a) suggests that to have a 66 per cent probability (ie two chances in three) of staying within a 2.0°C temperature increase by 2100, total cumulative emissions of CO_2 over the period 2011 to 2050 have to be limited to 1,000 gigatonnes (Gt – ie billion tonnes). At the current annual rate of GHG emissions from human activity of 41 Gt, this budget will be exhausted within 24 years. Figueres et al (2017) base their calculations on a much lower 'mid-range' carbon budget value of 600 Gt, equivalent to only 15 years of emissions at the current level. They emphasize that global emissions will have to peak in the near future and then drop sharply for us to have a realistic prospect of staying within even the 2.0°C limit. The longer this peak is delayed, the more precipitously future emissions will have to fall to stay with a 600 or 1,000 Gt limit. If we continue to emit GHGs at the current rate 'we would have to drop them almost immediately to zero once we exhaust the budget' (Figueres et al, 2017: 594). We must therefore act urgently to have a reasonable chance of keeping the temperature rise within an acceptable level. This cannot be left to future generations. As former President Obama observed,[2] 'We are the first generation to feel the effect of climate change and the last generation who can do something about it.'

Climate modelling suggests that the prospects of staying within a 1.5°C limit are very low, especially as average global temperature is already 1.1°C above the pre-industrial average. A recent forecast by the Met Office (2018) suggests that the 1.5°C limit may be exceeded as early as 2022. In the remainder of this chapter we will use emission calculations and targets based on the more realistic 2.0°C limit. Recent research suggests that even achieving this higher limit by 2100 will be very difficult. It will require both a phasing out of all energy-related emissions by 2050 and, in the later decades of this century, the absorption of GHGs already in the atmosphere, what are now called 'negative emissions' (Anderson, 2015). It may be necessary to capture as much as 810 billion tonnes of CO_2 from the atmosphere by 2100, although the technologies[3] that will be needed to do this are uncertain and unproven (*Economist*, 2017a). A recent study by EASAC (2018: 1) concluded that 'negative emission technologies may have a useful role to play but, on the basis of current information, not at the levels required to compensate for inadequate mitigation measures.' Placing heavy reliance on the future removal of GHGs from the atmosphere is therefore a risky strategy and best avoided by making deep cuts in emissions as soon as possible. This presents governments with a major challenge, particularly as they must reconcile pursuit of ambitious climate change targets with economic development and social welfare goals.

Governments are not all expected to cut emissions by a similar margin. Global efforts to address the climate change problem are underpinned by the principle of 'Common but Differentiated Responsibilities' (CBDR). This was a key element in the so-called Rio Declaration, made at the 1992 United Nations Conference on Environment and Development in Rio de Janeiro (the so-called 'Rio Earth Summit'). It required countries to share responsibility for dealing with environmental problems but recognized that their contributions should reflect differences in their levels of economic development. The IPCC (2014a: 5) elaborated on this fundamental principle: 'Countries' past and future contributions to the accumulations of GHGs in the atmosphere are different, and countries also face varying challenges and circumstances, and have different capacities to address mitigation and adaptation.' Developed countries, with relatively high annual emissions per capita and which have emitted most of the GHG already in the atmosphere, are expected to achieve much deeper reductions than developing countries. This will help to narrow the wide international differences that currently exist in average annual CO_2 emissions per capita, ranging (in 2014) from 45.42 tonnes in Qatar to 0.04 tonnes in Burundi (World Bank, 2017b). It also recognizes the need for less developed countries to be given the opportunity to expand their economies and cut their levels of poverty. They are, nevertheless, being strongly encouraged to pursue less carbon-intensive forms of development and to help with this they are receiving transfers of knowledge, technology and finance from richer nations.

Prior to 2015, efforts to achieve international agreement on climate change mitigation made limited progress. The COP21 conference in Paris in December of that year finally managed to secure the support of 195 countries for a new climate change accord. The key elements of this accord, which by December 2017 170 countries had ratified, were a common commitment to keep the increase in average global temperature 'well below 2°C between 1850 and 2100', a five-yearly review of countries' efforts to meet their carbon reduction targets, and a pledge by wealthier nations to financially support climate change initiatives in the developing world. The success of the COP21 negotiation was partly attributable to the freedom it gave countries to set their own carbon reduction targets. This 'bottom up' process involved countries submitting Intended Nationally Determined Contribution (INDC) documents indicating how and by how much they were planning to cut their GHG emissions by 2030 and 2050. Countries were encouraged to make ambitious but realistic carbon reduction commitments. When aggregated, the emission reductions promised in these INDCs were not large enough to keep the increase in average global temperature below 2°C by 2100. They would result in the increase in average global temperature exceeding the limit by 1.4°C (UNEP, 2016). This also assumes that countries will meet their INDC promises; they are under no legal obligation to do so. The agreement merely allows for a review after five years of every country's performance in trying to reach its self-declared carbon reduction target. The European Commission submitted a single INDC on behalf of all EU member states. Individual EU countries have, nevertheless, set long-term targets for reducing their emissions. The UK and Germany, for example, are aiming the cut their carbon emissions by 80 per cent between 1990 and 2050. The UK is one of the few countries in the world to enshrine its target in legislation, in the 2008 Climate Change Act.

The UK is also one of the few countries to include in its carbon reduction targets emissions from the international movement of people and goods to and from the country. This goes beyond the rules laid down in the Kyoto Protocol for the measurement of a country's total GHG emissions. These rules are based on a territoriality principle which confines the calculation to emissions arising from the country's land mass (Bastianoni, Pulselli and Tiezzi, 2004). Emissions from international transport services and the so-called 'embodied' emissions in imported goods are thereby excluded. This works to the advantage of countries which are large net importers, such as the UK, as they do not need to declare the embodied emissions in the imported products they consume, but to the disadvantage of large net exporters, particularly China, which have to shoulder the carbon burden associated with the manufacture of products consumed elsewhere. Several authors and organizations have highlighted this fundamental unfairness in the Kyoto approach to allocating emissions between countries and have proposed that it be done on the basis of consumption rather than production (eg Afionis et al, 2017). Little progress has so far been made, however, in getting political agreement for this change.

National governments can pull a range of domestic policy levers to cut carbon emissions. These broadly fall into four categories: regulatory, investment-related, advisory and fiscal. Regulatory initiatives can take various forms, such as lowering speed limits on motorways or imposing fuel economy standards on new vehicles. High levels of government investment, either directly or via public–private partnerships, will be required to decarbonize national economies. Much of it will go into the transformation of the energy supply system from fossil fuel to renewables and, in some countries, nuclear power. Investment in physical assets needs to be accompanied by expenditure on intellectual and behaviour-influencing activities such climate-related research and development, education, advice and advertising. Governments have an important role to play in encouraging citizens to adjust to low-carbon lifestyles and consumption patterns.

Under the fiscal heading, governments can use the tax system to penalize the use of higher-carbon products and services and offer grants and subsidies to incentivize a shift to lower carbon consumption. Governments can also use a market-based mechanism to achieve a similar effect. Emissions trading involves imposing a 'cap' on carbon emissions from particular sectors and creating a market within which businesses in these sectors can buy and sell 'allowances' to emit more than their prescribed limit. The world's first and largest Emission Trading Scheme is that set up by the EU in 2005, which operates at a continental rather than national level and involves 31 countries. The scheme currently covers 11,000 businesses in high-energy sectors which collectively account for around 45 per cent of GHG emissions in the EU. The sectoral 'caps' are lowered through time, which in theory should put pressure on companies to decarbonize their operations. The pressure comes mainly from the price of emission allowances, however, and this has fluctuated widely since 2005 between a peak of just over €30/tonne of CO_2 to below €3. In early 2018 it was around €9/tonne of CO_2. To put this price into perspective, the High-Level Commission on Carbon Pricing has calculated that, to be consistent with the climate goals of COP21, the price of a tonne of CO_2 would have to be US $40–80 by 2020 and $50–100 by 2030 (Carbon Pricing Leadership Coalition, 2017). The International Energy Agency (2016) estimates that by 2040 the price of a tonne of CO_2 (in 2013 prices) will have to be set at around $140 (€104) in developed countries and $125 (€93) in China, Russia, Brazil and South Africa to create the economic conditions needed to keep the increase in average global temperature within 2°C. The unilateral imposition by some countries or regions of emissions trading conditions capable of driving the carbon price up to such high levels would risk causing carbon-intensive industries to relocate to other parts of the world. To minimize this so-called 'carbon leakage' it would clearly be desirable for emissions trading to be widely, if not universally, adopted around the world. At present, this seems a very distant prospect.

In the meantime, various international funding mechanisms exist to transfer resources, technology and know-how from developed to developing countries to assist the latter with their climate change mitigation efforts. The largest and longest established is the Global Environmental Facility (GEF), set up just before the 1992 Rio Earth Summit to facilitate and help to fund environmental initiatives involving collaboration between countries and international organizations such as UN agencies, the World Bank and others. GEF funding, which is provided by 39 donor countries and leverages around $5 of additional finance for every $1 spent, is 'available to developing countries and countries with economies in transition to meet the objectives of the international environmental conventions and agreements' (GEF, 2017). To date it has supported around 790 climate change mitigation projects which have collectively reduced global GHG emissions by 2.7 billion tonnes.

Unlike the GEF, which addresses the full spectrum of environmental problems, the Clean Development Mechanism (CDM) relates specifically to climate change. It was created as part of the 1997 Kyoto Protocol to transfer resources from developed countries to the developing world through the trading of carbon credits. CDM-approved projects in developing countries are granted Certified Emission Reductions (CERs) that can be bought by developed countries to help them meet their carbon reduction obligations (1 CER = 1 tonne of CO_{2e}). In theory this mechanism should be mutually beneficial to both categories of country as it gives poorer countries a revenue stream for climate change mitigation and wealthier ones a cost-effective means of supplementing their domestic decarbonization efforts. In practice, however, the scheme has been widely criticized on several grounds (eg Subbarao and Lloyd, 2011). Critics complain that, among other things, many of the projects are not 'additional' and would have happened anyway without CDM support, the scheme's operating costs are too high, and it is subject to high levels of fraud.

Out of the 2007 and 2008 UN Climate Change Conferences emerged Nationally Appropriate Mitigation Actions (NAMAs) which developing countries can voluntarily register with the UNFCCC as a means of cutting GHG emissions and helping them meet their climate change commitments (Lütken et al, 2011). Some NAMAs, called 'unilateral', are undertaken without external support, but most do result either in a transfer of resources from international development agencies/donor countries or the generation of internationally tradable carbon credits on a similar basis to the CDM. NAMAs vary enormously in the nature of the activity, its geographical scale and sectoral coverage, reflecting the freedom given to countries to define them in a way that meets national and local needs.

One of the major problems with these three schemes is that they have so far provided relatively little financial support for the transport sector in general and freight transport in particular. This a matter of serious concern because, as discussed in the next section, the main growth of freight movement and related CO_2 emissions over the next few decades will occur in the developing world.

Carbon emissions from freight transport

According to the IPCC, transport as a whole, both passenger and freight movement, emitted 7.0 Gt of CO_{2e} in 2010, 15 per cent of the total for all human activity (Sims et al, 2014). Of this figure, 6.7 Gt were energy-related CO_2 emissions and the remainder emissions of other GHG gases, such as methane and HFCs. CO_2 emissions therefore accounted for 96 per cent of all GHG emissions from transport operations. Sims et al (2014) predicted on a business-as-usual (BAU) basis that transport emissions could reach 12 Gt CO_{2e} by 2050. According to the IPCC (2014a), staying within a 2°C global temperature increase will entail limiting GHG emissions from all human activity to 20 Gt by 2050. If transport emissions continue to grow at their current rate, they alone would account for 60 per cent of total allowable emissions in 2050. While transport is deemed to be an essential human activity, it is very unlikely to be considered important enough to claim such a large share of all permitted emissions. Sims et al (2014: 556) reckon that a 'reduction in total CO_{2e} emissions of 15–40 per cent could be plausible compared to baseline activity growth in 2050.' Even reductions at the upper end of this range would leave transport with a 36 per cent share of the emissions ceiling for 2050, more than double its 2010 share of GHG emissions. If it retains its current share, transport GHG emissions will have to drop from 7 Gt in 2010 to 3 Gt by 2050.

In its fifth Assessment Report, the IPCC did not distinguish personal from freight movement in its projections of future GHG emissions from the transport sector. In the same year, however, Guérin, Mas and Waisman (2014) singled out freight transport as one of the most difficult economic activities to decarbonize. Its baseline forecast would see freight's contribution to total anthropogenic GHG emissions grow from 7 per cent to 16 per cent between 2012 and 2050. In a low-carbon world of 2050 in which many other sectors had achieved 'deep decarbonization', the distribution of freight would then account for roughly one tonne of GHG out of every six emitted.

The International Transport Forum (2017b) has estimated that transport as a whole emitted 7.8 Gt of CO_2 (rather than CO_{2e}) in 2015 and forecast that, on a business-as-usual basis, this figure will rise to 13.3 Gt by 2050. This analysis differentiates between the movement of people and goods and, in the case of freight transport, distinguishes 'surface freight' by road and rail, from sea and air services. It predicts that total freight-related CO_2 emissions will increase from 3.2 Gt in 2015 to 5.7 Gt in 2050 (Figure 1.1) and sees freight's share of total transport CO_2 emissions going up slightly from 40 per cent to 43 per cent over this period. This contrasts with an earlier

Figure 1.1 Forecast growth in total freight movement worldwide between 2015 and 2050 (in tonne-kms)

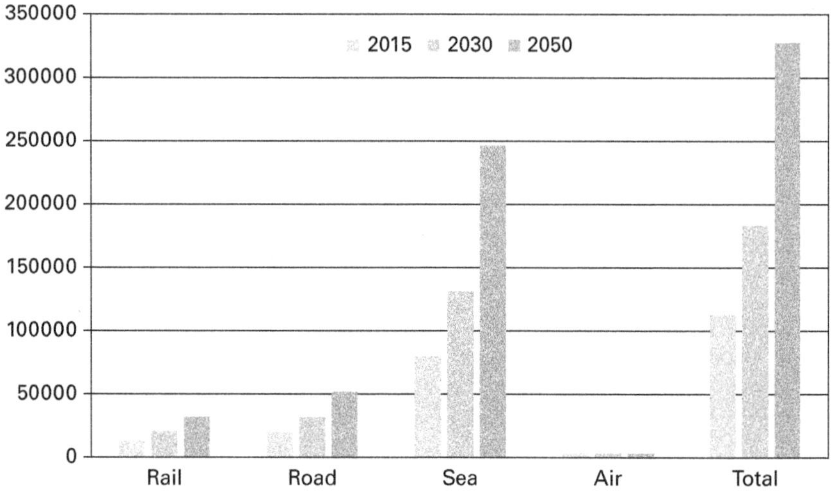

DATA SOURCE International Transport Forum (2017b)

prediction (International Transport Forum, 2015a), based on an assumption of higher future GDP growth, that the freight share of emissions would overtake that of passenger transport by 2050. Nevertheless, if freight movement were to emit 5.7 Gt of CO_2 by 2050 it alone would account for around 30 per cent of the 20 Gt limit on all emissions by human activity in that year.

In the wider modelling of GHG emissions and related global temperature increases, the comparison of baseline (BAU) projections with targets for specific years in the future has been supplemented by a GHG emission budgeting approach, as recommended by Meinshausen et al (2009). It is, after all, the accumulated stock of GHGs in the atmosphere which will determine the probability of future average temperature rises rather than the level of emissions in any given year. Already attempts have been made to estimate the total amount of emissions that the transport sector as a whole might be allowed to emit by 2050 as part of an economy-wide effort to stay within 1.5°C or 2.0°C limits (Gota, Huizenga and Peet, 2016). Viewing future freight-related emissions cumulatively and recognizing the need for them to stay within a tight 2050 budget should be creating a new sense of urgency among logistics managers and freight transport planners. The arguments of Figueres et al (2017) and others that total emissions must peak soon and then drop sharply apply at least as strongly to freight transport as to human activity in general.

This, however, presents a formidable climate change challenge for the freight transport sector, for basically two reasons. First, the demand for freight movement is expected to triple from around 112 trillion tonne-kilometres[4] in 2015 to 329 trillion in 2050 (International Transport Forum, 2017b). Second, it will be difficult to wean the freight sector off its very heavy dependence on fossil fuels over this period. In its baseline projection, the International Transport Forum anticipates a decline in the average carbon intensity of freight transport worldwide from 28 gCO_2/tonne-km in 2015 to 17 gCO_2/tonne-km in 2050, but this 40 per cent reduction is vastly exceeded by the three-fold growth in total tonne-kms over this period. If the forecast volumes of freight traffic materialize, CO_2 emissions per tonne-km will have to plunge to a small fraction of their current level by 2050 to keep freight-related emissions within a tight carbon budget. In its 'low-carbon scenario', the International Transport Forum 'combines the most optimistic scenarios for CO_2 for all modes and sectors' (p.61) to obtain an average freight carbon intensity value of 8 gCO_2/tonne-km. This would result in total freight-related CO_2 emissions declining by 12 per cent by 2050, even allowing for a three-fold growth in total tonne-kms (Figure 1.2). A 12 per cent absolute reduction in these emissions will fall well short of the COP21 requirements, making it necessary to constrain the future growth in freight demand. Similarly, at a European level, in the absence of policy

Figure 1.2 Projected growth in CO_2 emissions from freight transport 2015–2050: baseline and 'low carbon' scenario

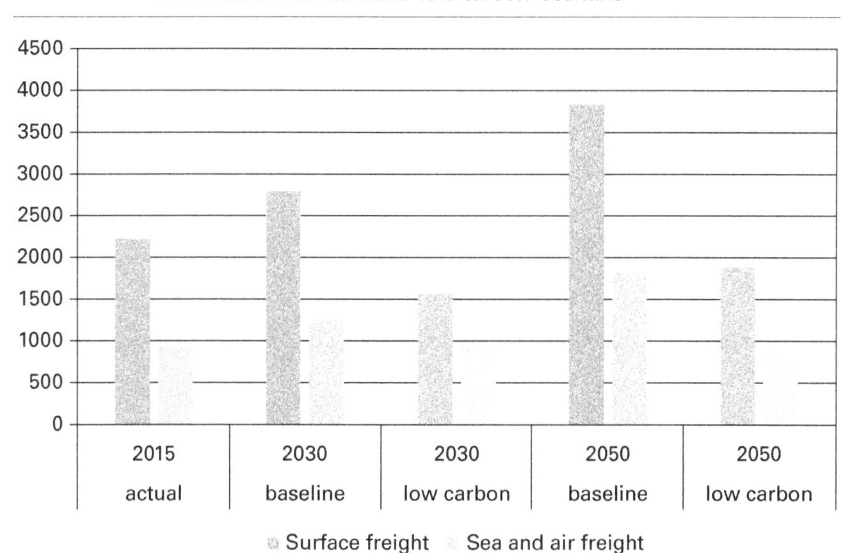

DATA SOURCE International Transport Forum (2017b)

initiatives to restrict freight traffic growth, carbon intensity would have to drop to unrealistically low levels. In 2011 the EU set a policy objective of cutting total CO_2 emissions from passenger and freight transport by 60 per cent between 1990 and 2050 (European Commission, 2011). Allowing for the increase in freight traffic since 1990 and current forecasts of its future growth to 2050 (European Commission, 2014), the carbon intensity of freight movement in Europe would have to fall to one-sixth of its 2015 level by 2050, what Smokers et al (2017: 2) call a 'factor 6' improvement in 'carbon productivity'. This will be extremely difficult to achieve even over a 35-year period.

Applying the concept of carbon budgeting makes this an even more daunting carbon reduction challenge, because it is not simply a matter of reaching a freight CO_2 target in 2050 that is 60 per cent below the 1990 figure. One must also take account of the freight transport emissions that will accumulate in the atmosphere over the intervening years. Figure 1.3 shows two emission reduction profiles for European freight transport which

Figure 1.3 Meeting the EU 2050 carbon reduction target for freight transport with differing emission reduction profiles

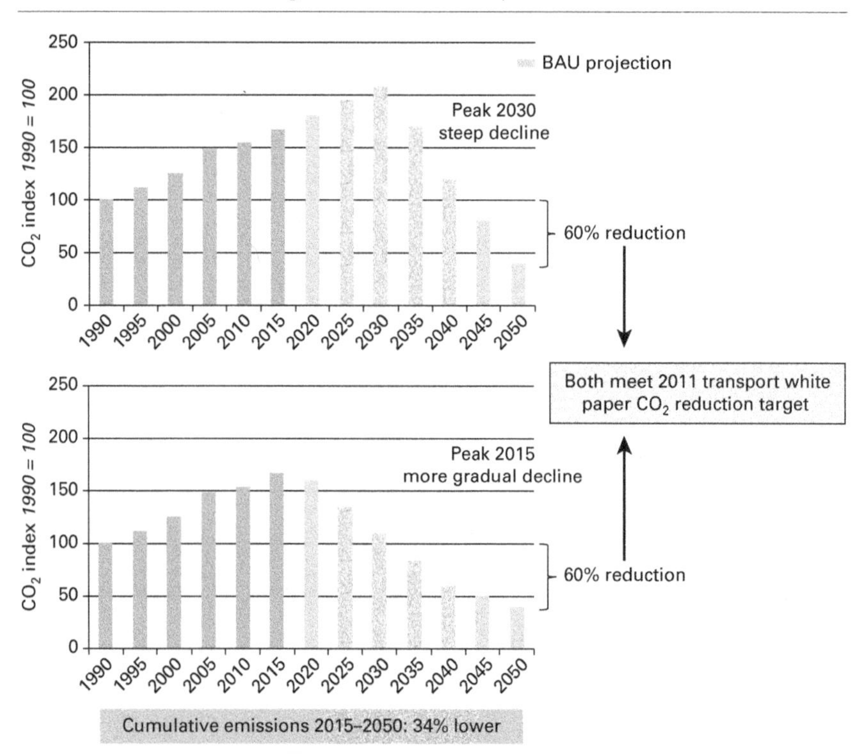

both achieve the required 60 per cent reduction by 2050 but differ markedly in their cumulative emissions. In the first case, emissions do not peak until 2030 and then drop quite sharply. In the second they peak now and decline more gradually. The latter profile puts a third less CO_2 into the atmosphere over the next 30 years, illustrating the need to cut emissions urgently. Carbon budgeting is now conveying this sense of urgency across the wider community of climate change analysts, advisers and policy makers, but has yet to be fully appreciated by logistics planners and managers. As 2050 is perceived to be a long way off, it is naturally tempting to see carbon reduction as a long-term project, much of which can be delegated to future generations of managers.

In the emerging markets, where the projected growth in freight movement is much higher than in Europe, the downward pressure on carbon emissions per tonne-km may have to be even greater. It is forecast that three-quarters of the global growth in road and rail tonne-kms between 2015 and 2050 will occur in Asia Pacific countries (International Transport Forum, 2017b). Africa is expected to experience the fastest rate of freight growth by surface modes, by 3.7 times over 35 years, though as this will be from a very low base, only around 5 per cent of the additional tonne-kms will be generated on this continent. Although Europe and North America will increase their surface freight tonne-kms by, respectively, 100 per cent and 50 per cent, they will together account for only 15 per cent of the global increase and see their share of total road and rail freight movement drop from 36 per cent in 2015 to 24 per cent in 2050. The share held by Asia Pacific countries is projected to rise from 53 per cent to 66 per cent. This means that the worldwide distribution of land-based freight transport activity is likely to change radically over the next few decades, shifting more of the responsibility for logistical decarbonization onto governments and businesses in the developing world. This does not absolve governments and businesses in richer countries of their obligation to cut freight-related emissions, particularly as the freight and emission intensities of their economies are already much higher than those of the emerging markets.

Despite being considered, in climate change terms, a problem sector, freight transport does not feature prominently in the INDC statements submitted by national governments to COP21. Only 13 per cent of these statements explicitly mention the movement of freight (Gota, 2016). In many countries, particularly in the developing world, logistics is not as yet on the climate policy radar. Even among the small group of countries whose INDCs did make reference to freight, there appears to be limited appreciation of the range of decarbonization policy options available to control freight-related emissions. By far the most frequently mentioned option is the transfer of freight to lower-carbon transport modes. While modal shift is unquestionably an important source of CO_2 savings, many other decarbonization measures, offering at least as great a carbon mitigation potential, are scarcely mentioned. These other measures are discussed in Chapters 3, 5, 6 and 7.

Although there is a need for strong freight decarbonization efforts in developing countries to counter high rates of freight traffic growth, the CDM, GEF and NAMA programmes have as yet done little to reinforce them. Of around 5,000 projects registered with the CDM, only 0.4 per cent relate to transport and of these 53 transport-related projects only two explicitly focus on the movement of freight. When compared to its share of GHG and other emissions, transport is also grossly under-represented in the GEF and NAMA funding schemes and, again, the freight share of this transport finance is very small. Only four (5 per cent) of 79 GEF initiatives targeted on the transport sector specifically apply to freight while just 16 per cent of transport NAMAs fall into the 'freight and logistics' category. The lack of CDM funding for transport-related projects has been attributed to 'methodological intricacies, costly data requirements, complex procedures and weaker capacity in the sector' (Asian Development Bank, 2013: iii). A similar diagnosis can be offered for the transport sector's inability, as yet, to secure much GEF and NAMA funding. More generously funded sectors, such as energy, have found it easier to quantify the environmental and social benefits and to establish effective measurement, reporting and verification (MRV) systems. The relative neglect of freight transport within the underfunded transport sector can be largely attributed to its low public policy profile. An analysis of long-term carbon reduction plans for transport in 60 countries revealed that they referred three times more to passenger-related policy measures than freight measures, despite the fact that freight accounts for 40 per cent of transport's CO_2 emissions (Gota et al, 2016).

Carbon footprinting logistics buildings

Very little data is available on GHG emissions from the buildings and terminals in which goods are stored, handled and transshipped. Governments collect data on emissions from buildings but seldom disaggregate it by building type and purpose. There are therefore no macro-level emission data sets for warehouses or other logistics buildings comparable to those compiled by governments and international organizations for freight transport. This may partly explain the observation by Ries, Grosse and Fichtinger (2017: 6486) that 'the environmental impact of warehouses due to heating, cooling, lighting and material handling... has largely been overlooked in the literature, with only a few notable exceptions.'

On the basis of available data, it is possible to make some general observations about the relative contribution of storage and handling to total GHG emissions from logistics. The World Economic Forum/Accenture (2009) estimated that 'logistics buildings' were responsible for 13 per cent of the logistics sector's total emissions. The UK Warehousing Association (2010) claims that warehousing is responsible for 3 per cent of total CO_2 emissions in the UK. This country's domestic freight transport emitted around 6.8 per cent of total emissions in 2012, suggesting that warehousing represented 30 per cent of the UK's total logistics emissions. On the other hand, estimates quoted by AEA (2012) (using employment as a surrogate for energy use) and Baker and Marchant (2015) suggest that warehousing's share of UK logistics emissions is much lower at around 11 per cent, closer to the WEF/Accenture figure. Ries, Grosse and Fichtinger (2017: 6497) estimate that in the United States, 'warehousing-related CO_2 emissions only represent about one-fifth of transport-related emissions'. Rüdiger, Schön and Dobers (2016) estimate that logistics facilities in Germany account for 1 per cent of the country's CO_2 emissions, a significantly lower figure than those quoted for the UK and US. This percentage can be expected to vary widely from country to country, reflecting variations in the average size, age, design and energy efficiency of the buildings, the degree to which material handling operations are mechanized and the carbon intensity of the electricity used in these facilities. Efforts to carbon footprint logistics activities performed in buildings are complicated by the fact that they are often undertaken in premises whose main function is retailing, wholesaling or production rather than logistics. The way in which governments compile business statistics usually makes it very difficult to analyse the nature and scale of these in-house logistics operations, let alone calculate their carbon emissions. This statistical problem is exacerbated by a blurring of the distinction between logistics and manufacturing premises as more value-adding activities, such as product customization, are carried out in what are classified as warehouse buildings (McKinnon, 2009).

Macro-level analysis of future warehousing-related emissions lacks both reliable baseline data and a forecasting model comparable to those developed for freight transport emissions. When predicting a BAU trend for warehousing emissions it is important to recognize that they are fundamentally different from transport emissions in two respects. First, the capital assets producing them have a much longer lifespan. Much of the current stock of logistics buildings will still be in use in 2050. The application of new, more energy-efficient design principles and technologies is relatively slow by comparison with technical upgrades in the transport sector, particularly trucking. This makes it all the more important that the design and construction of logistics buildings minimizes 'whole-life emissions' (Rai et al, 2011). In the short to medium term, buildings can be renovated and retrofitted with carbon-reducing systems and devices. According to WEF/Accenture (2009: 20) 'the potential savings from retrofits are larger than from up-scaling the technologies used in new builds.' Second, as logistics buildings are powered mainly by electricity they will directly benefit from the decarbonization of grid electricity, to a much greater extent than the freight transport system, at least over the next 10–20 years. They can also micro-generate their own power from on-site wind turbines and solar panels, hastening the transition to carbon neutrality. This shift to low-carbon energy is likely to offset increases in the energy intensity of warehousing as operations become increasingly automated and diversify into a broadening range of value-adding activities. Ries, Grosse and Fichtinger (2017), for example, suggest that in the United States, automation of storage and handling operations might increase warehousing-related CO_2 emissions by an average of 12 per cent. Overall, the WEF/Accenture (2009) study estimated that, mainly by the use of energy-saving measures, GHG emissions from logistics buildings could be reduced by a quarter and gave the feasibility of this happening an almost 100 per cent rating. There are many ways in which energy can be conserved in warehouses and their emissions reduced (Dhooma and Baker, 2012; Evans et al, 2014; Baker and Marchant, 2015). Combining these energy conservation measures with the decarbonization of electricity will offer a pathway to zero-carbon warehousing operations in some countries by the 2030s.

Decarbonization frameworks

Given the magnitude of the carbon reductions that logistics planners and managers will have to deliver over the next few decades, it is important that decarbonization is approached systematically and in a way that fully exploits all the opportunities. No single technology, software tool or business practice, available today or in prospect, has the potential to cut emissions by the required amount. There are, fortunately, a multitude of things that businesses can do to reduce the carbon footprint of their logistical operations. This offers flexibility and diversity, but also makes the development of a decarbonization strategy more complex. It is difficult to decide on the right mix of initiatives and how they should be co-ordinated. Frameworks exist to help managers and policy makers conceptualize the various options and formulate coherent decarbonizatin strategies for their companies or jurisdictions. This section of the chapter reviews these frameworks.

Kaya Identity

At a country level, what determines the amount of greenhouse gas released by human activity? According to the so-called Kaya Identity, there are basically four drivers: population, affluence (expressed in terms of Gross Domestic Product (GDP)), energy use and the carbon content of the energy (Kaya, 1990). It suggests that anthropogenic GHG emissions are a function of population multiplied by three critical ratios:

$$\text{GHG emissions} = \text{population} \times \text{GDP} / \text{population} \times \text{energy} / \text{GDP} \times \text{GHG} / \text{energy}$$

This decomposition of human-induced climate change into a few key parameters simplifies its analysis and makes it easier to determine appropriate policy options. For example, in the absence of controls on population and income levels, governments must aim to cut energy consumption relative to economic growth and encourage a switch to lower-carbon energy sources.

A similar analytical framework can be used at a sectoral level to model the relationship between GHG emissions and the level of logistical activity. In the case of freight transport, this generally involves substituting tonne-kms for population and confining the calculation to energy consumed by freight vehicles:

$$\text{GHG emissions} = \text{tonne-km} \times \text{GDP} / \text{tonne-km} \times \text{energy} / \text{tonne-km} \times \text{GHG} / \text{energy}$$

This might be called a 'freight identity' equation. The key ratios in this case are:

- Freight transport intensity of the economy: *tonne-kms of freight movement per $bn of output.*
- Energy intensity of the freight movement: *energy used to move one tonne one kilometre.*
- GHG intensity of the energy: *amount of GHG emitted per unit of energy consumed in the freight sector.*

These ratios were quantified for 15 countries by Guérin, Mas and Waisman (2014) in their analysis of 'pathways to deep decarbonization'. The research teams estimated their values in 2010 and the extent to which they were likely to change by 2050. In all 15 countries, except India,[5] it was predicted that each of the three ratios would decline, effectively decoupling freight demand from GDP, energy from freight movement and CO_2 emissions from freight-related energy use. As these trends would be mutually reinforcing, the net effect would be a substantial drop in the average carbon intensity of freight transport.

The study does not differentiate between freight transport modes or explicitly assess the extent to which switching modes could cut carbon emissions. The allocation of freight among transport modes can, however, be incorporated within a decompositional model (Figure 1.4).

This involves creating separate freight identity equations for each of the main freight transport modes and calibrating their respective intensity values accordingly. The modal GHG figures can then be aggregated to obtain an estimate of total GHG emissions from all freight movements within a country. Compiling separate freight identity equations for each mode is advantageous in two respects. First, it permits the modelling of modal split as a discrete decarbonization option and, second, it recognizes that the rate and extent of the decoupling of the key ratios varies by transport mode. This latter point is particularly significant because the longer-term carbon benefits accruing from a modal shift will depend on the relative speed at which the competing modes decarbonize. If, for example, a country's trucking sector is forecast to cut its carbon emissions per tonne-km more rapidly than the railways, the net carbon savings from a modal shift policy will diminish through time.

The application of this Kaya-based reasoning suggests that there are essentially five ways of decarbonizing freight transport:

1. *Cutting GDP*: few governments would countenance suggestions that economic growth be reversed to cut GHG emissions from logistics. This is a broader political issue that goes well beyond logistics and hence the scope of this book.
2. *Lowering freight transport intensity*: reducing the amount of freight movement generated by each billion dollars of GDP.
3. *Shifting transport mode*: increasing the proportion of freight moved by modes with a lower carbon intensity.[6]

Figure 1.4 Kaya Identity applied to CO_2 emissions from freight transport

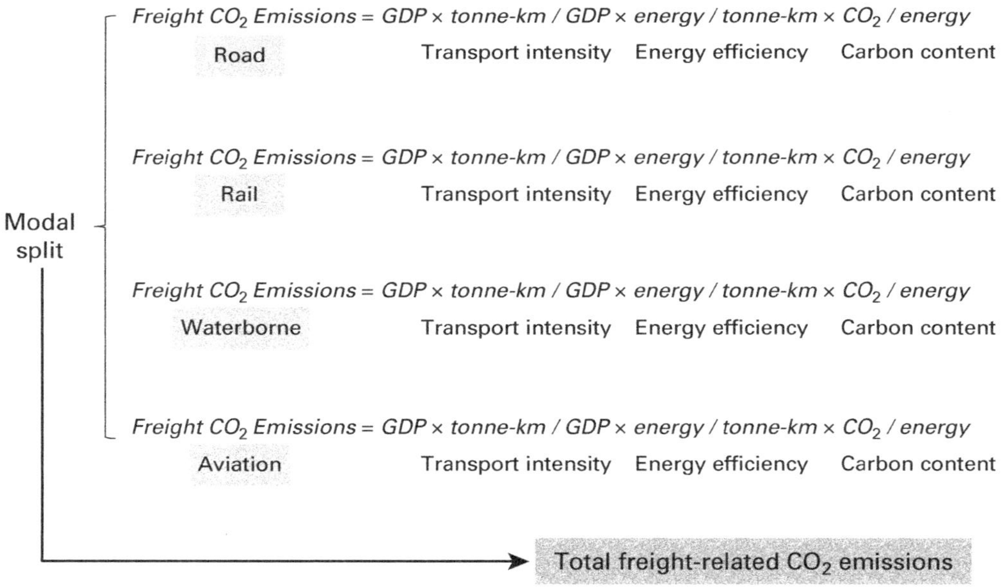

Freight CO_2 *Emissions* = *GDP* × *tonne-km* / *GDP* × *energy* / *tonne-km* × CO_2 / *energy*

Road — Transport intensity — Energy efficiency — Carbon content

Freight CO_2 *Emissions* = *GDP* × *tonne-km* / *GDP* × *energy* / *tonne-km* × CO_2 / *energy*

Rail — Transport intensity — Energy efficiency — Carbon content

Freight CO_2 *Emissions* = *GDP* × *tonne-km* / *GDP* × *energy* / *tonne-km* × CO_2 / *energy*

Waterborne — Transport intensity — Energy efficiency — Carbon content

Freight CO_2 *Emissions* = *GDP* × *tonne-km* / *GDP* × *energy* / *tonne-km* × CO_2 / *energy*

Aviation — Transport intensity — Energy efficiency — Carbon content

Modal split → Total freight-related CO_2 emissions

4. *Improving energy efficiency*: increasing the amount of freight movement per unit of energy consumed.
5. *Switching to lower-carbon energy sources*: such as biofuels or electricity generated by 'renewables'.

ASI and ASIF Frameworks

Options 2–5 of the 'freight identity' are closely aligned with the ASIF framework devised by Schipper and Marie (1999) and adopted by the IPCC (Kahn Ribeiro et al, 2007; Sims et al, 2014) in its modelling of the decarbonization of transport operations. ASIF is a framework for assessing opportunities for reducing the level of transport *Activity*, altering the modal *Structure* of the transport system, reducing the energy *Intensity* of the transport operation and cutting the carbon content of the *Fuel*.

Frequent reference is also made in the literature to the Avoid–Shift–Improve approach to greening transport. The similarity between this ASI acronym and ASIF is a regular source of confusion. Although the two frameworks use different words, there is little to separate them conceptually. Avoid involves reducing unnecessary transport Activity, Shift entails altering the modal Structure, and Improve reduces the energy Intensity of transport. The only difference between the two schemes is the inclusion of a separate Fuel parameter in ASIF to differentiate energy efficiency from the level of emissions per unit of energy. ASI usually subsumes this fuel variable in the Improve category.

The ASI and ASIF frameworks are applicable to all forms of transport. For example, Gota et al (2016) used the ASI one to classify the decarbonization actions listed in 450 sustainable transport reports. Almost two-thirds of them (61 per cent) fell into the 'improve' category, with the remainder split evenly between 'avoid' and 'shift'. Gota (2016) has made a similar classification of transport-related carbon mitigation measures mentioned in the INDCs submitted by 156 countries prior to the COP21 climate change conference in Paris. In the 13 per cent of INDCs that explicitly referred to freight transport, by far the most frequently mentioned carbon mitigation measure for freight was modal shift. Almost half (48 per cent) of the measures fell into the 'shift' category, far exceeding improvement-related actions such as increasing fuel economy (15 per cent), decarbonizing fuel (4 per cent) and raising vehicle load factors (4 per cent). This finding is unsurprising, as politicians and policy makers around the world have long seen modal shift to rail, and to a lesser extent waterborne transport, as a panacea for many freight transport problems. As we will see in later chapters, however, modal shift is only one of many ways of decarbonizing freight transport operations.

Green Logistics Framework

A framework has been devised more specifically for the greening of freight transport operations. It was originally developed in the course of a UK research project in the early 1990s (McKinnon and Woodburn, 1996), applied in an EU-funded research project called REDEFINE in the late 1990s (Netherlands Economic Institute, 1997) and more recently refined (Piecyk and McKinnon, 2010; McKinnon, 2015). It bears a close resemblance to the ASIF framework, though sub-divides some of its components and distinguishes statistical aggregates, parameters and determinants (or drivers). The latest version also incorporates storage and materials handling to give it a broader logistical perspective.

The framework maps the complex interrelationship between the economic output of a country or company and the logistics-related environmental effects and costs. This relationship is defined with respect to seven key parameters. As the main concern here is decarbonization, the version of the framework in Figure 1.5 focuses on a single externality, namely GHG emissions.

Modal split: indicates the proportion of freight carried by different transport modes. Following this split, subsequent parameters need to be calibrated for particular modes. As road is typically the main mode of freight transport within countries, the rest of Figure 1.5 has been defined with respect to this mode.

Handling factor: this is the ratio of the weight of goods in an economy to freight tonnes-lifted. The surveys used to calculate the tonnes-lifted statistics record the weight of goods loaded onto vehicles[7] at the start of a journey. As the typical product supply chain comprises several journeys, the same consignment can be recorded several times. Because of this multiple counting, the tonnes-lifted figure is much greater than the weight of goods produced or consumed. Dividing tonnes-lifted by the production or consumption weight yields the handling factor which can be regarded as a crude measure of the average number of links in a supply chain (McKinnon, 1989a).

Length of haul: this is the mean length of each link in the supply chain and essentially converts the tonnes-lifted statistic into tonne-kms.

Figure 1.5 Map of the relationship between economic output and GHG emissions from logistics

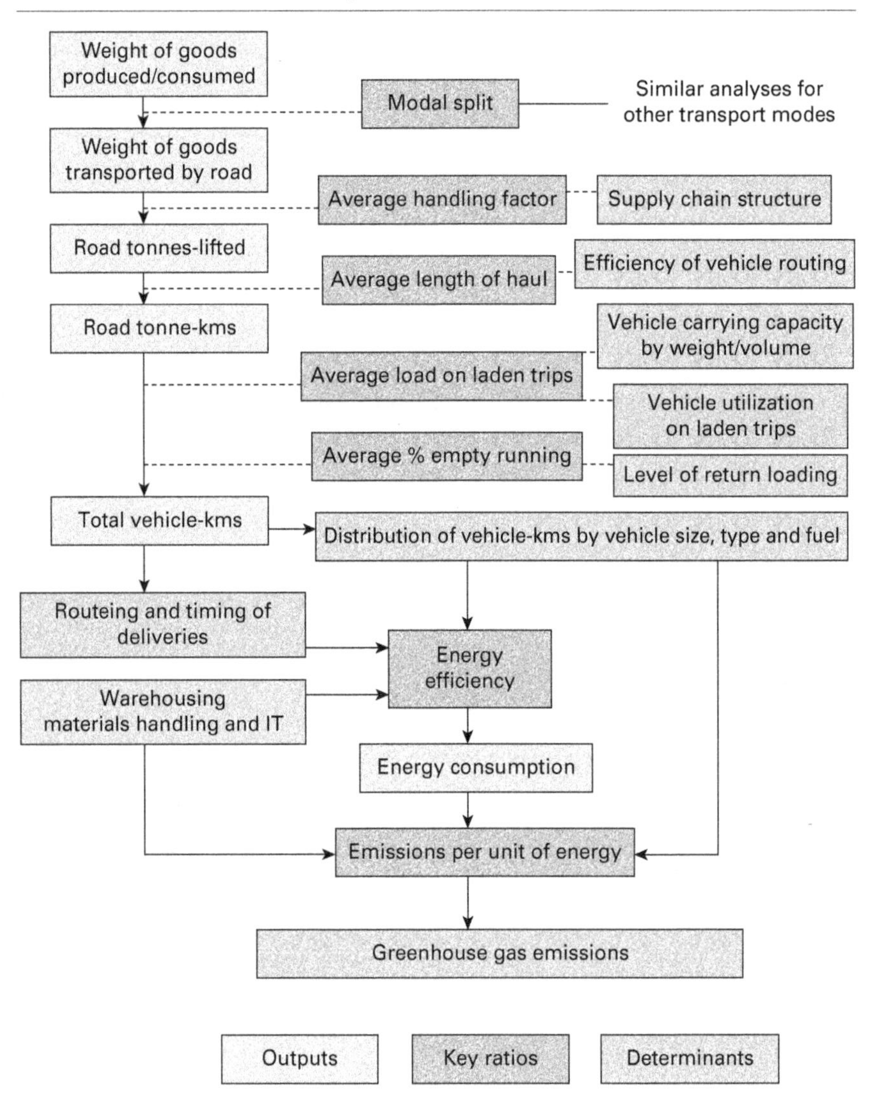

Empty running: freight vehicles travel empty on some journeys for commercial, operational or regulatory reasons. The extent of this empty running is expressed as the percentage of total vehicle-kms run empty.

Load factor (on loaded trips): this is normally measured solely in terms of weight, though in the case of lower-density products volumetric measures are often more appropriate.

Energy efficiency: defined as the ratio of distance travelled to energy consumed. It is a function mainly of vehicle characteristics, driving behaviour and traffic conditions.

GHG emissions per unit of energy: this depends mainly on the carbon content of the fuel, burned either in the vehicle or, in the case of electricity, at a power station. Where possible these emissions should be assessed on a well-to-wheel (WTW) basis, ie measuring emissions across the upstream supply chain as far back as the original source of the energy.

These are the seven parameters that companies and governments need to influence to decarbonize logistics. Figure 1.6 shows how they correspond to the more general ASI and ASIF frameworks.

For companies developing low-carbon strategies for logistics it is encouraging that there are so many decarbonization levers to pull and that, in most cases, their combined effect is mutually reinforcing. Governments can also take advantage of the range and diversity of these parameters when devising carbon reduction policies for the freight transport sector. They should also be aware of potential conflicts when trying to manipulate some of these parameters at a macro level. There is a major conflict, for example, between efforts to improve the energy and carbon efficiency of road haulage and efforts to shift freight from road to lower-carbon models. Making trucking more carbon efficient by raising load factors and fuel efficiency reduces its unit cost, making it harder for competing, greener modes to increase their share of the freight market. This issue is more fully discussed in Chapters 5 and 6.

Figure 1.6 Interrelationship between decarbonization frameworks for logistics

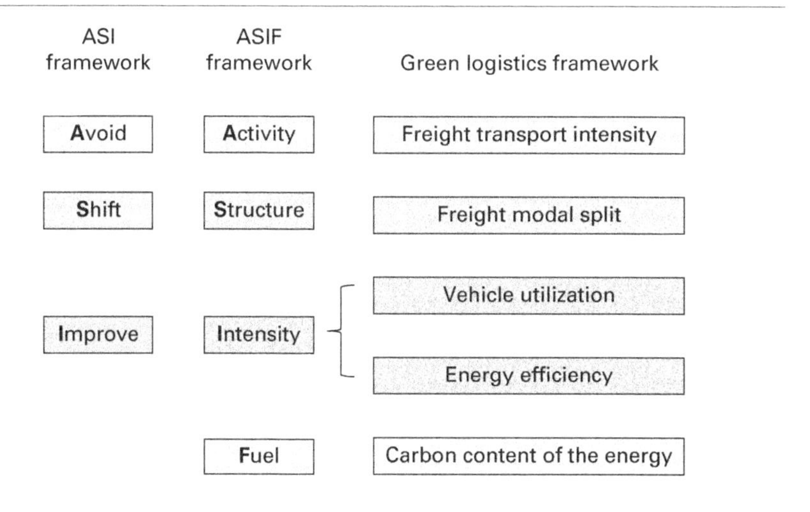

In this book the seven parameters have been reduced to five by combining *handling factor* and *average length of haul* into a 'freight transport intensity' parameter and by combining *empty running* and *load factor* into an 'asset utilization' parameter. This suggests that there are basically five ways in which logistics can be decarbonized, making up the logistics decarbonization framework:

1. *Reducing the demand for freight movement*: within the bounds of logistics management this involves reducing the freight transport intensity of economic activity.
2. *Shifting freight to lower-carbon transport modes*: taking advantage of the wide variations in carbon intensity between modes.
3. *Improving asset utilization*: using vehicle and warehouse capacity more effectively.
4. *Increasing energy efficiency*: reducing energy consumption relative to freight tonne-kms and warehouse throughput.
5. *Switching to lower-carbon energy*: reducing the carbon content of the energy used in logistics.

The remainder of the book is structured around these five activities.

TIMBER Framework

A distinction can be drawn between decarbonization levers that companies can directly control themselves and external forces, which can also affect logistics-related emissions, but over which they have little or no influence. The external developments can reinforce or inhibit companies' 'internal' efforts to cut their logistics-related carbon emissions. It is important that companies take account of these extraneous factors when drawing up a logistics decarbonization strategy. After all, they affect a business's ability to achieve its declared carbon reduction targets. Where they are well aligned with its internal efforts, it may be possible to set more ambitious targets. Where they have a counteracting effect, it may be necessary to intensify the internal initiatives and/or revise the targets downwards. It can also be possible for companies to gain additional leverage from external trends by, for example, taking advantage of new government schemes or upskilling staff to operate new energy-saving technologies.

The management literature has emphasized for decades that environmental scanning is an integral part of strategy development. When using the term 'environmental', authors are generally referring to the 'business environment' rather than the physical environment, though this is subsumed within the wider business definition. Companies are encouraged to use frameworks such as PESTEL to ensure that the scanning is comprehensive and covers all relevant factors in the political, economic, social, technological, environmental and legal domains.

A similar six-category framework has been developed[8] for the external factors influencing the level of carbon emissions from a company's or country's logistics system, which has been given the acronym TIMBER:

Technology: this includes advances in transport, warehousing and materials handling technology.

Infrastructure: this is predominantly transport infrastructure, comprising networks and terminals and covering all the main transport modes, but can also include energy and communication infrastructures.

Market: changes in the structure of the logistics services market, the way logistics services are traded and the nature of the demand for these services.

Behaviour: this can apply at industry and employee levels and, at the latter, includes skill levels and certification programmes.

Energy: comprising the nature of electricity generation, the availability of alternative fuels and the carbon intensity of the range of fuels used.

Regulation: at multinational, national and local levels. It can also be extended to cover fiscal policy measures and become a more general label for public policy intervention.

Each of the TIMBER factors can exert an influence on one or more of the five parameters in the logistics decarbonization framework outlined earlier. Figure 1.7 plots the main interactions between them, showing how each factor influences at least two sets of levers while regulation can affect them all.

Summary and preview

The scientific evidence that global warming is well under way and that human activity is its main cause is now overwhelming. Momentum is building in government circles nationally[9] and internationally to take the necessary actions to control the GHG emissions responsible for climate change. This will require deep reductions in emissions over the next few decades to create a sustainable low-carbon economy and society for future generations. These

Figure 1.7 TIMBER framework – relationship between external factors and key freight parameters

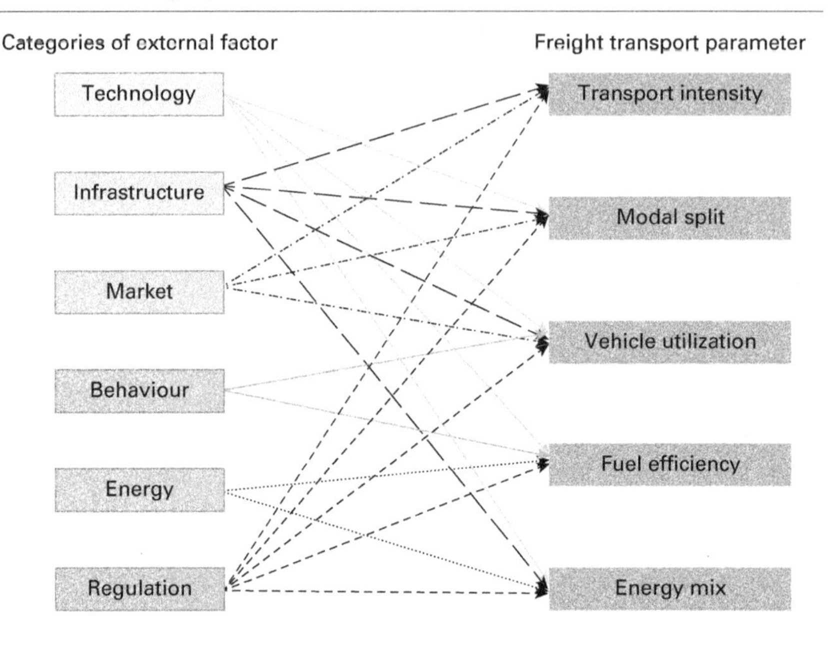

reductions will be difficult to achieve in the logistics sector because of the high growth in demand for its services and its heavy reliance on fossil fuels. It is anticipated that, on a business-as-usual basis, the total volume of freight could triple between 2015 and 2050, with around three-quarters of this growth occurring in the Asia Pacific countries. For freight-related CO_2 emissions to stay within the bounds of a 2°C scenario, it will be necessary both to restrain the growth of freight traffic and drive down its carbon intensity to a small fraction of its current level. This will require much stronger government intervention and the allocation of much more climate finance to mitigate efforts in the logistics sector. These efforts should target warehousing and materials handling activities as well as transport, as they probably account for 10–12 per cent of logistics' global carbon footprint. It is likely, however, to prove easier to decarbonize logistics buildings and terminals than the freight movements that interconnect them.

This book explains how logistics can be decarbonized mainly from a corporate perspective, though much of its content is also relevant to public policy makers. Chapter 2 outlines a 10-stage procedure that can be used to formulate a decarbonization strategy for logistics. It considers how logistics-related emissions can be quantified, carbon reduction targets derived, and appropriate actions selected from the large range of decarbonization options available to logistics managers and planners. The next five chapters examine these options in detail under the five headings listed earlier in this chapter. Chapter 3 addresses fundamental questions about the future trends in freight transport demand. For example, can it be decoupled from economic growth, will it be reduced by a dematerialization of the economy, and how much new freight traffic will be generated by the adaptation of our built environment to climate change? In Chapter 4 we focus on the freight modal split option and make a realistic appraisal of its likely contribution to decarbonization. Chapter 5 examines the extent to which improvements to the utilization of logistics assets can yield both CO_2 and cost reductions. The subsequent two chapters are based on the premise that energy use in the logistics sector will have to be transformed to achieve the deep reductions in GHG emissions required between now and 2050. The transformation will involve both quantum improvements in energy efficiency and a major switch away from fossil fuels to low- and zero-carbon energy. Chapter 6 reviews both these energy options for the road freight sector, while Chapter 7 does the same for the movement of freight by sea, air and rail. The concluding chapter adopts a national perspective on logistics decarbonization, using the United Kingdom as a case study. It illustrates how the TIMBER framework can be used both to review a country's prospects of meeting its freight-related carbon reduction targets and to make companies aware of external trends that influence the carbon intensity of their logistics operations but over which they have little or no direct control.

Notes

1 When the other GHGs are expressed in terms of the global warming potential of CO_2, the notation CO_{2e} (ie CO_2 equivalent) is often used.

2 Obama, Barack (2014) We are the first generation to feel the effect of climate change and the last generation who can do something about it (Twitter, 23 Sept) [online] https://twitter.com/barackobama/status/514461859542351872?lang=en [accessed 5 February 2014].

3 UN Environment (2017) [online] https://www.unenvironment.org/ [accessed 5 February 2018].

4 The tonne-kilometre is the standard measure of freight activity representing the movement of one tonne over one kilometre.

5 India's transport intensity was expected to rise between 2010 and 2050, though this would be offset by a moderate reduction in the energy intensity ratio and an

exceptionally large drop in the carbon intensity of the energy used by the freight sector.

6 Carbon intensity is measured by gCO_{2e}/tonne-km.

7 The term 'vehicle' here applies to all freight transport modes. It is a generic term for trucks, vans, rail wagons, vessels and aircraft.

8 This framework was developed in the course of a study of logistics decarbonization in thirteen countries undertaken by Kuehne Logistics University for Unilever and Kuehne+Nagel.

9 With the exception of the United States, where in 2017 President Trump announced that the country would be withdrawing from the Paris Accord on Climate Change.

Formulating a decarbonization strategy for logistics 02

Much of the responsibility for developing and implementing carbon reduction plans rests with business. Many companies now have such plans in place at a corporate level, but relatively few have as yet formulated a strategy for cutting their logistics-related emissions. Their corporate social responsibility (CSR) statements often present examples of carbon-saving measures being applied by the distribution function, but as isolated initiatives and not part of a well-articulated decarbonization scheme. How should a company develop a decarbonization strategy for its logistics operation?

Several frameworks have been proposed for the development of a climate change strategy at the corporate level. One of the most widely cited is that by Hoffman (2006), which divides the process into eight steps (Figure 2.1).

On the basis of work for and discussions with companies operating logistics systems, I have developed an alternative framework which is more closely tailored to the needs of this sector. Most of the activities are similar, though they are differently sequenced and defined more specifically with respect to logistics. This logistics decarbonization framework comprises a 10-stage procedure with each stage labelled with a word beginning with the letter 'C'. This 10C approach can be applied both by companies whose main activity is transport/logistics and those for whom it is an ancillary activity, such as manufacturers and retailers (Figure 2.2).

This chapter outlines the 10 stages of the process in sequence.

Corporate motivation and leadership

A very strong business case can be made for decarbonizing logistics. Several surveys have enquired about companies' reasons for wanting to green their

Figure 2.1 Hoffman (2006) framework for climate-related strategy development

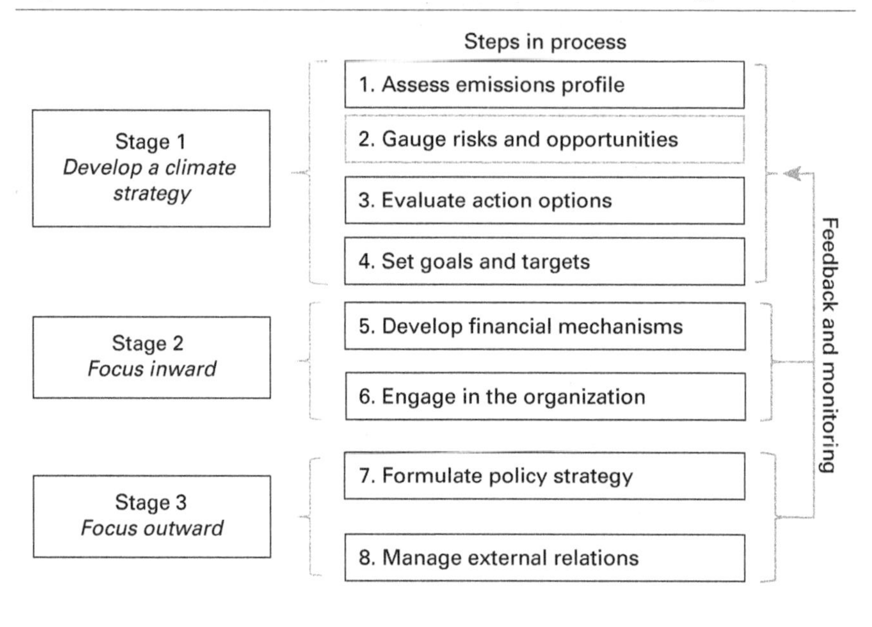

Figure 2.2 Developing a decarbonization strategy for logistics: 10C approach

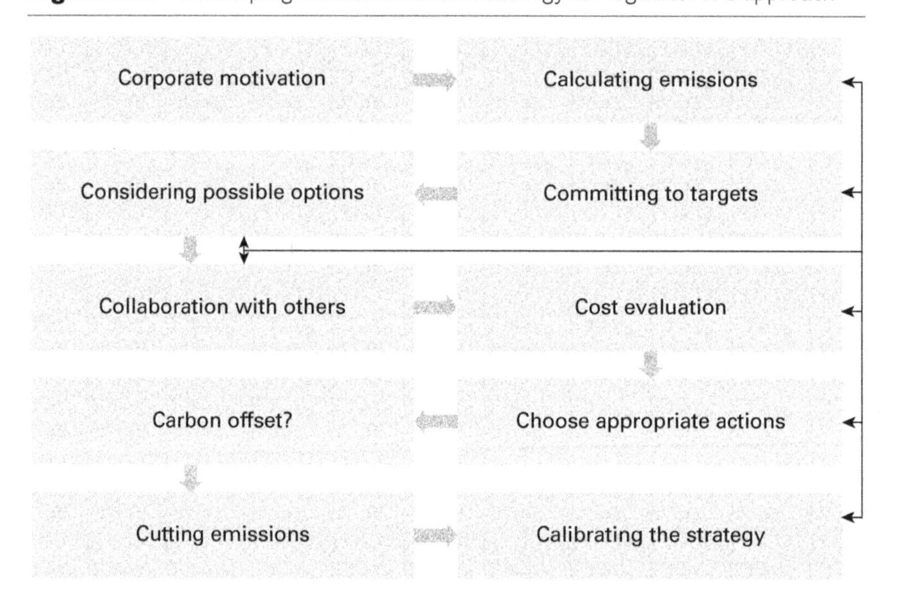

logistics operations and supply chains and found that 'corporate image, competitive differentiation, cost saving and compliance with government regulation' figure prominently (McKinnon, 2015: 17). In the case of carbon emissions, there is a longer-term existential threat to business as we currently know it if future generations suffer the full impact of unrestrained global warming. Within a shorter time horizon, companies are coming under increasing pressure from customers, shareholders and their peers to reduce their carbon footprints. They can also anticipate tightening government regulatory and fiscal policy on climate change and may gain competitive advantage by being proactive and decarbonizing while it can still be done voluntarily. These arguments carry just as much force for the logistics function as for the business as a whole.

One of the central tenets of logistics strategy development is that it should be consistent with the broader goals of the business (Copacino and Rosenfield, 1985). Ideally it should be closely integrated within the general strategy development process to ensure that the plans for logistics are carefully co-ordinated with those of other functions (Fabbes-Costes and Colin, 2007). What applies to logistics strategy as a whole applies equally to a decarbonization strategy for logistics. Efforts to decarbonize logistics must be consistent with the wider sustainability goals of the business and have the backing of senior directors/executive officers. It also helps to have senior logistics/operations managers championing the cause of carbon reduction.

In the case of logistics service providers, whose core activities are moving, storing and handling products, corporate and logistical decarbonization strategies should by definition be well aligned. In companies for which logistics is 'non-core', carbon trade-offs will have to be made between different functions to minimize the carbon footprint of the business as a whole. These resemble the general inter-functional cost and service trade-offs that constantly impact on the way logistics is managed. Similar trade-offs exist in the management of carbon emissions. Minimizing these emissions at a corporate level may cause logistics to have a larger carbon footprint than if, in carbon terms, it were optimized separately. For example, sourcing materials over longer distances from lower-carbon suppliers may more than offset the additional emissions from the longer freight hauls.

In modelling these intra-organizational trade-offs it is helpful to conceive of logistics-related decision making as being conducted at four levels (McKinnon and Woodburn, 1996):

Strategic level: decisions relating to fixed assets requiring longer-term capital investment and determining the numbers, locations and capacity of factories, warehouses, shops and terminals.

Commercial level: decisions about the sourcing of inbound supplies, the subcontracting of production processes and the marketing of finished products which establish the geographical pattern of trading links between a company and its suppliers, distributors and customers.

Operational level: decisions relating to the scheduling of production and distribution that translate the trading links into discrete freight flows and influence the speed at which inventory rotates at the various nodes in the supply chain.

Functional level: decisions affecting the management of logistical resources on a daily basis. Within the constraints imposed by decisions made at the previous three levels, logistics managers still have discretion over the choice, routeing and loading of vehicles and practices within storage and handling facilities.

Most of the carbon-reducing logistics-related decisions that companies have adopted are at the lowest tier in this hierarchy. This is revealed, for example, by surveys of companies belonging to the Logistics Carbon Reduction Scheme in the UK (FTA, 2016). Functional-level initiatives such as training in fuel-efficient driving, the installation of aerodynamic profiling and the use of load-matching schemes are now common in many developed countries. Their uptake is driven much more by a desire to save money than to cut carbon emissions but, whatever the main motivation, the CO_2 savings are a very welcome co-benefit. These carbon reductions at the functional level, however, are often offset by the carbon impacts of decisions made at the higher-level operational, commercial and strategic levels, relating for example to just-in-time delivery, inventory centralization or wider sourcing of supplies, which tend to make logistics systems more transport-intensive and hence more environmentally intrusive (McKinnon, 2003).

Financial calculations usually show that these higher-level decisions yield a net increase in corporate profitability after allowance is made for the additional logistics costs. After all, logistics costs, typically representing only 3–10 per cent of a manufacturer's or retailer's total costs (A T Kearney, 2008; Establish Inc, 2010), can appear a relatively small item on a balance sheet dominated by the cost and revenue contributions of production, procurement, marketing and merchandising activities. Much less analysis has been done of the relative carbon impacts of logistics-related decisions at different levels in the hierarchy and of the cost-carbon trade-offs at each level. Available evidence suggests that in businesses whose core activities are production or retailing the logistics share of total carbon emissions is also relatively small, often between 3 and 8 per cent (Punte and Bollee, 2017). Cutting emissions from these core activities gives companies greater leverage on their corporate carbon footprints and therefore, quite understandably, receives greater priority and managerial attention. It is, nevertheless, good practice to assess the degree of alignment between a company's efforts to decarbonize its logistics operation and its corporate carbon reduction goals and to instil in senior management an awareness of the implications for logistics-related carbon emissions of higher-level strategic, commercial and operational decisions.

When embarking on a decarbonization programme, logistics managers can gain support, encouragement and morale from industry-wide green freight schemes. Over the past decade there has been a proliferation of such schemes around the world, some focusing on particular countries, regions, sectors or transport modes, others with a global reach (Table 2.1). They typically offer advice, training and benchmarking services to their members, accredit companies meeting certain environmental standards and provide a forum within which managers can share experiences and exchange ideas. Among the longest-established and most successful national/regional schemes are the US SmartWay programme, the Dutch Lean & Green programme and the UK Logistics Carbon Reduction Scheme.

At a global level, the Smart Freight Centre has assumed the main role in co-ordinating green freight initiatives worldwide, harmonizing the measurement of logistics-related carbon emissions and raising the profile of green issues among governments and international organizations. It is also promoting the concept of 'smart freight leadership' and has identified a group of corporate 'leaders' whose efforts to cut logistics emissions

Table 2.1 Environmental organizations promoting the decarbonization of freight transport/logistics

Global	Smart Freight Centre (SFC), Global Logistics Emission Council (GLEC), Climate and Clean Air Coalition (CCAC), Global Fuel Economy Initiative (GFEI), Partnership on Sustainable Low Carbon Transport (SLoCaT), Carbon War Room, Rocky Mountain Institute, Partnership for Clean Fuel and Vehicles, Clean Cargo Working Group, Sustainable Shipping Initiative, World Ports Climate Initiative, EcoTransIT
Regional	Green Freight Asia, Clean Air Asia, Northern Corridor Green Freight Strategy, Network for Transport Measures (NTM), ECO Sustainable Logistics Chain Foundation (ECOSLC)
Multi-country	Lean & Green, Ecostars
Country-specific	SmartWay (US), SmartWay (Canada), Transporte Limpio (Mexico), Logistics Carbon Reduction Scheme (UK), Objectif CO_2 (France), TK Blue Agency (France), China Green Freight Initiative, Green Freight India, Japan Green Freight Partnership

set a good example to other businesses (Punte and Bollee, 2017). Having reviewed the annual reports and websites of 145 multinational corporations belonging to one or more international green freight programmes, the SFC selected 37 as potential smart freight leaders on the basis of their emission reduction targets, measurement–reporting–verification (MRV) systems, past record in cutting emissions and supply chain collaboration. So newcomers to the process of logistics decarbonization can join a community of like-minded businesses and tap into accumulated industry know-how on how to cut emissions quickly and cost-effectively.

Calculating logistics-related emissions

Only by quantifying the amounts of GHG emitted by different activities and processes can managers assess the relative level of emissions from different activities and processes and discover where the carbon reduction efforts should be focused. Regularly monitoring trends in logistics emissions also allows them to measure the impact of their carbon reduction strategy. In several countries, such as the UK and France, it is now mandatory for companies above a certain size to measure and report their carbon emissions. In addition to complying with legal obligations to submit aggregate emissions data, it is desirable for companies to compile disaggregated data on individual logistical activities.

Early attempts to carbon audit logistics operations proved very frustrating as companies lacked guidance on the measurement, analysis and reporting of emissions. Many devised their own methodologies and often incorporated into their calculations published emission factors that were inappropriate for their operations.

Standards agencies, industry bodies and governments then responded by providing advice on how the calculations should be done and the results reported. Unfortunately, their efforts were not co-ordinated and the resulting proliferation of methods, calculators, standards and emission factors left managers unsure which to adopt. Inventories of these support tools revealed a disconcertingly large choice available to logistics managers (Auvinen et al, 2014). This diversity would not have been a problem had the numerous carbon auditing schemes been based on a consistent set of principles. Regrettably this was not the case. Efforts have therefore been made in recent years to harmonize carbon accounting in the logistics sector. The European Committee for Standardization (CEN) has prepared standards for the 'calculation and declaration of energy consumption and GHG emissions of transport services (freight and passengers)' (CEN, 2012). In 2016, the Global Logistics Emissions Council (GLEC), published its 'framework for logistics emissions methodologies'. This document 'provides a much more closely harmonized basis for the calculation of emissions from freight transport chains across modes and global regions than existed previously' (GLEC, 2016: 9). Rather than adopt a clean slate approach, the GLEC has built its framework as much as possible on existing methodologies that are already widely used by particular modes and sectors, such as those of the Clean Cargo Working Group for container shipping, EcoTransIT for rail and both EN 16258 and the US SmartWay system for trucking (Table 2.2). So good progress is being made in the pursuit of clear and consistent global standards for the measurement and reporting of carbon emissions from logistics.

The carbon auditing of logistics operations is underpinned by a series of principles set out in the Greenhouse Gas Protocol of the World Business Council for Sustainable Development (WBCSD) and World Resources Institute (WRI). This seminal document recommended that the data collected on carbon emissions be relevant, complete, consistent, accurate and transparent (WBCSD/WRI, 2004). The 'tenets' of the GLEC framework also include two other criteria: simplicity – in the sense of making the results easy to comprehend – and flexibility, allowing the methodology to be 'useable by all business models and transport modes' (GLEC, 2016: 11). When judged against one or more of these seven criteria much of the emissions data collected and

Table 2.2 Existing carbon calculation methodologies adopted by GLEC framework

Transport mode	Base methodology adopted
Road freight	US SmartWay
	European Committee for Standardization (EN16258)
Rail freight	EcoTransIT
	US SmartWay
Shipping	International Maritime Organization (IMO)
	Clean Cargo Working Group (BSR)
Inland waterways	IMO
	US SmartWay
Air cargo	International Air Transport Association (IATA)
Transhipment centre	Green Efforts project
	Fraunhofer Institute for Material Flow and Logistics

SOURCE GLEC (2016)

reported in the logistics sector is deficient. This is largely because the diversity of logistics activities, their wide geographical spread and the high propensity to outsource them greatly complicates the carbon auditing task.

When establishing a carbon measurement system for logistics a company needs to resolve a series of issues.

1. Which GHG emissions to measure

The UN Kyoto Protocol identified six GHGs, which have widely varying global warming potentials, measured relative to that of CO_2. CO_2 accounts for 90–95 per cent of all GHG emissions from the logistics sector and so it is common for companies only to measure and report emissions of this gas. Indeed, the current GLEC framework only relates to CO_2 emissions. Companies whose logistics operations are temperature controlled and/or make significant use of natural gas should also monitor so-called 'fugitive emissions' of HFC and methane, particularly as they have high GWP values. The leakage of these gases from refrigeration equipment, pipelines and vehicles can account for a substantial proportion of some companies' total GHG footprint. For reporting purposes, these other GHGs are expressed in terms of CO_2 taking account of their respective GWPs and given the designation CO_2 equivalent or CO_{2e}. Freight transport operations also emit significant quantities of 'black carbon', the tiny dark soot particles that result from incomplete combustion of diesel fuel in trucks and heavy fuel oil (HFO) in ships (Comer et al, 2017). While in the air these particles directly absorb heat warming the atmosphere but this effect is 'short-lived' as they soon fall to the ground. When emitted in the polar regions, black carbon darkens the ice cover, reducing its albedo and causing a longer-term warming effect. Research suggests that, although it is not officially a global warming gas, black carbon is second only to CO_2 in warming the atmosphere (Bond et al, 2013). Over 20 years, it has a GWP around 3,200 times that of CO_2. There is an urgent need therefore to incorporate black carbon into the carbon auditing of logistics operations. The Smart Freight Centre has outlined a methodology that can be used to calculate and report black carbon emissions from logistics (Greene, 2017).

2. Where should the boundary be drawn around the logistics system?

This is a multidimensional boundary, with primarily four dimensions:

Organizational This relates to the division of emissions among participants in the value chain and usually reflects the allocation of financial responsibility.

Whoever pays for a logistics activity should be assigned the related CO_2 emissions. The Greenhouse Gas Protocol established three 'Scopes' to differentiate emissions for which a company is directly responsible (Scope 1), from those arising from the electricity it purchases (Scope 2) and from those released 'indirectly' by suppliers of purchased materials and outsourced services (Scope 3) (WBCSD/WRI, 2004). Businesses conforming to the original Protocol were required to measure and declare their Scope 1 and 2 emissions, but for them the inclusion of Scope 3 emissions was optional. Since the release of the Accounting and Reporting Standard for Scope 3 emissions in 2011, companies have been encouraged to comply also with the Scope 3 Protocol for which declaration of these indirect emissions is also mandatory. Often the majority of a company's logistics emissions are categorized as Scope 3. Where a company outsources its logistics, almost all the related emissions will be in Scope 3. Likewise, the carbon footprint of a logistics provider subcontracting much of its transport to independent carriers may comprise mainly Scope 3 emissions. Extending the organizational boundary to include Scope 3 emissions complicates the data collection exercise, but guidance is available 'to help companies prepare a true and fair Scope 3 GHG inventory in a cost-effective manner, through the use of standardized approaches and principles' (WBCSD/WRI, 2011: 4).

Activity It is generally accepted that the core logistical activities of transport, warehousing and materials handling should be included in the carbon calculation, but there is less agreement about the inclusion of related IT, business travel, equipment maintenance etc. The best advice is to keep the range of activities as comprehensive as possible and preferably inclusive of all the activities for which the logistics department has responsibility. This permits a more thorough analysis of the carbon trade-offs that may have to be made during the decarbonization process, where emissions from one activity may have to rise to achieve a more-than-offsetting reduction in total logistics emissions.

System There are five nested levels of system boundary around a logistics operation[1] (Figure 2.3).

To date, almost all measurements of carbon emissions from logistics have been enclosed within system boundary 1 (SB1) and relate to the direct emissions from the vehicles, handling equipment and warehouses. In the case of vehicles powered by onboard fuel these emissions are described as 'tank to wheel' (TTW). At the SB2 level, the boundary is extended to embrace emissions from the energy supply chain, making a 'well to wheel' (WTW)

Figure 2.3 System boundaries around measurement of freight transport emissions

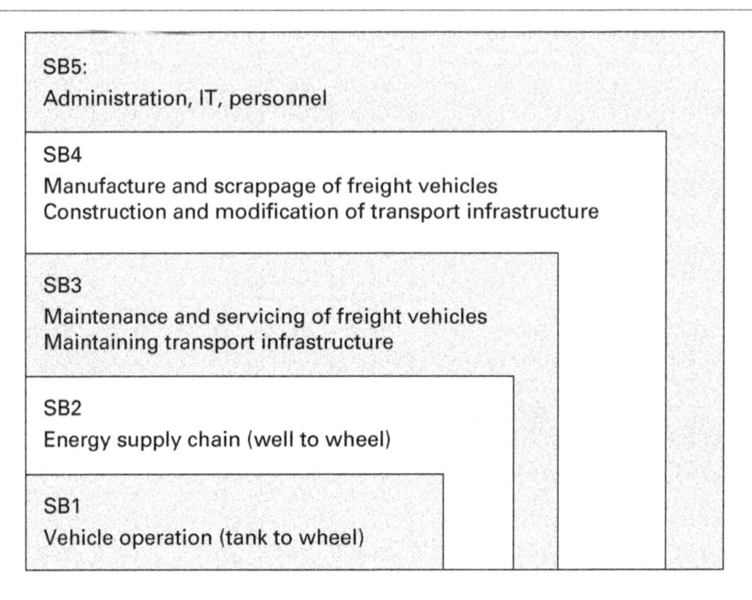

Adapted from NTM (Network for Transport Measures)

assessment. SB3 also includes the servicing and maintenance of vehicles, materials handling equipment and logistics buildings, while SB4 pushes the boundary out further to include emissions from the manufacture of the vehicles and equipment, the construction of logistics property and the eventual scrappage, dismantling and demolition of these assets. SB5 brings emissions from related office functions and the activities of staff within the perimeter of the calculation. Data limitations currently confine most companies' carbon auditing to levels SB1 and SB2, though there are some exceptions. For example, the French logistics business Geodis includes in its carbon footprint calculation the construction-related emissions embodied in its warehouses. UPS (2016a) declares CO_{2e} emissions from business travel and its employees commuting to work. More organizations, such as DEFRA (2017) and NTM (2017), are now publishing tables of WTW as well as TTW emission factors, facilitating the transition to SB2 accounting. Research is at a much earlier stage in the calibration of logistics accounting at the SB3, SB4 and SB5 levels. Only when this process is complete will companies get a truly holistic view of their logistics carbon footprint and be able to manage critical trade-off between operational and life-cycle emissions. For example, it is currently difficult for companies to tell if it is better over a 10-year planning horizon to run and maintain existing trucks more carbon-efficiently or to accelerate their replacement with new, more fuel-efficient vehicles, given the amounts of carbon embodied in the new trucks and carbon emissions associated with the scrappage of the old ones.

In its original system boundary framework, NTM included emissions from the construction and maintenance of transport infrastructure in, respectively, SB4 and SB3. Responsibility for these emissions rests with infrastructure providers, construction companies and governments, rather than the users of the transport networks. The system boundaries can therefore be redefined for different stakeholders in what might be called the macro-logistics system of a country or region (Havenga, Simpson and Bod, 2013).

Geographical A company with extensive operations may wish to disaggregate its logistics-related emissions by country or region to meet local carbon reporting requirements and/or devolve responsibility for decarbonization to national or regional management teams.

3. What approach to carbon measurement should be adopted?

Broadly speaking there are two approaches.

Energy-based approach Since almost all CO_2 emissions from logistics are energy-related, the simplest and most accurate way of calculating them is to record energy use and employ standard emission factors to convert energy values into CO_2. This approach is preferable and available to logistics providers and companies with in-house logistics operations which directly consume energy and so can access the necessary energy data to calculate their Scope 1 emissions. Companies which outsource their logistics and for whom the emissions are in Scope 3 are often unable to get this energy data. Many logistics companies are not prepared to disclose their energy usage as this would compromise commercial negotiations with their clients. In the case of trucking operations, for example, fuel can account for a third of total costs and be the biggest item on the balance sheet. In the absence of energy/carbon transparency, companies contracting out their logistics have little choice but to adopt the alternative activity-based approach.

Activity-based approach This involves multiplying an index of the level of logistical activity by an industry-standard emission factor. As freight transport accounts for the vast majority of logistics-related CO_2 emissions (around 90 per cent on average), the dominant activity index is the tonne-kilometre.

There are now numerous published data sets of tonne-km-based emission factors for different transport modes. Company records, ERP systems[2] and delivery manifests can provide the necessary data for tonne-km calculations. Obtaining distance data for rail and waterborne transport can be more problematic, though the EcoTransit[3] environmental assessment tool can be used for this purpose. Fewer data sources exist for emissions generated by warehousing and terminal operations and permit much less differentiation by activity, scale and type (Rüdiger, Schön and Dobers, 2016; Baker and Marchant, 2015).

Companies are always encouraged to adopt the energy-based approach whenever possible and should regard the activity-based approach as a second-best, fall-back option. If, however, a company is able to apply the energy-based approach and discovers that it yields a higher carbon value than an activity-based calculation using industry standard emission factors, what incentive would it have for reporting the more accurate company-specific energy-based value? Until all companies are required by law to adopt the energy-based approach, which in turn would require full energy/carbon transparency across the supply chain, there will always be a strong temptation to declare an average activity-based emission value if it is lower. It is important, therefore, that published, industry-standard emission factors for logistical activities are as accurate and consistent as possible and sufficiently disaggregated to permit a good match with particular types of distribution operation.

4. What carbon emission factors should be used?

The main European sources of carbon emission factors for freight transport operations were reviewed by McKinnon and Piecyk (2010) and the EU COFRET project. They found significant variations in the emission values for particular transport modes and considered them attributable to several factors, mainly:

Method of derivation: some are based on the actual operations of a sample of companies while others rely on lab-based emission testing of vehicles running on standard duty cycles.

Loading assumptions: emission factors are highly sensitive to assumptions about vehicle loading. These assumptions vary between data sets and are often not explicitly stated.

Geographical context: most data sets are country-specific and, because of national differences in the primary energy mixes, traffic conditions and operating practices, should not be extrapolated internationally. As most countries do not have their own dedicated sets of emission factors, however, this has become quite a common practice. For example, the values derived by DEFRA (2017) for UK transport operations are widely applied in other countries.

A company's choice of carbon emission factors for its logistics operations can be influenced by numerous factors including the range of transport modes it uses, the density of its products, the required degree of data disaggregation by equipment type and the geographical extent of its operations. The near-universal adoption of the tonne-km as the denominator in activity-based emission factor ratios tends to favour companies transporting higher-density products as they generate fewer vehicle-kms and use less energy per tonne-km than those moving low-density freight. Emission factors based on the cubic volume of freight transported would give a fairer representation of the carbon intensity of firms moving low density products. Very little volumetric data is currently available for carbon accounting purposes, however, and varying the activity metric in the emission factor calculation would undermine the comparability of carbon intensity measurements in the logistics sector. GLEC (2016: 26) acknowledges that there are 'pros and cons' over the choice of metric but bases its framework on weight as this is the most used measure in carbon intensity calculations.

5. How should carbon emissions be allocated between orders/consignments?

Where different types of traffic and the consignments of different shippers share the same vehicle, a method must be found to allocate emissions between them. There are basically four allocation problems in logistics:

Between freight and passengers Just over half of all air cargo moves in the 'bellyholds' of passenger aircraft (Airbus, 2016), making it necessary divide the emissions between the people and the cargo. Both IATA (2014a) and ICAO (2014) have devised methodologies for splitting these aviation emissions. Kristensen and Hagemeister (2012) have undertaken similar work for the International Maritime Organization on the division of emissions between passengers, cars and freight vehicles on roll-on roll-off (Ro-Ro) ferries.

Between freight consignments Where freight belonging to different shippers is combined in the same truck, rail wagon, ship or plane, it can be difficult to divide emissions between them because the consignments may differ in their weight, density, cube and handling characteristics. The allocation exercise is even more complicated where the consignments are picked up and/or dropped off at different points on a multiple-stop journey. Under these circumstances, there is no single, objective method of allocating emissions. As the UK Department for Transport (2010: 30) explained: 'In practice, the mathematics required to definitively allocate carbon emissions from a specific multi-drop trip quickly become almost unworkably complex, requiring far more data than is likely to be available.' Its report goes on to outline four practical ways of allocating emissions between consignments on a mixed-load, multiple-drop round in the road freight sector. At a European level, EN 16258 gives companies standard options for dividing emissions generated by shared loads among individual consignments (CEN, 2012). At a global level, the World Economic Forum (2010) proposed a standard system of consignment-level carbon reporting for shipping and airfreight companies.

Between empty and loaded trips Empty running typically accounts for between a quarter and a third of truck-kms. When vehicles are repositioned empty to collect their next load, who should take responsibility for emissions from the kilometres run empty? How should they be allocated between the carrier and the shippers responsible for the previous and following trips? Again, there is no objective answer to these questions, but guidance is provided by CEN (2012).

Between products or unit loads stored and handled in a warehouse Allocating total warehouse emissions to individual units of throughput presents a formidable challenge as emissions per unit will vary with, among other things, the time they spend in storage, the amount of space they occupy, the nature of the handling they receive and the degree of temperature control they require. The more heterogeneous the product range, the more difficult the allocation exercise becomes. There is currently little demand for warehouse CO_2 data disaggregated to a pallet, case or product level, but, as discussed under the next heading, this would be a prerequisite for the carbon labelling of consumer goods.

6. To what extent should logistics-related emissions be disaggregated?

The simple answer to this question is to disaggregate the carbon data to the level (i) dictated by regulators or customers, or (ii) needed to manage the carbon reduction process effectively, whichever is lower. As the carbon calculation drills down from the corporate level through business unit, branch plant, depot, customer, delivery, consignment to individual product, the required time, cost and effort steeply escalate. Some organizations (eg Carbon Trust, 2006; PCF Project Germany, 2009) and individuals (eg Goleman, 2010) nevertheless argued that supply chains should be carbon audited at a product level to permit carbon labelling, in the expectation that this would encourage consumers to switch to lower-carbon products. Methodologies were devised for carbon footprinting supply chains at a product level (eg British Standards Institution, 2011; International Standards Organization, 2013) and some companies applied them to collect enough data to put carbon labels on a sample of their products. Most found this to be an extremely costly exercise. A large UK retailer and a global food manufacturer reported the resulting carbon auditing costs to be as high as £20,000–25,000 per stock-keeping unit (SKU). Tesco, whose CEO in 2008 committed the retailer to putting carbon levels on its entire product range of around 70,000 SKUs by 2012, claimed to have got this cost down to an average of £2,000–3,000, though this would still have entailed to total expenditure of £140–210 million (McKinnon, 2012a). By the time Tesco abandoned this exercise in 2011, it had only carbon audited 112 products, auditing them at a rate that would have taken 510 years to cover the 2008 product range (Smithers, 2010). Most businesses now accept that supply chain auditing at a product level is a task of 'labyrinthine complexity' (Lynas, 2007a). Market research has also revealed that the carbon labelling of products would induce only a marginal shift towards lower carbon consumption (Upham, Dendler and Bleda, 2011), making this a very expensive means of cutting carbon emissions. I discuss the case for and against carbon auditing and labelling at a product level in greater detail in McKinnon (2010c).

Committing to carbon reduction targets

Having quantified the carbon emissions from different logistical activities at appropriate levels within the business, the next step is to set targets for reducing them. There is a fundamental distinction between *absolute* targets to cut a company's total logistics-related emissions and *intensity-based* targets which aim to reduce emissions relative to the level of logistics activity. Most companies fear that the pursuit of absolute reductions will constrain the growth of their business and therefore prefer to express the target as a decline in the carbon intensity of their logistics operations. The problem with intensity-based targets is that even quite large reductions in carbon intensity can be wiped out by increases in the level of activity, with the result that the absolute level of emissions continues to rise. As explained in the previous chapter, keeping the increase in average global temperature within an environmentally acceptable limit will require massive reductions in total carbon emissions. In recognition of this planetary imperative, governmental GHG reduction targets are defined in absolute terms. There is therefore a mismatch between the absolute targets of government and the intensity-based targets of business.[4]

The Science-Based Targets (SBT) initiative aims to bridge this gap by giving companies advice on how to set absolute carbon reduction targets consistent with the climate science. Within 18 months of its launch in November 2016, 200 businesses had joined this initiative; they had a combined market value of $4.8 trillion (equivalent to the valuation of the Tokyo stock exchange at the time) and were collectively emitting annually as much GHG as South Korea.[5] It is accepted that the potential for decarbonization and its relative cost-effectiveness vary by sector and so the SBT initiative has proposed a methodology for deriving sector-specific GHG reduction targets (CDP/WRI/WWF, 2015). Its differentiation of sectors draws heavily on key publications of the IPCC and International Energy Agency. Unfortunately, freight transport is lumped into a residual 'other transport' sector for which the SBT analysts could find 'no activity information' in the IPCC and IEA reports (IPCC, 2014b; International Energy Agency, 2014). Modelling of this more general sector was based on monetary surrogates. It is to be hoped that this rather crude 'climate science-based' method of emission target setting for freight transport will be refined as more macro-level activity data becomes available.

Setting a science-based carbon reduction for logistics is complicated by its pivotal role as a support service to many other sectors. As illustrated in Figure 2.4, these sectors vary in their potential to reduce GHG emissions (R) (height of the line) and the abatement cost per tonne of GHG saved (C) (slope of the line) in each of the sectoral diagrams.

These variations should be taken into account when determining the carbon reduction targets for each sector. The targets for sectors A to G will in turn influence the height and slope of the carbon abatement cost line for logistics, making it difficult to derive a meaningful target for its future emissions. The modelling of these sectoral interdependences is particularly important for logistics but is analytically challenging.

Few national governments currently set GHG reduction targets for individual sectors. In the absence of sector-specific targets, it is usually assumed

Figure 2.4 Impact on logistics of sectoral variations in the level and cost of carbon abatement

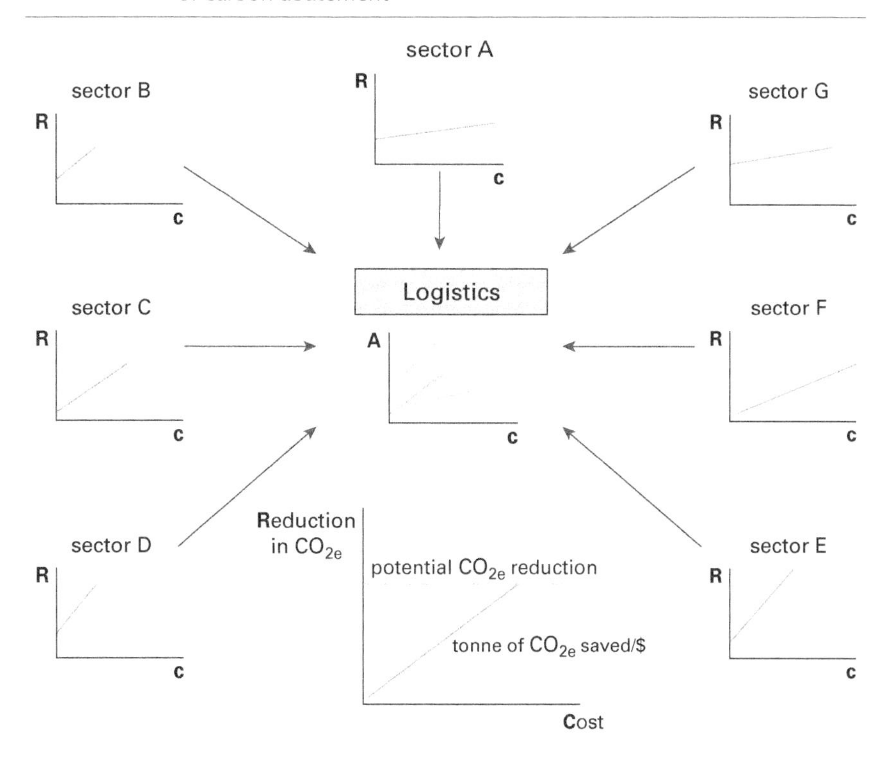

that economy-wide targets apply at the sectoral level. In the UK, for example, the road freight sector is expected to conform to the national target of an 80 per cent reduction in GHG emissions between 1990 and 2050, decreed in the 2008 Climate Change Act. On the other hand, the 2011 EU White Paper on Transport (European Commission, 2011) conceded that transport (both passenger and freight) would be more difficult to decarbonize than other sectors and so set it a lower target (60 per cent reduction between 1990 and 2050) than the EU economy as a whole (80–90 per cent). An analysis by McKinsey & Co (2009) of sectoral contributions to meeting a carbon reduction target of 35 per cent by 2030 (against a 1990 baseline) found that the transport sector would require 37 per cent of the carbon abatement budget to save only 10 per cent of the CO_2. This demonstrates that targeting at a sectoral level should be sensitive to variations in the scope and cost of carbon abatement. It may be inappropriate therefore for a company simply to apply a government's national, economy-wide target to logistics.

Industry associations can have a better understanding of the potential for carbon reduction in their sectors, though some see it as their role to lobby against government carbon reduction plans to 'protect' their members against any associated tax increases or regulations. Those that take a more enlightened view of climate change and are prepared to work with governments can make an important contribution to the decarbonization of logistics. This can include the derivation, promotion and monitoring of industry-specific carbon reduction targets. For example, in 2011, the Logistics Carbon Reduction Scheme in the UK set a target for its 50 members, who at the time operated 45,000 trucks and vans, of reducing the carbon intensity of their freight transport operations by 8 per cent over five years (FTA, 2011). This target was subsequently endorsed by the government and achieved at an industry level in 2015 (FTA, 2016):

> This collective approach to target setting is beneficial in that it gains the endorsement of industry peers, helps to build up momentum for carbon reduction across the logistics industry, ensures greater consistency in targeting and demonstrates to government that business is serious about meeting its climate change obligations (McKinnon and Piecyk, 2012: 636).

It also discourages companies from indulging in what has become known as 'competitive greenery' where they try to outbid each other with excessive carbon reduction claims which often command little confidence.

Setting carbon reduction targets for the logistics sector is complicated by its close interdependence with many other sectors. For example, geographical patterns of production and trade are likely to change over the next 35 years to reflect spatial variations in the rate at which electricity decarbonizes, the climate changes, water reserves are depleted, population migrates etc. As the servant of other economic and social activities, freight transport will have to adjust to these external forces. It is possible that to help other sectors meet their carbon targets and adapt to climate change, freight volumes will have to rise, even more than predicted in Chapter 1. Setting an absolute carbon reduction target for freight transport in isolation might therefore prove unrealistic. Indeed, it could be counterproductive if it resulted in quantitative controls being imposed on logistical activity which prevented other sectors from attaining their GHG reduction or climate adaptation goals. It is not yet possible to estimate by how much meeting these wider goals will inflate the rates of freight traffic growth factored into current forecasting models.

Nor is it clear how the business community would react to the setting of an absolute carbon reduction target for the freight sector as a whole. If a consensus emerged that it could be achieved entirely by reductions in carbon intensity, an enhancement of current decarbonization efforts might suffice. Research is underway in several countries to see how far phased deployments of a broad range of technological and operational measures might take us along the decarbonization pathway. The carbon reductions required of the logistics sector by 2050 may be so deep, however, that more radical action may be needed. A tight cap may have to be imposed on total logistics emissions and mechanisms put in place to allocate carbon credits among sub-sectors of the logistics market, transport modes, carriers, regions etc. This is still a distant prospect, but setting 'science-based' targets for absolute reductions in freight-related GHG emissions puts us on a trajectory that leads in this direction.

For a more detailed discussion of the derivation of carbon reduction targets for logistics, readers should consult McKinnon and Piecyk (2012).

Considering possible options

At this stage in the strategy development process the objective is to compile a long list of measures likely to cut carbon emissions regardless of their financial implications. Their relative cost-effectiveness is assessed at a later stage. The Hoffman framework focuses attention on 'technological solutions'. In the logistics sector, they can be supplemented by a broad range of behavioural and operational measures.

To ensure that the list is constructed systematically and that no major carbon reduction opportunities are missed, reference can be made to various classificatory frameworks outlined in the previous chapter, particularly the green logistics one which is the most detailed and most customized to logistics operations.

There has been a tendency on the part of both analysts and managers to see advances in vehicle technology and a switch to low-carbon energy as the main, or even sole, means of decarbonizing freight transport. This reflects a more general view that there is a 'technological fix' to the climate change problem. In the case of the road freight sector, much attention has been focused on the inner core of Figure 2.5, what I call a 'rectangular onion' diagram.

This places heavy reliance on truck manufacturers, biofuel suppliers and the wider decarbonization of electricity. While new vehicle designs and low-carbon power sources will certainly make major contributions to future decarbonization in this sector, they alone are very unlikely to deliver the deep cuts in CO_2 that will be needed over the next few decades. That will probably require companies to exploit carbon savings across all the levels of Figure 2.5. Interventions at the inner four levels, relating to the performance,

Figure 2.5 Scoping the decarbonization of freight transport

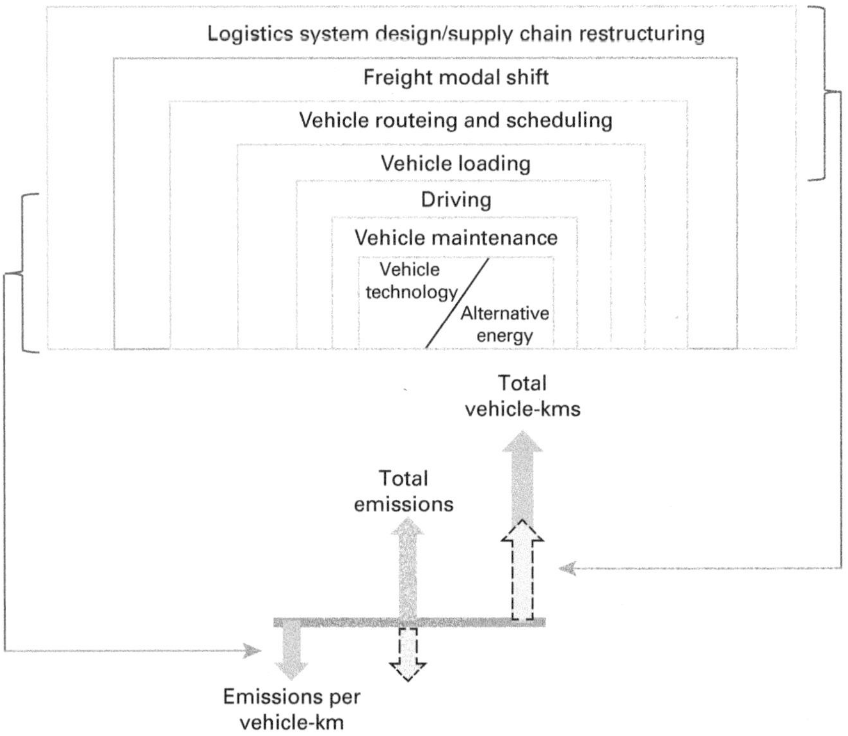

Logistics system design/supply chain restructuring
Freight modal shift
Vehicle routeing and scheduling
Vehicle loading
Driving
Vehicle maintenance
Vehicle technology
Alternative energy
Total vehicle-kms
Total emissions
Emissions per vehicle-km

energy use, maintenance and operation of vehicles, have the effect of reducing carbon emissions per vehicle-km. These reductions, however, may be more than offset by the forecast growth in total vehicle-kms, leaving net road freight emissions significantly higher than in the base year. Actions are therefore required at the outer layers to reduce total vehicle-kms by a sufficient margin to achieve absolute reductions in emissions overall. This will require more efficient vehicle routeing, a freight shift to lower-carbon modes, a redesign of logistics systems and a reconfiguration of supply chains to reverse the long-term growth in demand for freight movement. This illustrates the need to 'cast the net wide' when reviewing the options for cutting logistics-related emissions.

The options can be classified with respect to the logistics decarbonization framework outlined in Chapter 1:

1 *reducing the demand for freight transport*;

2 *shifting freight to lower-carbon transport modes*;

3 *improving asset utilization*;

4 *increasing energy efficiency*;

5 *switching to lower-carbon energy*.

Table 2.3 lists specific measures that can be taken under each of these headings. These lists are not exhaustive, but they give a sense of the huge variety of things that companies can do to minimize the amount of carbon emitted by logistics operations. Most of the measures are fairly generic and applicable to most companies' operations. They vary, however, in their potential for cutting carbon emissions, their feasibility and their relative cost-effectiveness.

In an influential report, the World Economic Forum and Accenture (2009) assessed the carbon abatement potential and feasibility of 13 key supply chain decarbonization measures. They used a range of data sources and methodologies to estimate, at a global level, the potential savings in GHGs that these 'commercially viable' measures could offer. It was claimed that they could, collectively, cut emissions of CO_{2e} by 1,400 billion tonnes in the medium term, representing around half of the emissions generated by the 'logistics and transport sector' at the time. Their analysis suggested that 'around 60 per cent of the potential carbon abatement originates from the sector's own emissions. Others come from the broader supply chain and can be achieved through changed logistics and transport configurations' (p.4). Feasibility was defined as the likelihood of a measure being successfully implemented and depended on such things as operational complexity, the rate of technological innovation and adoption, management capability and, of course, cost-effectiveness.

Table 2.4 shows how the 13 measures were ranked on the basis of a composite abatement–feasibility index. Those measures over which logistics managers have a high degree of control are shaded and comprise three of the five top-rated measures. Clean vehicle technology is considered to be the largest source of CO_{2e} reductions, followed closely by 'de-speeding logistics', in other words, reducing the speed at which products flow through the supply chain. Both of these sources of emission savings are discussed at length in Chapters 6 and 7 of this book. Also highly rated are the optimization of logistics networks and training and communications, which are discussed in Chapters 3 and 6. The majority of the measures on the list require wider structural change to patterns of procurement, production and retailing, showing how the carbon intensity of logistics is intimately related to other business processes. Although this first major study of supply chain decarbonization was pitched at a macro level and is broad in its functional

Table 2.3 Examples of carbon-reducing interventions in the road freight sector

1 Reduce number of links in the supply chain:	**5 Reduce empty running:**
Dis-intermediation – bypassing agencies/nodes in the supply chain	Use load-matching services (online freight exchanges/web-based procurement)
Greater vertical integration of processing – reduce intermediate journeys between processing plants	Promote collaborative initiatives – both vertical and horizontal collaboration
	Explore backloading opportunities during purchasing negotiations
2 Reduce the average length of haul:	Source supplies on ex-works basis to increase backloading opportunities
More localized sourcing of inbound supplies	Incorporate the planning of backloading into vehicle routeing and scheduling software
Decentralize processing, storage and distribution operations	Use telematics to increase 'visibility' of the fleet and help exploit backloading opportunities
Move production/storage/distribution facilities into more transport-efficient locations	Consolidate return of handling equipment (roll-cages/dollies) in a fewer vehicles
Industry swap arrangements – to minimize delivery distances (as in oil refining sector)	Maximize use of returning shop delivery vehicles for collection of packaging material
Use computerized vehicle routeing and scheduling (CVRS) to reduce vehicle-kms	Relax delivery schedules to accommodate more backloads
Incorporate predictive analytics into CVRS to further improve routeing efficiency	Improve the reliability of loading and off-loading operations to build confidence in backloading schedules
Use real-time telematics with CVRS to dynamically reroute	Increase the ratio of trailers to tractors (ie the 'articulation ratio') to create more flexibility for backloading

3 Promote transfer of freight to lower-carbon modes	**6 Reduce exposure to traffic congestion**
Send greater % of freight by rail	Reschedule deliveries to inter-peak periods and evening/night
Send greater % of freight by waterborne services	Extend opening hours of premises for collections and deliveries
Relocate production facilities/warehousing to be adjacent to alternative transport network	Introduce unattended delivery systems for out-of-hours/off-hours delivery
Invest in rail siding and/or rolling stock	Use real-time telematics with CVRS to dynamically reroute vehicles
Develop/invest in equipment to facilitate intermodal transfers	
Reschedule distribution operations to match timetables of the alternative mode	**7 Improve fuel efficiency**
Link modal choice to inventory management using synchromodality principle	Develop fuel management programme
Apply for any available government modal shift incentives	Appoint fuel champion
	Train drivers in the techniques of fuel-efficient driving (eco-driving)
4 Increase vehicle payloads on laden trips:	Use telematics/onboard devices to monitor driving performance
Relax just-in-time replenishment schedules to permit greater load consolidation	Regularly debrief drivers on fuel performance
Increase use of primary consolidation (at expense of adding an extra link to the supply chain)	Give drivers financial and other incentives to drive more fuel-efficiently Install smart cruise control
Give hauliers/transport departments more advanced warning of traffic demands	Reduce the vehicle replacement cycle to accelerate adoption of more fuel-efficient vehicles Prioritize fuel efficiency as a vehicle purchase criterion

(*continued*)

Table 2.3 *(Continued)*

Promote collaborative initiatives – both vertical and horizontal collaboration	Use vehicle with stop-start system
Shift from dedicated contracts with 3PL to shared-user contracts/ network services	Lightweight the vehicles
Adopt 'vendor-managed inventory' (VMI) arrangement with suppliers	Improve aerodynamic profiling of vehicles
Expand the use of 'nominated day' delivery systems	Upgrade vehicle maintenance/undertake more preventative maintenance Adopt vehicles with automatic transmission
Replace the monthly 'order – invoice' cycle with a system of rolling credit	Ensure effective tyre management/inflation of tyres to fuel-efficient level
Use more 'space-efficient' handling equipment	Use supersingle tyres
Minimize the amount of secondary and primary packaging	Use low 'rolling-resistance' tyres
Stack loads to greater height (within warehouse slot height constraints)	Install vehicle speed limiters and set them at lower maximum speeds
Right-size vehicle size and weight to the loads being transported	Fit anti-idling devices
Use longer and/or heavier vehicles when justified by load size/weight	Adopt more energy-efficient forms of vehicle refrigeration
Make greater use of double-deck vehicles (where infrastructure permits)	Reduce pre-loading time for refrigerated vehicles
Switch from powered- to fixed-deck double-deck trailer	**8 Reduce emissions per litre of fuel consumed:**
Use compartmentalized vehicles to increase load consolidation opportunities	Increase use of hybrid and electric vehicles

Increase storage capacity at delivery points – to permit delivery of larger loads	Increase use of gas vehicles – preferably with biomethane
Use online procurement platforms ('freight exchanges') to increase opportunities for load consolidation	Increase use of dual-fuel vehicles (only with minimal methane slip)
Deploy load optimization software (including agent-based systems)	Increase % blend with environmentally-sustainable biofuel
Use telematics to improve management of the vehicle fleet	Recharge vehicle batteries with low-carbon electricity
	Use lower-carbon energy in refrigeration equipment
	Minimize refrigerant gas leakage from vehicles

Table 2.4 Rating of 13 supply chain decarbonization opportunities

	GHG abatement potential	Feasibility index	Combined score
Vehicle technology	175	0.8	140
Slowing down supply chain	171	0.8	137
Low-carbon sourcing of agricultural produce	178	0.6	107
Optimizing logistics networks	124	0.8	99
Improved training and communication	117	0.8	94
Improved packaging design	132	0.7	92
Low-carbon sourcing in manufacturing	152	0.6	91
Energy-efficient logistics buildings	93	0.9	84
Freight modal shift	115	0.7	81
Reverse logistics/recycling	84	0.6	50
Increase in home delivery	17	0.5	9
Reduced congestion	26	0.3	8
Near-shoring of production	5	0.7	4

DATA SOURCE World Economic Forum / Accenture (2009)

scope, it is still of relevance to individual companies drawing up carbon reduction plans for their logistics.

Several online calculators have been constructed to give companies an indication of typical CO_2 savings from a range of carbon-reducing interventions in the road freight sector or the shifting of freight between modes (Table 2.5). Most base their estimates on industry default values, but also allow users to insert company-specific data and thereby tailor the calculation to particular fleets and operations. These tools can be used to compute the total emission reductions accruing from either a single measure or a combination of measures, making allowance for any interdependences between them.

Collaborative decarbonization

It is at this step in the procedure that companies should look beyond their corporate boundaries to explore opportunities for logistical collaboration

Table 2.5 Examples of online carbon calculators for freight transport operations

Calculator	Organization	Transport mode	Website
SRF Optimizer	Centre for Sustainable Road Freight	Road freight	http://bit.ly/2Eklr6T
Carbon Intervention Modelling Tool	Freight Transport Association	Road freight	http://bit.ly/2nWeQKE
EcoTransIT World	IVE	All modes	http://www.ecotransit.org/
Clean Fleet Management Toolkit	Clean Air Asia	Road freight	http://bit.ly/2nZjQxm
Truck Carrier Partner Tool	US SmartWay	Trucking	http://bit.ly/2EzdfnB
CSX Carbon Calculator	CSX	Rail freight	http://bit.ly/2dKkU1U
Carbon Emissions Calculator	JF Hillebrand	International freight services	http://bit.ly/2HaTxgi
Sustainable Performance Monitor	UN ECLAC / University of Applied Sciences, Bremen	Ports/freight terminals	http://spm-terminals.com

with other businesses. The pursuit of deep carbon reductions in the logistics sector will almost certainly demand a high level of inter-company collaboration in the sharing of logistics assets, pooling of information and aggregation of loads. This will make a reality of something that has been discussed, researched and promoted for several decades, but is still the exception rather than the rule. For example, an executive survey by Accenture back in 2004 found that 'collaborating with multiple partners' was the most important supply chain challenge facing businesses. Since then there have been well-publicized examples of companies engaging in logistical collaboration, particularly in the fast-moving consumer goods sector (eg ECR, 2008; Muylaert and Stoffers, 2014). Many of these collaborations can be described as 'opportunistic' in the sense that they often emerge from chance encounters between logistics managers. As collaboration gathers momentum one might expect companies to develop a more systematic approach to seeking out potential collaborators and cultivating relationships. Mounting pressure to decarbonize logistics may hasten this move to a situation where exploiting collaborative links becomes a key component of business strategy. Where companies have maximized their internal carbon efficiencies and are still falling short of their carbon reduction targets, they may have little choice but to work with suppliers/distributors in the vertical supply chain (ie vertical collaboration) and businesses at the same level in the supply chain (ie horizontal collaboration) to close the gap. The reported levels of CO_2 reduction that companies have been able to achieve as a result of logistical collaboration, discussed in Chapter 5, suggest that this can be a very effective way of reinforcing internal decarbonization efforts.

Cost evaluation

Companies naturally wish to select and prioritize those measures that yield large carbon savings and are relatively cost-effective to implement. It is now generally accepted that within logistics systems there is a close and direct correlation between energy use, CO_2 emissions and cost. Indeed, some senior logistics managers argue that in their experience almost everything they do to cut carbon saves money. The close correlation between cost savings and carbon reductions is well illustrated by Walmart, which doubled the efficiency of its truck fleet between 2005 and 2015 through a combination of improved 'loading, routeing and driving, as well as through collaboration with equipment and system manufacturers on new technologies' (Walmart, 2017: 63). On an annual basis this cut the company's truck operating costs by roughly $1 billion in 2015 and its CO_2 emissions by 650,000 metric tons. Even self-financing carbon reduction measures like this still need to be differentiated by the scale of investment required and the rate and level of return measured both in economic and carbon terms.

By conducting a marginal abatement cost (MAC) analysis it is possible to compare decarbonization options in terms of the amount of CO_2 likely to be saved and the cost per tonne of CO_2 saved. In a MAC graph, economic costs and benefits are measured on the vertical axis and CO_2 savings along the horizontal axis (Figure 2.6).

Figure 2.6 Comparison of the marginal abatement costs of decarbonization initiatives

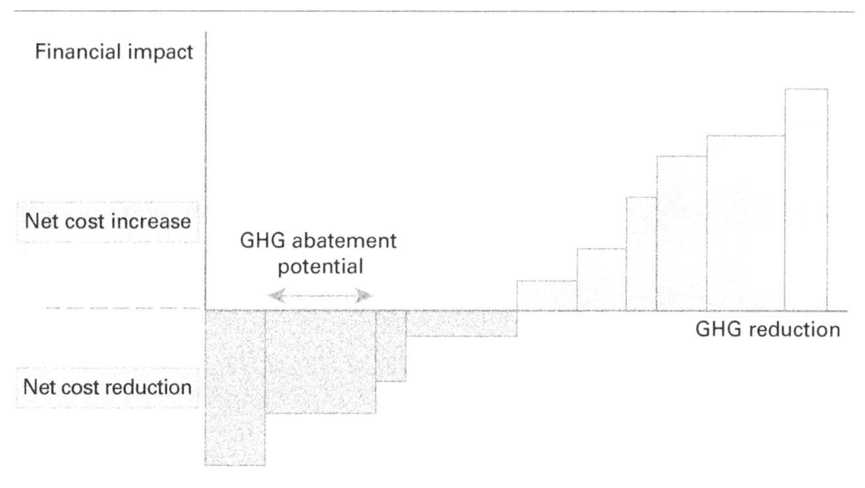

Each intervention is represented by a bar of varying width and height/depth. The width of the bar indicates the amount of CO_2 likely to be saved. Where the bar is above the X-axis the cost per tonne of CO_2 saved is positive, which means that carbon reductions will come at a price. Where it is below the X-axis, a negative cost is incurred, in other words the measure saves money as well as cutting carbon. Analyses of this type in the freight transport sector typically show that many carbon-reducing interventions fall into this category. These measures are often described as the 'low-hanging fruit' as it is easy for companies to make a financial case for implementing them. Indeed, this is often little more than the application of good business practice. In the longer term, however, it is very unlikely that harvesting all this 'fruit' will achieve the deep carbon reductions that will be expected of the logistics sector. Once it is exhausted, economic sacrifices will have to be made at both business and consumer levels. We may then enter a period when greater investment is required in decarbonization schemes and logistical cost trade-offs rebalanced to prioritize carbon reduction (Figure 2.7).

At a later stage, even more draconian measures may be required which squeeze profit margins, investment returns and share prices. For the foreseeable future, however, the so-called 'green-gold' measures should be used to get companies onto a carbon reduction trajectory that will ultimately lead to low-carbon logistics.

While a MAC analysis is an effective means of comparing carbon-reducing measures, it can be difficult for individual companies to perform, particularly smaller businesses. They often lack practical experience of the measures and

Figure 2.7 Likely phases in the future decarbonization of logistics

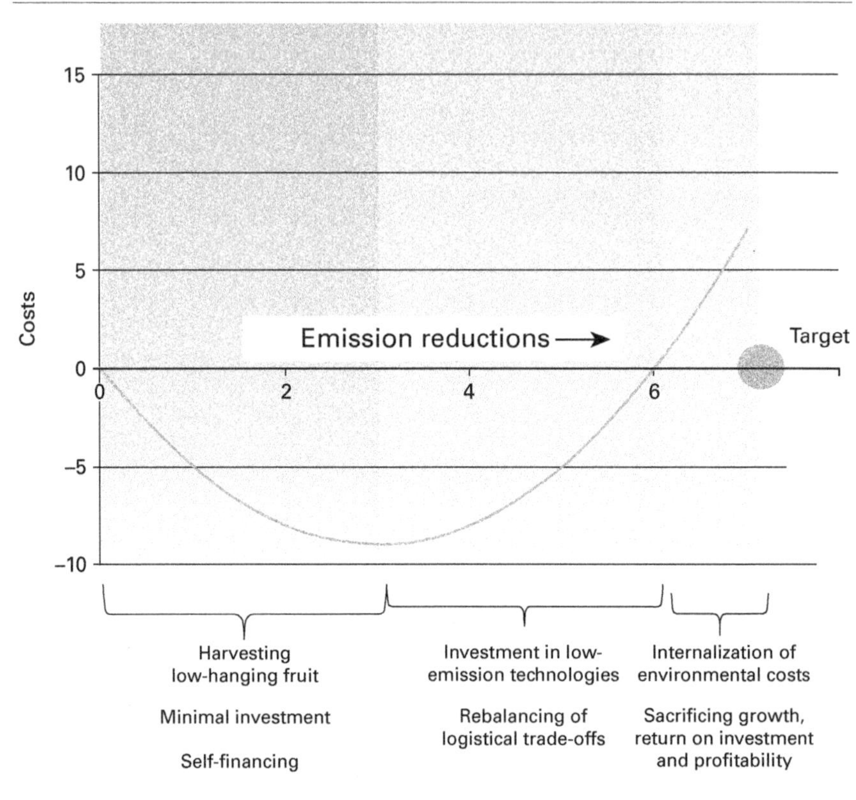

SOURCE Based on graph in Tavasszy (2017)

so are unable to calibrate the analysis with carbon and cost data. There is a need, therefore, for industry data sets that pool knowledge of the relative cost-effectiveness of a range of carbon-reducing measures. There are still relatively few data sets of this type in the public domain. Many of the larger businesses that have undertaken MAC analyses internally consider the results to be commercially confidential and so are naturally reluctant to share them. Several academic and consultancy studies, however, have gathered data from various sources to perform similar analyses for public dissemination (eg Atkins, 2010). Most of these cost analyses have been restricted to new technologies and technical upgrades. Quantifying the carbon cost-effectiveness of 'softer' initiatives such as changes to business processes, operating practices and inter-company relationships is much more difficult and usually relies on company case study data. Unlike a technical intervention, such as retrofitting a fuel-saving device, which has a tangible and discrete impact, modifications to a business practice are context-specific and can be applied to varying degrees. One of the online calculators listed in Table 2.5, the SRF Optimizer, provides estimates of the average cost-effectiveness of a range of decarbonization measures, including some of these managerial initiatives.

In addition to cutting emissions and saving money, some interventions yield other social and environmental co-benefits. For example, truck drivers who have been trained to be more fuel efficient also tend to drive more safely, thereby reducing accident rates and insurance premiums. It is important, therefore, when appraising the value of a carbon reduction measure, to take account of these wider benefits. They can either be expressed in terms of environmental and social welfare metrics or converted into monetary values. Monetary valuations of the environmental and social effects of transport, compiled for the European Commission, can be found in Ricardo-AEA (2014). Once monetized, the co-benefits can be summed to produce a single index of their total economic, environmental and social contribution, in keeping with the 'triple bottom line' view of sustainability. Although the monetary values ascribed to the social and environmental externalities have little or no financial significance at present, this will change if, as many environmental NGOs and politicians have been advocating, governments were to fully internalize the external costs of transport in higher taxes.

Arguably, first in line to be internalized should be the external cost of GHG emissions, reflecting the gravity of the global warming problem. Monetary values are currently assigned to these emissions in various ways. In countries and regions with 'cap and trade' systems, market trading of carbon credits determines the price of a tonne of CO_2. In January 2018, this price was around €9 in the world's largest carbon market, the European Emissions Trading Scheme (ETS). As explained in Chapter 1, this figure is deemed to be much too low to give companies the financial incentive they need to get emissions down to a level consistent with a 2°C global temperature rise limit. Carbon prices will need to reach levels of US $50–100 by 2030 to induce the necessary change in commercial behaviour (Carbon Pricing Leadership Coalition, 2017). It is very unlikely that emissions trading schemes will deliver carbon prices of this magnitude in the foreseeable future. Nor do they currently cover all economic sectors. The World Bank, Ecofys and Vivid Economics (2017) found that across 42 national and 25 sub-national carbon pricing schemes, only 15 per cent of carbon emissions were covered. The only logistical activity currently included in the ETS is air cargo operations within the EU. Carbon valuations based on emissions trading also have the disadvantage of being relatively volatile, making it difficult to build a financial case for investment in GHG reduction projects. The imposition of carbon taxes held at a more stable level or gradually rising in line with a declared long-term plan would provide a firmer basis for appraising these investments, though very few governments or jurisdictions have so far introduced such taxes (Carbon Tax Center, 2017).

If logistical activities were subject to a carbon tax, this would offer a fiscal basis for assigning a monetary value to carbon emissions. In the absence of such taxes, governments still need to put a price on carbon emissions when appraising public policy decisions that have a bearing on climate change. There is a substantial literature on the economics of valuing GHG emissions, much of it undertaken or commissioned by governments. As knowledge of the subject has grown, valuation methods have been refined. The UK government, for example, used to estimate a social cost of carbon, based on the damage that would be done by GHG emissions from a project or policy during its lifetime. Today it uses 'a hybrid methodology for producing short-term traded carbon values... using a market-based approach based on futures prices' (DECC, 2015: 3). This predicts the likely price for GHG emissions per tonne in an emissions trading market under conditions of tightening carbon abatement to meet future GHG reduction targets. The government quotes low, central and high estimates up to 2030. In 2015, the central value for a tonne of CO_{2e} emissions was £5.94, which would rise to £78.5 (in real terms) by 2030. It is anticipated that these values will ramp up steeply during the 2020s. Such figures could be used as a benchmark for companies appraising logistics investments that have an impact on GHG emissions. Increasing numbers of businesses are now factoring an internal carbon price into their financial modelling and planning. According to the Carbon Disclosure Project (CDP, 2017), over 1,300 businesses were doing this in 2015. It argues that 'internal carbon pricing has emerged as an important mechanism to help companies manage risks and capitalize on emerging opportunities in the transition to a low-carbon economy' (p.5).

Choosing appropriate actions

How does a company select a manageable set of carbon-reducing measures from the long list of possible options in Table 2.3? It can do this by systematically comparing them against a range of criteria, such as the following:

- *Carbon abatement potential*: in other words, how big a carbon saving is the initiative likely to deliver?
- *Cost-effectiveness*: what will be the economic cost or benefit per tonne of CO_2 saved over a given period?

- *Level of investment*: how much capital investment will be required and what will be the likely returns and payback period?
- *Ease of implementation*: to what extent will internal systems and procedures have to be modified and staff retrained?
- *Interrelationship with other measures*: often groups of measures are mutually reinforcing and therefore best implemented together.
- *Synergies with other business goals*: is the measure likely to bring wider co-benefits to the company?
- *Availability of government support*: are any tax incentives available to help defray some of the cost?
- *Experience of other businesses*: how widely and successfully has the measure been applied across the industry?
- *Opportunities for collaboration*: can the initiative be introduced jointly with other companies to leverage higher carbon and cost savings?
- *Timescales*: how long would it take to implement and scale up the measure to achieve the desired carbon and cost savings?
- *Risk profile*: how does the measure affect the reliability of the logistics operation and wider business continuity?

A numerical scoring system can be used to formalize this process. This involves assigning the criteria different importance weightings and then rating each decarbonization option against each criterion. Depending on the length of the 'long list', the exercise can involve a good deal of data collection, though use of online calculators can make the job more manageable. It also requires a significant amount of expert judgement on the part of individuals or teams tasked with developing the logistics decarbonization strategy.

In estimating the combined impact of a set of interventions, one must exercise caution. Most interventions are mutually reinforcing and so their carbon savings are cumulative. This does not mean that the percentage savings they yield can simply be added together. This is illustrated by the worked example in Figure 2.8, in which CO_2 emissions from a road-based distribution operation are reduced in six different ways.

A modal shift to rail reduces the tonnage carried by road. Improved routeing cuts the average distance this tonnage is moved by road in loaded vehicles. An increase in the level of backloading cuts the proportion of empty running and hence total vehicle-kms. Higher fuel efficiency reduces fuel consumption per vehicle-km while a switch to lower-carbon fuels cuts average CO_2 emissions per litre of fuel used. Each successive saving has to be

Figure 2.8 Worked example of carbon-reducing interventions applied to a road-based distribution operation.

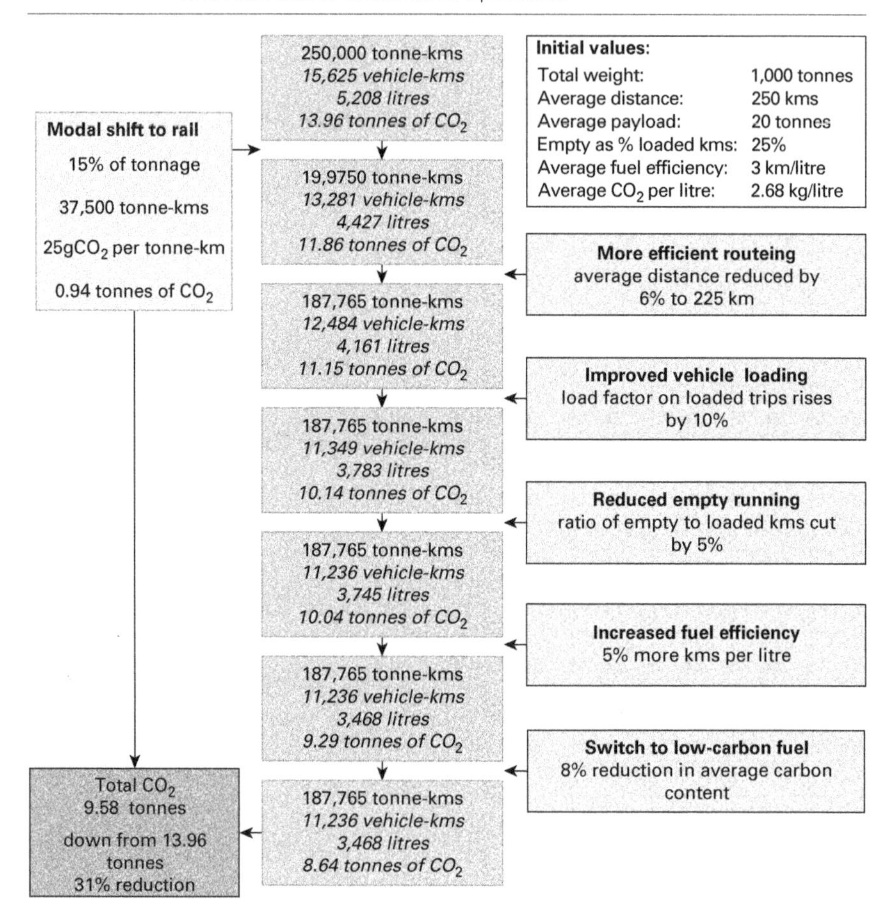

calculated relative to the declining baseline. For example, the improvement in fuel efficiency is applied, not to the original number of vehicle-kms, but to the much lower number of vehicle-kms after modal shift, load consolidation and improved backloading have taken effect. If the percentage improvements from each intervention were factored together and applied to the initial set of parameters, the combined carbon saving in this example would be 42 per cent. This significantly overestimates the actual saving of 31 per cent that would be achieved when the percentage impact of each intervention is applied to the reducing totals of tonne-kms, vehicle-kms and fuel consumed.

The calculation must also take account of the counteracting effects of some interventions. For example, lowering the average speed of a truck (the so-called 'downspeeding' option) reduces emission savings from improved aerodynamic profiling. Retrofitting a vehicle with an aerodynamic kit also increases its tare weight and is therefore in conflict with efforts to 'light-weight' the fleet. In calculating the emission savings from streamlining freight vehicles, one must 'net off' the effects of downspeeding and increasing tare weight. Most online calculators allow for the negative interaction between some carbon-reducing initiatives, using industry default values to make a realistic assessment of their combined impact on CO_2 emissions.

Carbon offsetting

Carbon offsetting involves one business paying another to save GHG emissions on its behalf. This is generally done on a market basis through an exchange in which carbon credits are traded. Typically the money raised by carbon offsetting is used to finance afforestation, renewable energy and resource conservation projects which cut GHG emissions. Many of these projects are located in the developing world where a given amount of carbon saving can be 'purchased' relatively cheaply. This offers companies a means of meeting a carbon reduction target for logistics which exceeds the level of carbon savings they are able to make themselves within a given time period. For example, a business may calculate that by fully exploiting all its logistics decarbonization options over the next five years, and allowing for anticipated growth in sales volumes over this period, it will only be able to achieve 80 per cent of the target reduction in GHG emissions. It can then buy carbon credits equivalent to 20 per cent of the required GHG savings. Alternatively, companies can compare the marginal costs of decarbonization and carbon offsetting to find the most economical way of meeting the target. As discussed above, however, MAC analysis usually reveals that the marginal costs of decarbonization in the logistics sector are negative, making this financially a more attractive option than offsetting.

There is a danger, however, that carbon offsetting is regarded as an 'easy option'. It can allow companies to buy their way out of their climate change obligations rather than make a serious effort to cut their own emissions. When companies are setting their own carbon reduction targets for logistics it is often better to calibrate them with respect to the savings they can make themselves rather than pitch them at a higher level and have to rely on some offsetting. Where the targets are set externally and a gap exists, carbon offsetting may be a necessity, but it should be seen as a last resort. This is not to deny that the money raised by carbon offsetting schemes is usually put to good use in supporting low-carbon economic development in other parts of the world. To ensure that it is used for *bona fide* purposes, it is always advisable to use an accredited carbon offsetting agency and programme. It is also good practice, as UPS (2016a) does, to provide details of the offsetting projects that have been sponsored and the resulting amounts of GHG 'retired'.

One sector of the logistics market where carbon reduction plans rely heavily on carbon offsetting is air cargo. ICAO, the UN agency for civil aviation, has committed international aviation as a whole to carbon-neutral growth beyond 2020. This will be difficult to achieve, given the forecast growth of air traffic. Air freight tonne-kms, for example, are predicted to grow by around 4 per cent annually between 2016 and 2036 (Boeing, 2016; Airbus, 2016). It is estimated that, up to 2040, operational improvements, new technology and a switch from kerosene to alternative fuel will take the international aviation sector roughly 60 per cent of the way to carbon neutrality (ICAO, 2016). What ICAO calls 'market-based measures', mainly carbon offsetting, will be needed to close the gap. For this purpose, it has established a programme called CORSIA (Carbon Offsetting and Reduction Scheme for International Aviation), which will initially be voluntary but will become mandatory for most countries after 2027. The decarbonization of airfreight operations is discussed in Chapter 7.

Cutting emissions and assessing the impacts

Having selected a package of carbon-saving measures that is well suited to the logistics operation and is likely to yield the required reduction in carbon emissions, the company then has to plan its implementation. Synergies will exist between some measures, making it sensible to implement them as a group. Improving the aerodynamic profiling of trucks, for example, involves retrofitting different types of fairings and this is best done in a single operation. Training in eco-driving needs to be supplemented with driver monitoring and incentive schemes to ensure that the improvements in driving behaviour are maintained. The company also needs to phase the introduction of the measures in line with budgetary considerations and other pressures on the business. It is desirable to trial the measures in particular locations or on particular fleets prior to a full roll-out. Lower-level operational measures which impact solely on the logistics department will generally be easier and quicker to implement. Higher-level, strategic initiatives need cross-functional support and will require more time to plan and co-ordinate. In some cases, carbon offsetting may have a role to play in the short to medium term to reduce a company's logistics-related carbon commitments until its logistics decarbonization plan is fully implemented.

Companies need to establish a system for measuring the impact of the various carbon-reducing measures. The carbon accounting system developed at the second stage of the framework should have sufficient granularity to assess the effect on emissions of applying specific measures. Where several measures are implemented simultaneously it can be difficult to isolate their individual impact, particularly where synergies exist. The review process should also collect cost data to permit analysis of the relative cost-effectiveness of the various measures that have been implemented. In larger businesses, standard accounting systems should be able to furnish managers with much of the necessary data, though smaller carriers may need to undertake additional fuel and cost monitoring.

Calibrating the strategy

Analysis of the effects of individual carbon-reducing measures and the strategy as a whole on energy use, carbon emissions and costs will help companies build up their knowledge of the decarbonization process. In the light of this practical experience, they may wish to go back and revise their carbon reduction targets or the package of measures they selected. The targets may have been set too high or too low, or the timescales over which they are to be met may need adjustment. Estimates of the cost-effectiveness of the measures made may have to be revised, with some possibly being dropped and others given greater priority. This recalibration of decarbonization procedures is represented by the addition of feedback loops to Figure 2.2, shown as broken lines. The development and refinement of the carbon reduction strategy then becomes an iterative process, trialling and reviewing different options and ultimately converging on a set of measures that best meets the company's operational, financial and environmental requirements.

Summary

The 10C approach outlined in this chapter offers guidance on how a company can plan the decarbonization of its logistics in a systematic and comprehensive manner. It has highlighted many of the issues and pitfalls that managers typically encounter along the way, but should also have provided assurance that the task is manageable and likely to be rewarding in both corporate and personal terms. Thanks to a combination of academic/consultancy research and the experience of a number of pioneering companies we have built up a reasonable understanding of how the decarbonization of logistics should be organized. The challenge is now to get more and more companies to embark on the process. As the recent 'We Mean Business'[6] initiative demonstrates, momentum is clearly gathering at a corporate level to achieve deep carbon reductions. It is important for logistics managers, departments and providers to make their contribution. The next five chapters explain how this can be done by deploying the five sets of decarbonization options listed on pages 48 and 49.

Notes

1 These boundaries were classified by the Swedish transport and environment organization Network for Transport Measurement (NTM) and elaborated on by McKinnon and Piecyk (2010) and Piecyk (2015).

2 Enterprise Resource Planning software supports integrated, cross-functional management within business, spanning procurement, production, logistics, sales etc.

3 EcotransIT World (2018) EcoTransIT World – a sustainable move [online] http://www.ecotransit.org/ [accessed 6 February 2018].

4 One very notable exception is Deutsche Post DHL Group, the world's largest logistics business, which in March 2017 committed to reducing 'all logistics-related emissions to net zero by the year 2050'. Details of how it plans to make its operations carbon neutral can be found at www.dhl.com/en/press/releases/releases_2017/all/dpdhl_commits_to_zero_emissions_logistics_by_2050.html

5 Science-Based Targets (2018) 200 Companies commit to science-based targets, surpassing expectations for corporate climate action [online] http://sciencebasedtargets.org/2016/11/16/200-companies-commit-to-science-based-targets-surpassing-expectations-for-corporate-climate-action/ [accessed 6 February 2018].

6 We Mean Business Coalition (2018) [online] https://www.wemeanbusinesscoalition.org/ [accessed 6 February 2018].

03 Reducing freight transport intensity

The logistics sector will unquestionably be difficult to decarbonize. This is mainly because the demand for freight movement is steadily rising and is predicted to remain on a steep upward path for the next few decades. If the growth forecasts outlined in Chapter 1 prove accurate, the carbon intensity of freight transport will have to plunge to a small fraction of its current level for this sector to stay within carbon limits consistent with a 1.5–2°C global temperature increase. If the growth rate is more subdued than expected or can be managed downward, the required reductions in emissions per tonne-km could be set at a more realistic level. This chapter examines the various forces driving freight traffic growth. How are they currently evolving and what could be done to weaken or reverse them? The chapter begins by looking beyond the remit of a logistics manager at the total amount of stuff that has to be transported. Is it true, as some commentators have suggested, that the economy is dematerializing and there will be less stuff to move in the future? The next two sections discuss other ways in which the growth of freight traffic might 'decouple' from economic growth, first at a national level and then internationally. One issue seldom discussed in the context of freight-GDP decoupling is the logistical impact of climate change adaptation. For example, climate-proofing infrastructure and settlements and relocating populations at risk of rising sea levels will entail the movement of large amounts of material. The fourth section gives examples of just how much freight traffic this might generate. Having charted a course through the more contentious and speculative aspects of the subject, the chapter ends with a review of one of the least controversial and most widely adopted means of cutting vehicle- and truck-kms – optimizing the routeing of freight flows.

A dematerializing economy?

The level of logistical activity is a function of the amount of material that must be moved, handled and stored across the global economy. This mass of material could be substantially reduced without hindering future improvements in living standards. The word 'dematerialization' has 'often been broadly used to characterize the decline over time in weight of the materials used in industrial end products' (Herman, Ardekani and Ausubel, 1990). On a related theme, Allwood et al (2011: 362) examine ways of reducing 'material demand' by improving 'material efficiency', including making products last longer, modularization, component re-use and 'designing products with less material'. In this section we review a range of processes that can reduce the physical quantity of goods required to deliver a given amount of consumer value.

Waste minimization

Waste is endemic at all levels in the supply chain and across all economic sectors. For example, it is estimated that between 37 and 50 per cent of post-harvest crop biomass is wasted at various points along the supply chain and at final point of consumption (Alexander et al, 2017). This range rises to 42–61 per cent if 'consumption in excess of nutritional requirements is included as a loss'! The construction sector also has a poor record for squandering resources. In the UK, for instance, 15 per cent of the materials delivered to building sites 'go straight to tip' due to poor ordering practices and product damage: 60 million tonnes per annum (Sharman, 2017). In any waste management scheme, top priority should be given to minimizing waste at source in adherence with the 'reduce, re-use, recycle' principle (WRAP, 2012). Wastage also occurs along the supply chain where, for example, products are damaged or temperature control fails. The disposal or recycling of waste generates additional transport, much of which is avoidable.

Recycling

It is not simply waste that is recycled. Many end-of-life products can be re-used, recycled or remanufactured in ways that impose less burden on the logistical system than the manufacture and distribution of new products. Recycling one tonne of steel, for example, saves 1.1 tonnes of iron ore,

633 kg of coal, and 54 kg of limestone (Planet Ark, 2017). It saves 2 kg of GHG for every 1 kg recycled. Reducing the weight of materials used, however, does not guarantee a reduction in freight tonne-kms as account must also be taken of the distances that new and recycled products are moved. As Weetman (2016) observes, in moving from a linear to a circular economy 'geography matters' (p.22). For recycled products, 'closed regional and local loops are intuitively the most attractive as they are based on close proximity between points of production and use' (Ellen MacArthur Foundation and McKinsey, 2014: 40). Recent research on the circular economy has shown that there is huge potential for using materials more intensively, thereby reducing the amount of stuff passing through logistics systems. It is estimated that in Europe 'after the first use cycle, we recapture only 5 per cent of the raw material value' (Ellen MacArthur Foundation and McKinsey Center for Business and Environment, 2015). This is because an 'open linear material take–make–dispose [approach] still vastly dominates supply chain logistics' (Ellen MacArthur Foundation and McKinsey, 2014: 42). If the principles of the circular economy become widely applied across business and society over the next few decades, the level of logistical activity might be significantly reduced.

Digitization

Media products such as books, newspapers, magazines, software and entertainment are increasingly being distributed digitally rather than in physical form, effectively replacing freight consignments with electrons. Supply chains for the timber, pulp, paper, chemicals, plastics and dyes used in the manufacture of the corresponding physical products are relatively transport-intensive. Of the 400 million tonnes of pulp, paper and paperboard produced globally in 2014, 130 million tonnes were used for printing,[1] much of this tonnage moved over long distances between paper mills, printing works, warehouses, shops and homes. In many countries, the demand for this paper has been declining as demand switches from newspapers and magazines to digital media (PwC, 2016). E-books have also captured a large share of the book market. In the United States this share rose from 2 per cent in 2009 to 36 per cent in 2014, though this has since declined to 32 per cent in 2016 (Milliot, 2016). Worldwide the e-book share appears to have peaked, suggesting that in this sector the future potential for dematerialization may be limited. In the music sector, however, digitization continues apace. Between 2006 and 2015, the share of music sales in the United States in physical formats (CDs, music videos etc) dropped from 83 to 32 per cent (Sisario and Russell, 2016). In early research on the environmental impact of this trend, Hogg and Jackson (2008) were pessimistic. They argued that 'digital formats have not contributed to dematerialization due to the growth in hardware' and from this drew the more general conclusion that 'the trend towards digitization is not necessarily a driver for dematerialization in environmental terms' (p.142). Combined with the miniaturization of electronic hardware, this trend is likely to have reduced freight transport demand, though it is not known how many tonne-kms of freight movement have already disappeared as a result of product digitization or might be eliminated over the next few decades. As this trend is confined to a few sectors, its net effect on the total carbon footprint of logistics is likely to be relatively small.

Miniaturization

This has been a long-term trend in the computing, electrical equipment and telecommunications sectors, where products are shrinking while their functionality increases. For example, flat-screen televisions typically have 20 per cent of the cube and 25 per cent of the weight of the cathode ray tube (CRT) equivalents that they have replaced. The substitution of tablet computers for laptops and desktops and the replacement of several devices with a smartphone are similarly reducing the amount of material that needs to be transported. Product 'downsizing' is, nevertheless, confined to specific sectors and has been partly offset by 'upsizing' in others, most notably the automotive sector where SUVs have increased their share of the vehicle parc and are around 20 per cent heavier than the average car (Cheah, 2010). These are examples of products being fundamentally redesigned, often to incorporate a new technology. Designers can also make incremental changes to products to reduce their weight and/or volume while maintaining their functionality. The cumulative effect of such changes both to the products and associated packaging can significantly reduce freight traffic.

Material substitution

This involves the substitution of lighter materials, such as aluminium (average density 2.7 g/cm^3), for heavier ones, such as steel (average density 8.0 g/cm^3). 'Lightweighting' has become a hot topic in the automotive sector as vehicle manufacturers struggle to meet tightening fuel economy standards. It will be discussed in Chapter 6 as a means of reducing the carbon intensity of trucking. It is, however, a trend with wider implications for the decarbonization of logistics because any lightening of the loads carried in freight vehicles translates into lower fuel consumption and fewer emissions per km travelled. The replacement of heavier metals by lighter ones and of metals and wood by plastics has contributed to a long-term decline in the average density of freight. This has been reflected in the increasing proportion of truck loads 'cubing out' before they 'weigh out'. Very little data is available, however, to track the nature and extent of this trend and to forecast its future course. The switch to lighter metal alloys, plastics, carbon fibre and radically new lightweight materials, such as graphene, will partly depend on the narrowing of current material cost differentials. Even simple, low-cost innovations, however, can save significant amounts of freight-related CO_2. For example, replacing the 25-kg concrete blocks typically inserted into the 3.2 million washing machines sold each year to stabilize them during spinning, with plastic containers that can be filled with water at point of use, would cut around 44,600 tonnes of CO_2 from the distribution operation (Harrabin, 2017).

Additive manufacturing

This family of production technologies, of which 3D printing is the currently the most common, contributes to dematerialization in several ways. Being an additive rather than subtractive process it uses fewer materials and produces less waste than conventional methods of moulding and assembly (Campbell et al, 2011). It also economizes on material by producing hollow, lightweight structures (PwC, 2014). Reverse flows of unwanted, unsuitable or damaged product are minimized as on-demand manufacturing at point of use permits a close matching of product specifications to customer preferences (Kewill, 2013). So there is little dispute that additive manufacturing uses fewer materials than traditional forms of manufacturing to make a given set of products. But this set of products is unlikely to be fixed; additive manufacturing is extending product ranges. It is, for example, giving the 'maker movement', composed of individual inventors and designers, a new means of supplying their products directly to customers (Waller and Fawcett, 2014). The resulting product diversification, particularly into items that are inherently more highly customized and less easy to distribute through conventional supply chains, will expand the total quantity of goods produced and material required (McKinnon, 2016b). This need not result in much more freight movement, though, because the logistics intensity of additive manufacturing is relatively low.

Indeed it is through its logistical characteristics rather than its use of materials that additive manufacturing will have its greatest impact on future freight transport trends. By replacing complex multi-link supply chains with much simpler ones delivering only 3D printing materials to the point of production (and, in many cases, final demand) it could substantially reduce freight traffic. The reduction could be large if there were mass uptake of 3D printing at the consumer level and home production became the norm for a broad range of household products. As I discuss at length elsewhere (McKinnon, 2016b), this seems very unlikely for the foreseeable future.

What then are the prospects of additive manufacturing transforming production processes further up the supply chain? Until recently the prevailing view was that at a corporate level, additive manufacturing will find mainly niche applications, in sectors such as aerospace, automotive and medical equipment in the manufacture of prototypes and specialist components. If so, the net effect on logistics and freight traffic levels would be very modest. This view is based on the assumption that there are few economies of scale in additive manufacturing, certainly by comparison with conventional batch production. Janssen et al (2014: 11), for example, observed that 3D printing costs 'do not decrease significantly with an increase in scale' and because of this, 'mass production with additive manufacturing is not profitable now, nor expected to be in the near future'. While this may have been true of earlier layer-based forms of 3D printing, new types of additive manufacturing, such as 'digital light synthesis' being used by Adidas to mass-produce the soles of training shoes (*Economist*, 2017b), are making it economical for a widening range of products and businesses. So additive manufacturing may now make a more significant contribution to the decarbonization of logistics than previously thought.

Attempts to forecast the overall impact of additive manufacturing on freight transport volumes have been at best unconvincing. Most, such as those by Birtchnell et al (2013), ING (2017) and Boon and van Wee (2017), have elicited expert opinion on the matter but to limited effect. For example, an academic survey of experts by Boon and van Wee (2017: 13) on the likely effects of 3D printing on transport 'led to a scattered array of answers ranging from large positive impacts to large negative impact' with a majority of experts anticipating 'either moderate or no impact'. ING (2017) supplemented their consultation with a very crude modelling exercise underpinned by unrealistic assumptions about the uptake of 3D printing and its impact on world trade. Its suggestion (in its 'accelerated scenario') that 3D printing will be responsible for 50 per cent of all manufactured output by 2040, creating a reduction in global trade volumes by 40 per cent lacks credibility.

Postponement

The postponement principle governs the production and distribution of many products. Originally articulated almost 70 years ago by Alderson (2006), it states that companies should delay customizing products or dispersing them to regional markets until they have a good idea of the likely demand. If properly applied, the principle minimizes inventory and the risk of over-supply. There is also an important transport dimension to postponement which is often overlooked or underplayed in much of the literature on the subject. Delaying the final packaging and customization of products until they are near their final point of consumption or use can reduce product mass, and hence tonne-kms, across the intermediate links in the supply chain. This reduces the total quantity of freight moved along the chain, and is well exemplified by changes in the global distribution of wine over the past 20 years. Traditionally it was bottled near the point of production and shipped by the case. Today much of the Australian, Latin American and South Africa wine consumed in Europe and North America is shipped in bulk in containerized tanks and bottled within the import market. The proportion of this New World wine exported in bulk rose from 23 per cent in 2001 to 50 per cent in 2013 (Rabobank, 2012). This practice increases the amount of product per cubic metre of ship space by a factor of roughly 2.0–2.5. It has been estimated that 'shipping wine from Australia in bulk (to the UK) reduces CO_2 emissions by 164g for each 75cl bottle, or approximately 40 per cent when compared to bottling at source' (WRAP, 2011: 2). Many opportunities for transport-reducing postponement are already being exploited, but a more concerted effort to concentrate weight- and volume-adding activities at later stages in supply chains, combined with localized sourcing of the final 'ingredients', could further economize on transport.

Some of these dematerialization processes, such as the elimination of waste, are universal, while others, such as miniaturization and additive manufacturing, are likely to be confined to particular sectors. Some rely on future advances in technology while others depend more on changes in business practice and consumer behaviour. In a virtuous scenario, a combination of these processes could drive down the total amount of stuff in the global logistics system, making decarbonization more manageable.

Decoupling freight traffic growth and economic growth at the national level

One of the most enduring trends in transport economics has been the relationship between economic growth (measured by gross domestic product (GDP)) and freight traffic growth (measured in tonne-kms). Most countries have exhibited a close correlation between these variables over many years. This may seem to be intuitively obvious and merely reflect the expectation that as countries get wealthier they produce and consume more material goods which then have to be moved. Research has shown that the relationship between GDP and tonne-kms is actually much more complex than this, but there is no denying their close interdependence.

The closeness of this relationship may suggest that the only way of restraining future increases in freight movement will be to slow economic growth. This is something that few governments would seriously contemplate, given the strong and universal commitment to continuing economic development. One of the few countries to do so was Morocco. In 2008 it published a plan to 'reduce the number of tonnes per km transported in Morocco by 30 per cent to decrease carbon emissions by 35 per cent over a five-year period' (Oxford Business Group, 2011: 107). This was not simply a climate change policy, however; it also had an 'overarching objective... to accelerate GDP growth by increasing added value with reduced logistics costs'. At a global level, climate change policy is predicated on the assumption that economies will continue to grow, people will get wealthier and poverty will gradually be eradicated. As Stern (2015: 131) argues, 'responsible climate policy can combine with economic growth and the battle against poverty. This combination can generate a new and sustainable pattern of development and prosperity.' This new, climate-sensitive development may require some redefinition of the traditional link between freight transport and economic growth, but it can be assumed that this growth will continue.

The rate of economic growth may, of course, vary and deviate from the trends factored into freight transport growth models. For example, between 2015 and 2017, the International Transport Forum (2017b) revised downwards its assumptions about the long-term GDP growth rates underpinning its global freight modelling exercise. Given the cyclical nature of economic activity over differing timescales it is possible that over the next few decades the global economy will grow at rates well below the historic average or experience long periods of stagnation. Freight traffic levels would then be correspondingly lower than predicted.

One development which could both restrain economic growth and depress demand for freight transport would be a steep rise in oil prices. It has triggered global recessions in the past and typically had a double impact on the freight sector, reducing freight volumes at the same time as inflating the cost of one of the sector's main inputs, fuel. Given the past volatility of the oil price, commentators are generally reluctant to make long-term projections of its future trend. The International Energy Agency (2017b: 94) suggests that the 'oil market can find a longer-term equilibrium in the range of $50–70/barrel (in real terms)', in other words around the current price of $55–60/barrel. The World Bank (2017a) anticipates the price possibly rising to $80/barrel by 2030. The UK government's central forecast sees the oil price rising gradually to $80 (in 2016 prices) by 2030 and remaining stable for the rest of that decade. These projections are well short of the peak global oil price of $145 reached in 2008, and it is unlikely that there will be another oil price 'shock' in the foreseeable future. It is more likely that the real cost of oil will decline as renewables, nuclear and shale gas capture an increasing share of global energy consumption. Owners of fossil fuel reserves may be inclined to 'offload' them more rapidly in anticipation of further erosion of their asset value in a low-carbon world. Over-supply might then further depress the oil price. In the end, it may be a carbon price rather than the oil price which financially restrains the growth of freight traffic.

Freight transport demand would also decline if GDP, as conventionally measured, was deliberately managed downward, as advocated by members of the so-called 'degrowth' school. Schneider, Kallis and Martinez-Alier (2010: 512), for instance, envisage 'sustainable degrowth' involving 'an equitable downscaling of production and consumption that increases human well-being and enhances ecological conditions at the local and global level'. They argue that 'we cannot count on dematerialization after 20 years of limited progress in relative terms and no progress at all in absolute terms' (p.516). Victor (2012) constructed a 'degrowth scenario' for Canada up to 2035, which cut the country's GHG emissions by almost 80 per cent. Figures have actually been put on the maximum 'material footprint' per capita consistent with long-term sustainability. This 'footprint' measures the total annual amount of material resources consumed by an individual and calculated on a life-cycle basis. Lettenmeier, Liedtke and Rohn (2014: 489) 'propose an amount of eight tons per person in a year for household consumption and two tons per person in a year for public consumption (eg education, health care, and public administration).' A limit of 10 tonnes would be well below the average material footprint per capita in many countries in 2008, the last year for which we have data (Figure 3.1).

Wiedmann et al (2015) note that the material footprint 'of nations reflects the increasing complexity and multi-country nature of global supply

Figure 3.1 Domestic material consumption per capita in selected countries 2010

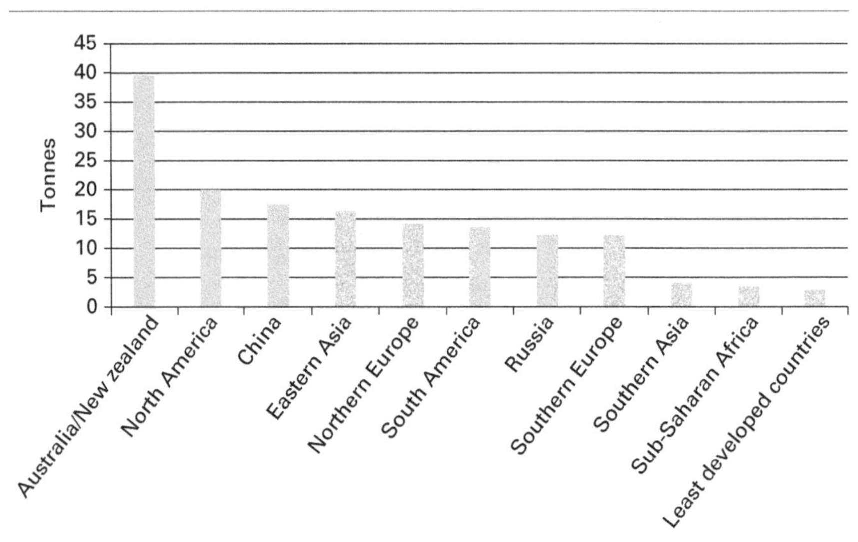

SOURCE UN Statistics Division (2017)

chains and is the appropriate indicator if the aim is to pinpoint the ultimate consumer responsibility of a country for impacts associated with raw material extractions worldwide' (p.5). A long-term effort to bring this footprint down to a more sustainable level would therefore have a major impact on global material flows. Degrowth and proposals to shrink material footprints, however, raise wider political issues which are beyond the scope of a book on logistics. For the purposes of this discussion, we will assume that economic growth continues and that any reduction in total freight transport demand will stem mainly from a decoupling of GDP and tonne-km trends.

Tapio (2005) examined the different forms of GDP/tonne-km decoupling and classified them into eight categories. If one assumes that GDP continues to grow, the main distinction is between a 'positive' decoupling where tonne-km growth rate slackens relative to that of GDP and 'negative' decoupling where it increases at a faster rate. The nature and extent of decoupling has been analysed in many countries such as Finland (Tapio, 2005), the UK (McKinnon, 2007b), Denmark (Kveiborg and Fosgerau, 2007) and Spain (Alises, Vassallo and Guzmán. 2014). In recent years, decoupling trends appear to have become more divergent. For example, over the period 2000–2015, freight tonne-kms across the EU as a whole grew almost exactly in line with GDP, but disaggregating the data by member state reveals that some countries, such as Bulgaria and Poland, generated more freight traffic per € billion while others, such as France and Ireland, experienced a

Figure 3.2 Percentage change in ratio of freight tonne-kms to GDP in a selection of EU Member States

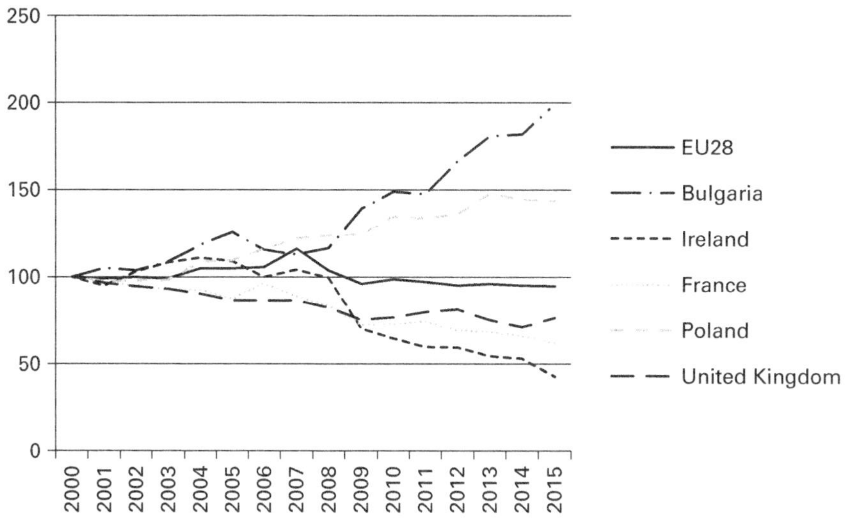

SOURCE Eurostat (2014)

pronounced reduction in their freight transport intensity (Eurostat, 2014) (Figure 3.2).

Those countries exhibiting the positive form of freight decoupling have shown how it is possible to enjoy economic growth without impairing economic performance. How then have they achieved this?

Several interrelated processes appear to have contributed to this decoupling:

Increasing share of national expenditure going on services rather than physical goods. As per capita income rises, people's consumption patterns change and they spend more on services and less on objects. Advanced economies also become more 'knowledge-based', specializing to a greater degree in high-end business services. Although difficult to quantify using available statistics, it is widely accepted that money spent on service activities generates less freight movement per $ billion than material consumption. Some commentators have extrapolated recent trends and suggested that in wealthier countries, demand for material goods will ultimately peak as consumers' physical demands are satisfied. The former head of sustainability in IKEA, for example, has hinted that we may be on the eve of 'peak home furnishings' (Farrell, 2016). Goldman Sachs (quoted by Gwyther, 2017) sees two trends possibly leading to 'peak stuff'. The first is the ascendancy of 'access over ownership', when people will more freely share assets rather than possess them. For example, the

International Transport Forum (2015b: 5) have estimated that 'nearly the same mobility can be delivered with 10 per cent of cars' using 'TaxiBots[2] combined with high-capacity public transport'. With the development of the 'share economy' people are showing greater willingness to share products, vehicles, accommodation etc and online platforms such as Airbnb and BlaBlaCar give them the means to do so. Widespread adoption of a sharing mentality, particularly by the younger generation of consumers, could significantly reduce the material content of modern life. The second trend is prioritization of 'experiences over possessions', assuming that the material intensity of these experiences is less than in the amount of stuff that would have been purchased for the same money.

While these trends are clearly evident in the developed world, there remains a huge, unsatisfied demand in lower-income countries for many of the basic material possessions now taken for granted in the West. At a global level, therefore, the concept of 'peak stuff' is rather fanciful.

Offshoring of manufacturing. The corollary of the transition to a service economy has been the transfer of production capacity to countries in the Far East, Latin America and Eastern Europe etc that have lower labour costs. When the production plants relocate to these countries, with them go their upstream supply chains, eliminating the inbound freight movements that used to bring in raw materials and components from several tiers of supplier. This has the effect of reducing the freight transport intensity of the offshoring country and increasing it the country to which the manufacturing and related supply chain activities migrate. In the territorially based Kyoto system of carbon accounting, the country offshoring the activities effectively offloads the freight-related emissions to the receiving nation. These emissions are not simply displaced geographically, however; they often increase overall as the less developed country is likely to run older, less fuel-efficient vehicles, and greater use is now made of long-haul transport to ship the finished products back to Europe or North America. So the negative decoupling of a single country's freight and GDP growth trends does not necessarily yield a net decarbonization benefit at a planetary level. This is part of a more general international transfer of responsibility for GHG emissions. Taking a production perspective on this transfer, Wei et al (2016: 2) estimate that, as a result of international trade, '3–9 per cent of responsibility for the increased atmospheric CO_2 concentration was shifted from the developed to the developing countries between 1990 and 2005.' Peters et al (2011) quantified the 'net emission transfers' on a consumption basis and found them increasing in the opposite direction, from developing to developed countries, from 0.4 Gt CO_2 in 1990 to 1.6 Gt in 2008. This revealed a growing 'spatial disconnect between the point of consumption and the emissions in production' (p.5).

Slackening of spatio-economic trends. Much freight traffic growth has been the result of goods being transported over greater distances. For example, between 1953 and 2015 the average length of freight hauls on the UK road network rose from 72 km to 110 km (Department for Transport, 2017d). On the Chinese road network the increase has been even more dramatic, up from 32 km in 1978 to 184 km in 2015, a lengthening of 4 km per annum (National Bureau of Statistics of China, 2016). Figure 3.3 illustrates the range of geographical processes that can lengthen the links in a supply chain (McKinnon, 1989b).

Figure 3.3 Spatial processes causing a lengthening of freight hauls

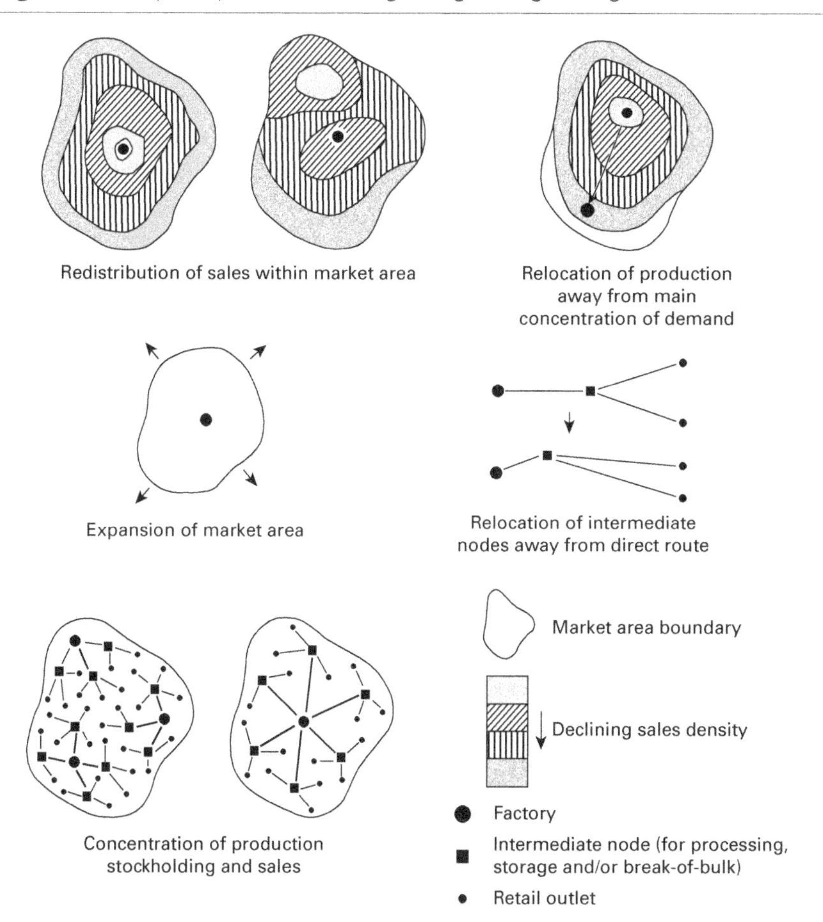

SOURCE McKinnon (1989b)

The two main drivers of freight traffic growth have been the expansion of market areas and the geographical concentration of business activity, two trends that are intrinsic to economic development (McKinnon and Woodburn, 1996). Market expansion has allowed companies to scale up production to more efficient levels and search more widely for suppliers and distributors. Centralization has also enabled companies to exploit economies of scale in production, storage and handling operations and to achieve deep reductions in inventory levels. Concentrating inventory in fewer locations allows companies to take advantage of the so-called 'square root law' and achieve substantial reductions in stock levels while maintaining the level of customer service (Maister, 1976; Oeser and Romano, 2016).

Other things being equal, centralization extends the average distance between the stockholding point and the customer, increasing the amount of freight movement. An analysis by a large Swedish manufacturer of the transport impact of inventory centralization at a European level found that reducing the number of stockholding points from eight to one increased tonne-kms by 34 per cent and CO_2 emissions by 42 per cent (Kohn, 2005). The company was, nevertheless, able to lessen these traffic and CO_2 impacts by increasing the degree of load consolidation across the centralized network, making more use of rail and cutting the number of emergency deliveries (Kohn and Huge-Brodin, 2008). This confirmed the conclusion of McKinnon and Woodburn (1994) that greater load consolidation within more centralized logistics systems can reduce the carbon intensity of the longer freight movements. From a transport standpoint, this can mitigate the negative environmental effects of centralizing logistics operations.

The spatial processes causing freight hauls to lengthen cannot continue indefinitely as centralization and the elongation of supply lines must ultimately reach their maximum extent. Within many developed countries the processes appear to be at an advanced stage, causing the rate of freight traffic growth to slow or even reverse. In the UK, for example, average length of haul peaked in 2000 at 123 km and has since declined by 11 per cent. At a continental level, however, the processes still appear to have some way to run. A large sample of logistics executives surveyed in 2013 predicted that the percentage of distribution centres in Europe serving the continental market would rise from 45 to 64 per cent, while the proportion supplying goods only within national boundaries would drop from 20 to 10 per cent. This enlargement of warehouse catchment areas suggests that the logistical integration of the EU is still well underway and partly explains why the amount of freight movement in the EU is projected to grow by 57 per cent between 2010 and 2050.

The ratio of freight traffic growth to economic growth remains high in low-income countries and shows little sign of decoupling. According to the International Transport Forum (2017b) it is 44 per cent higher in countries with a per capita income of less than $4,000 than in countries where the average income is over $40,000. The average value for this ratio in lower-income countries (1.18) indicates that freight tonne-kms are growing about 20 per cent faster than GDP. This is of particular significance because, as noted in Chapter 1, much of the forecast growth in freight traffic over the next few decades will be in countries at an earlier stage in their development.

Rising freight transport intensity is one of several trends in lower-income countries that combine to increase logistics emissions at a faster rate than economic growth. Figure 3.4 maps the interrelationship between these trends and shows how they contribute to higher GHG emissions.

As an economy develops and per capita income rises, patterns of consumption change, with people acquiring more consumer durables and trading up to higher-value products. Some of the new demand for processed food and manufactured goods is satisfied domestically as the country industrializes and reduces its dependence on the primary production of raw materials. These changes in economic structure and consumer demand alter the mix of commodities channelled through the country's logistics systems.

Figure 3.4 Mapping economic development – freight transport CO_2 relationship

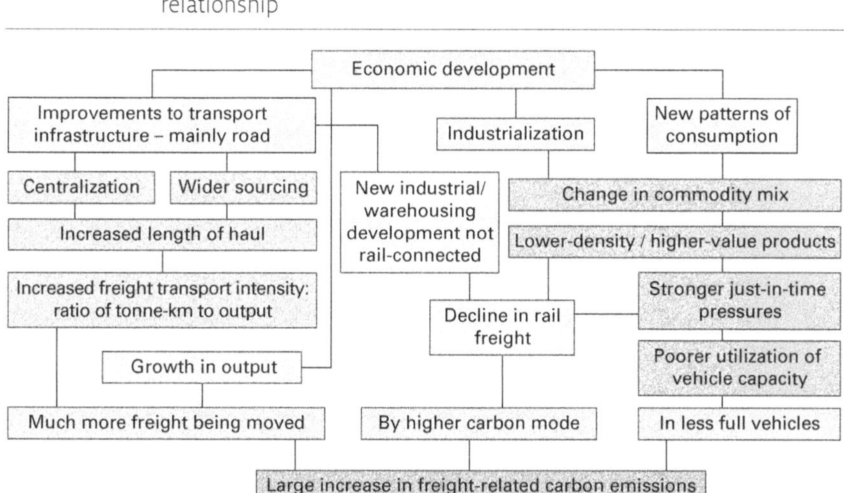

The proportion of higher-value, lower-density product moved, stored and handled steadily increases. Given the higher value of this inventory and rising customer delivery expectations, logistical operations become more tightly scheduled and experience stronger 'just-in-time' (JIT) pressures. As observed in developed countries, this can lead to companies prioritizing inventory reductions and rapid service over the efficient utilization of vehicle capacity, causing average vehicle load factors to drop.

Meanwhile, the upgrading of transport infrastructure induces a series of logistical changes. It makes it easier and cheaper for companies to serve wider areas from more centralized facilities. As discussed earlier, this promotes the concentration of production and inventory and enlargement of market areas which in turn lengthens freight hauls and increases transport intensity. Each increment in GDP then generates an even larger increment in tonne-kms (ie negative decoupling). The division of these tonne-kms among transport modes also changes because infrastructural spending tends to favour the road network over the railways and waterways. This strengthens road's position as the dominant transport mode and causes companies to locate new investment in factories and warehouses at points of high accessibility on the highway network. This erodes freight traffic from lower-carbon rail and waterborne services and causes longer-term logistical 'lock-in' to the use of trucking.

Expressed in rather crude terms, the combined effect of these trends is much greater movement of freight by a more carbon-intensive transport mode (trucking) in less full vehicles. On the basis of historical experience, the economic development process may therefore seem to be hardwired to inflate GHG emissions from the logistics sector. This need not be the case, however. Logistics in developing countries may be able to deviate from the evolutionary path it has followed in Western countries to chart a lower-carbon course. For this to happen, logistical decision making in developing countries will need to be shaped by government policies which, among other things, ensure more modally balanced investment in transport infrastructure and use taxation and road pricing systems to incentivize high vehicle loading and energy-efficient operation. These modal split and asset utilization issues are discussed in greater detail in Chapters 4 and 5.

Future trends in international trade and transport

At a national level the focus is on the binary relationship between domestic tonne-kms and GDP. In the case of cross-border freight traffic, there is a three-way relationship between global/regional GDP, international trade and freight tonne-kms. This can be decomposed into two ratios: the ratio of trade to GDP (both monetary variables) known as the 'trade multiplier' and the ratio of trade, expressed in monetary terms, to freight volumes, a physical measure we shall call the 'trade freight multiplier'.

Trade multiplier

The ratio of trade to GDP has been subject to wide fluctuations over the past 60 years. In the 1960s, it averaged around 1.5, meaning that the value of trade grew 50 per cent faster than GDP; between 1975 and 1985 the two variables increased more or less in parallel; and from 1985 to 2000, as the process of globalization gathered pace, the ratio rose to between 2.5 and 3.0, falling back to around 2.0 in the early years of this century (Hoekman, 2015). Since the 2007–08 global recession, the trade multiplier has averaged around 1.5. The volatility of this multiplier over the past few decades makes it very difficult to predict its course to 2050 and beyond. Numerous studies (eg Constantinescu, Matoo and Ruta, 2015; ECB, 2016) have tried to determine if a trade–GDP elasticity in the range of 1–1.5 is likely to be the 'new normal' or if there is a prospect of it returning to values twice as high. Most suggest that this is not simply a temporary aberration caused by the deep 2007–08 recession that is likely to be corrected within a few years. They identify several factors likely to cause a longer-term slowdown in international trade growth:

- Much of the increase in global GDP has been occurring in emerging markets which typically have a lower propensity to trade than advanced economies. This gap may narrow but probably not in the short term.
- The process of trade liberalization has slowed and the number of new restrictions on trade has multiplied. The Doha round of multilateral trade negotiations, launched in 2001, made little progress and was largely abandoned in 2016. The number of bilateral and plurilateral trade deals has also been reducing. Among G20 countries, 1,583 trade-restrictive measures were imposed between 2008 and mid-2016, around 1,200 of which are still in place (WTO, 2016). Hopes that the Trans-Pacific Partnership (TPP) trade deal, involving 12 countries and covering 40 per cent of global trade, would revive the growth in trade volumes were dashed in 2017 by the Trump administration's decision to withdraw the United States from the partnership agreement.
- The fragmentation of global value chains, which was very pronounced between 1990 and 2010 and a major driver of trade growth, appears to

be stabilizing and even showing signs of reversal. The 'offshoring' process resulted in value being added more incrementally in more production locations, causing a surge in the amount of 'intermediate trade' and related freight movement (Gereffi et al, 2001). Between 1990 and 2010 there was a 430 per cent increase in the value of 'intermediate trade' as global supply chains became increasingly complex. As the economies of emerging markets mature and their production capabilities diversify they can internalize more value-adding processes thereby reducing the need for intermediate trade and cross-border movement. It is also possible that some manufacturing capacity will 'reshore' from low-labour-cost countries to North America or Europe as cost differentials narrow, production is automated and political pressure to repatriate industrial activity mounts. This trend is discussed more fully below.

Much of the decline in the trade multiplier is likely to be attributable to longer-term structural processes 'running their course'. We may, therefore, be witnessing a permanent reduction in this ratio, bringing it closer to parity over the next few decades. This is reflected in the conclusion of the ECB (2016: 5) that 'the upside potential for trade over the medium term appears to be limited' and the recalibration of the OECD/ITF's global freight model with trade–GDP elasticity values in the range of 1.2 to 1.4.

Another possible explanation for the decline in this elasticity is that the long-term reduction in the real cost of international transport may be levelling off. Between 1930 and 2000, average ocean freight and port charges fell by two-thirds in real terms (Eddington, 2006), strongly promoting trade growth. ECB (2016: 10) cites 'lower transportation costs' as a 'factor that had previously contributed to global trade outpacing global output growth' but now provided 'waning support'. This is not borne out by the statistical evidence, however. The Baltic Dry Index, which measures the average cost of bulk shipping, peaked at 11,793 in 2008 before collapsing to 290 in 2012 and has been fluctuating around 1,200 since then (Rodrigue, Comtois and Slack, 2017). According to Drewry Consultants,[3] the average freight rate for shipping a 40-foot container dropped from almost $2,500 at the start of 2012 to $1,000 at the end of 2015. Their 'East-West Air Freight Price Index' dropped by roughly 12 per cent between May 2012 and August 2016. Both the ocean shipping and air cargo markets have had substantial excess capacity over the past decade and this has driven down freight rates. So the trade–GDP elasticity has declined despite the fact that moving goods internationally has become significantly cheaper. These low freight rates are unsustainable in the longer term, however, in both business and environmental terms. International transport costs are likely to rise as a consequence of a long-overdue rationalization of the shipping and air freight sectors, the imposition of market-based climate mitigation measures, and possibly rising energy prices. This could retard future trade growth.

The inhibiting effect of increasing transport costs will be partly offset by trade facilitation initiatives which 'expedite the movement, release and clearance of goods, including goods in transit' at international frontiers (WTO, 2017). The 2013 Bali Trade Facilitation Agreement, which is believed to have the potential to generate an extra $1 trillion worth of international trade, came into force in February 2017, when two-thirds of the World Trade Organization's member countries ratified it. It is estimated, for example, that it will boost global air freight by between 1.4 and 4.4 per cent in value terms (IATA, 2014a). With trade liberalization efforts stalling, the main institutional impetus for future trade growth may come from a reduction in the amount of 'logistical friction' at national borders.

Trade freight multiplier

The decline in the trade–GDP ratio over the past 20 years has been mirrored by a reduction in the ratio of container shipping volumes (measured in TEUs) to GDP. This ratio declined from an average of 4.2 in the 1990s to 3.2 between 2000 and 2010 and to 'essentially zero' in 2016 when container volumes stagnated (Rodrigue, Comtois and Slack, 2017). The trade freight multiplier is also sensitive to changes in the average value density of international trade measured in $/tonne. This average value density may vary over the next few decades as a result of interrelated changes in the spatial pattern of trade and mix of commodities that are internationally traded.

Changing geography of international trade

The financial transaction between two states, which registers as a monetary trade flow, gives no indication of their distance apart or the amount of freight movement generated. Scenarios constructed by the global freight model of the International Transport Forum (2017b) suggest that the 'average hauling distance' for international trade will lengthen by 12–15 per cent between 2015 and 2050, though no explanation is given of the spatio-economic processes likely to cause this increase.

There are, however, circumstances under which the 'average hauling distance' would shorten, thereby reducing the ratio of tonne-kms to trade. This could result from companies sourcing more of their products from suppliers in less distant countries, in other words opting for so-called 'nearshoring' strategies. This reversal of the long-term globalization trend is already occurring in some sectors, most notably that part of the clothing business known as 'fast fashion'. Companies such as H&M and Zara have been introducing new 'collections' at an increasing rate and rushing them into the market to gain competitive advantage. This has caused many retailers in this sector to switch much of the sourcing for their European shops from the Far East to countries such as Turkey and Morocco (Fernie and Grant, 2015). Future erosion of the labour cost advantage of the emerging markets combined with an increase in transport costs for the reasons mentioned earlier could compress the international supply chains of a broader range of products. Re-localization of in-house or subcontracted production capacity would have a similar effect. Already some global corporations, such as Unilever and P&G, regionalize much of their production thereby minimizing their dependence on transcontinental transport services, more effectively customizing products to local demand and reducing their supply chain reaction times.

The 'reshoring' of manufacturing operations from low-labour-cost countries to North America and Europe would also shorten supply chains. There has been much debate in recent years about the extent to which this is happening and likely to happen in the future. In 2011 the Boston Consulting Group (2011: 5) predicted:

> Within five years the total cost of production for many products will be only about 10–15 per cent less in Chinese coastal cities than in some parts of the United States... factor in shipping, inventory costs and other considerations, and the cost gap between sourcing in China and manufacturing in the US will be minimal.

In a more detailed economic analysis sponsored by BCG, Sirkin, Rose and Zinzer (2012) argued strongly that conditions were ripe for an 'American manufacturing comeback'. PwC (2013: 5) also reckoned that reshoring was 'becoming increasingly viable due to factors that have emerged since the onset of offshoring'. In surveys of senior manufacturing executives, the Boston Consulting Group (2015b) found that the proportion of respondents 'actively reshoring' rose from 7 per cent in 2012 to 17 per cent in 2015, with a desire to 'shorten the supply chain' identified as the main reason for doing so.

In contrast, another US consultancy firm, A T Kearney (2015: 1), concluded that 'the reshoring phenomenon, once viewed by many as the leading edge of a decisive shift in global manufacturing, may actually have been just a one-off aberration.' To back up this claim it devised a US Reshoring Index[4] and tracked it over the period 2004 to 2015. In only one year (2011) was this Index positive, suggesting net reshoring. Between 2011 and 2015, net offshoring became more pronounced. The European Parliamentary Research Service (2014: 1) could also find 'little evidence of reshoring from China' to the EU and concluded that 'significant mass returns of the manufacturing jobs that left developed countries from the 1980s onwards appear unlikely.' In a review of academic work on the subject, Wiesmann et al (2016: 36) found that 'the current research view seems to be that reshoring will not lead to a "re-industrialization" of Western economies'.

With opinions so sharply divided on this issue, it is difficult to judge the extent to which manufacturing capacity is likely to relocalize. PwC (2013: 5) advises companies to conduct a 'deep, disciplined analysis' and 'evaluate a combination of reshoring, onshoring, nearshoring and offshoring in order to right-shore and optimize supply chain networks.' While this sounds a rather bland statement, it does recognize the intricacy of this issue. Much of the discussion on reshoring underestimates the complexity of global value chains and is preoccupied with the location of final assembly plants. It is possible that more of the value in finished products will be added locally, particularly as the degree of automation increases (A T Kearney, 2017), but many of the upstream supply chains through which components and sub-assemblies are sourced will continue to span the globe.

These examples of globalization being reversed have a business rather than environmental motivation. Environmental pressure groups have been campaigning for a more general return to localized sourcing to cut transport energy use and emissions. Much of their attention has been directed at the lengthening of food supply lines with the accusation that many unnecessary 'food miles' are being travelled by products that could be obtained locally. This debate can be traced back to the famous 'well-travelled yoghurt pot' study by Böge (1995), which showed just how transport-, energy- and carbon-intensive the production and distribution of even a very basic foodstuff could be. Since then the subject has been extensively researched (eg Smith et al, 2005; Coley, Howard and Winter, 2011; Garnett, 2015) and expanded to include other categories of product, such as T-shirts (Rivoli, 2005). Pursued to a logical extreme, the argument for localized sourcing would undermine the whole basis of international trade, curtail economic development in poorer countries and deny consumers everywhere the product diversity, quality and value they have come to expect. The argument therefore needs to be bounded by certain assumptions about the economic, social and political sacrifices that would have to be made to secure the environmental benefits of localization. This takes it beyond the realms of logistics into a wider discussion of equity and welfare issues and the overall sustainability of consumption patterns and lifestyles.

There is one critical point that must be stressed as it has distinct relevance to logistics. This is the observation that minimizing the distance a product is transported does not necessarily minimize its overall carbon footprint when measured on a life-cycle basis. Transport usually accounts for a small fraction of all the embodied GHG emissions in a traded product, even in the case of products transported long distances. A much larger share of these emissions is associated with production operations and the carbon intensity of these operations can vary enormously from country to country. In the case of agricultural produce, this intensity varies with climate, soil type, farming practices, fertilization and storage methods. For all traded products, the average carbon content of electricity and the efficiency with which energy is used are key determinants of the amount of embodied GHG in a product. Figure 3.5 shows how widely the amount of GHG emitted per kWh of electricity varies internationally and the differing rates at which it has been declining in recent years.

In trying to minimize embodied GHG emissions in the products we consume it is generally more important to source them from locations where production-related emissions are low, even if this entails transporting them a long way.

This was confirmed for the international trade in food products by Smith et al (2005: ii) who concluded that 'the impacts of food transport are complex and involve many trade-offs between different factors. A single

Figure 3.5 Carbon intensity of electricity (gCO_2/kWh) in selected countries: 2010 and 2015

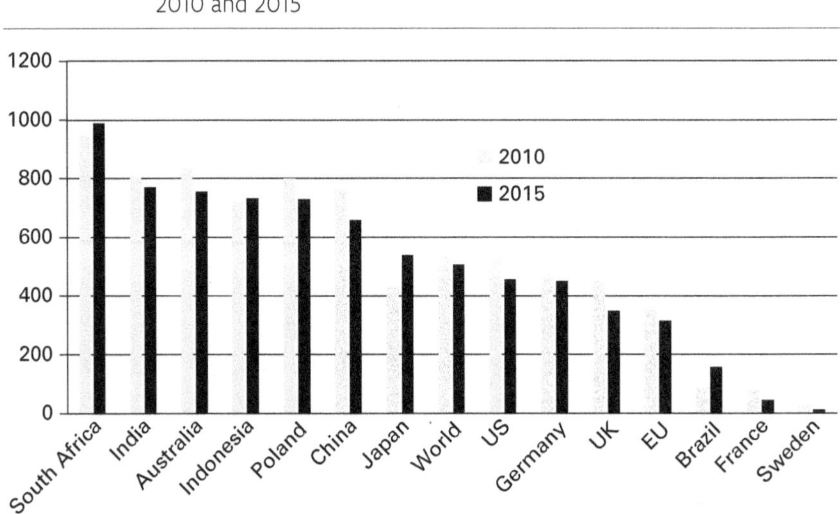

SOURCE International Energy Agency (2017b)

indicator based on total food kilometres travelled would not be a valid indicator of sustainability.' This was illustrated by a comparative analysis of the life-cycle emissions of agricultural products consumed in the UK but sourced from either the UK or New Zealand. The amount of CO_2 emitted per tonne of dairy produce, lamb and apples supplied to the UK market was, respectively, 52 per cent, 75 per cent and 30 per cent lower when sourced from New Zealand, despite the 18,000-km journey by refrigerated deep-sea container (Saunders and Barber, 2007). This reflected the 'less intensive production system in NZ than the UK, with lower inputs including energy' (p.vii). Edwards-Jones et al (2008), however, review several life-cycle analyses (LCAs) of agricultural produce sourced from different locations and find some of the evidence conflicting. They contend that these 'contradictory results emphasize the need to utilize similar system boundaries and methodologies when making comparisons between different food systems' (p.267).

On the basis of a partial equilibrium analysis of trade, input–output and environmental data, Cristea et al (2013) suggest that around 31 per cent of international trade (by value) yields a net reduction in carbon emissions because production-related emissions in the exporting country are lower than those in the importing country by a sufficiently large margin to offset the transport-related emissions. The remaining trade was responsible on average for an additional 158 g of CO_2 per $ of value. For many traded products, therefore, the pursuit of a localization strategy would be counter-productive in terms of climate change mitigation because it would inflate rather than reduce total emissions. Only by conducting a rigorous and consistent LCA can one determine, for any given product, what pattern of sourcing truly minimizes carbon emissions.

Changing commodity mix

The trade freight multiplier is sensitive to the value density of the traded commodities, ie their monetary value per tonne. An increase in the average value density of trade reduces the amount of freight movement (in tonne-kms) per $ billion of trade. This can occur as a result of several processes. For example, in the course of their economic development, countries exporting raw materials can acquire the capability to process them, thus increasing their value prior to sale. Many developing countries are currently 'moving up the value chain' in this way (Dicken, 2015). Another mega-trend likely to affect average value density is the contraction of the global market in fossil fuels. In 2016, the G7 nations committed themselves to phasing out the use of fossil fuels by the end of this century. Many countries will abandon fossil fuels sooner than this; Sweden, for example, plans to be fossil-fuel-free by 2040. The contraction of the global fossil fuel market will radically alter the composition of international trade in both monetary and physical terms. In 2016, approximately 4.2 billion tonnes of coal, oil and gas were moved internationally by sea, accounting for 41 per cent of all seaborne international trade (UNCTAD, 2017). By 2050 much of this maritime traffic will have been eliminated by the substitution of renewable and nuclear energy for fossil energy, coupled with major improvements in energy efficiency. In the intervening period, many countries will have replaced foreign-sourced oil and coal with domestically produced gas, some of it fracked and used as a lower-carbon 'bridging fuel' during the transition to zero-emission energy.

While the amounts of fossil fuel moved internationally will decline, switching electrical energy-generating capacity from fossil fuels to renewables and nuclear power will require the movement of vast amounts of material, much of it sourced over long distances, as the supply chains for wind turbines, solar panels and batteries are worldwide. Analysis of the material requirements for a wind turbine illustrates how much freight is likely to be generated. A three-megawatt (3MW) turbine requires an average of 1,140 tonnes of concrete in its base and 276 tonnes of steel (Crawford, 2009). Realizing the International Energy Agency's vision of wind power generating up to 18 per cent of global electricity by 2050 (International Energy Agency, 2013) would require the installation of 2,370 GW of wind turbine capacity, equivalent to 790,000 3MW wind turbines. Constructing and installing this number of turbines would require the movement of 900 million tonnes of concrete and 218 million tonnes of steel. To put these figures into perspective, they would represent, respectively, around 20 per cent and 15 per cent of total annual global demand for these products, though this demand and related freight movement would be spread over many years. The creation of a renewable energy infrastructure is causing a temporary surge in the amount of freight movement. This is well justified in transport terms, however, because once in place the new infrastructure will reduce, and ultimately eliminate, the need to move fossil fuels. It has been estimated, for example, that the installation of 180 GW of renewable energy capacity in Europe (enough to power 107 million households) replaces the refining and distribution of roughly 228 million tonnes of oil (European Wind Energy Association, 2008).

The logistics of adapting to climate change

Another future source of freight traffic, which has not yet been factored into the major freight forecasting models, is the adaptation of our built environment to climate change. Protecting infrastructure and settlements against extreme weather events and rising sea level will require the movement of large quantities of construction material. No attempt has been made to quantify the total amount of logistical activity this will generate, though the following examples give an indication of the scale of the task:

Coastal protection for the world's largest ports. Becker et al (2016) have estimated that to provide coastal defence for 221 large ports against a two-metre rise in sea level, 3,600 kms of coastal structures would be required comprising 148 million m^3 of cement, 125 million m^3 of quarry stone, 110 million m^3 of sand and 76 million m^3 of other materials. To put this volume of material into perspective it is equivalent to roughly 20 per cent of the carrying capacity of all the shipping containers in the world. Becker et al's calculations exclude roughly 3,100 other ports around the world. They conclude that 'adaptation will be a monumental task and will significantly tax global resource capacity.' It will become a gargantuan task when allowance is made for other forms of transport, energy and communication infrastructure and other elements of the built environment exposed to climate risk.

Displacement and resettlement of population. The World Bank (2017c) estimates that approximately 3 per cent of the global population, currently around 200 million people, live in areas less than five metres above the sea and therefore at high risk from a rise in sea level. Much of this population will have to be relocated to new settlements and building these settlements will be very transport-intensive. The construction sector already accounts for a substantial amount of freight movement: for example, it is responsible for almost a quarter of road freight tonne-kms in the UK (Department for Transport, 2017c). Using a recent assessment of the 'material footprint of the planet' (Zalasiewicz et al, 2017) it is possible to get an order-of-magnitude indication of how much material may have to be moved to rehouse populations displaced from low-lying land. It is estimated that the total weight of buildings in urban and rural areas is around 17.4 trillion tonnes. If just 0.1 per cent of this stock of buildings had to be abandoned and reconstructed elsewhere the weight of materials in these buildings would be equivalent to the total tonnage of freight moved in the EU in 2015 and this would not allow for any wastage of materials in the rebuilding process.

Installation of air conditioning. Increases in average temperature and the frequency of extreme heat will encourage citizens and businesses to install air conditioning. This growth in demand will be reinforced by rising incomes. It is estimated that the ownership of air conditioning equipment in warm climates goes up by 2.7 per cent for every $1,000 increase in household income (Davis and Gertler, 2015). The annual growth rate for sales of air conditioners is already 10–15 per cent per annum in countries such as India, Brazil and Indonesia. The Lawrence Berkeley National Laboratory (LBNL) has forecast that 700 million more air conditioning units will be installed worldwide between 2015 and 2030 (Mooney and Dennis, 2016). On the basis of a life-cycle analysis conducted by Shah, Debella and Ries (2008), each air conditioning unit installed in the United States generated an average of 122 tonne-kms of freight movement. If the air conditioning supply chains of other countries were similarly transport-intensive, admittedly a rather crude assumption, then the LBNL forecast would translate into 86 billion tonne-kms of freight movement, roughly half the total amount of freight transport in the UK in 2015 (Department for Transport, 2017d).

It is ironic that the movement of materials for climate proofing and adaptation will increase freight transport emissions and conflict with mitigation initiatives in the logistics sector. As the three examples above illustrate, adaptation has the potential to generate a huge amount of new logistical activity. This could substantially inflate the future growth of freight movement in pursuit of the very worthy goal of protecting communities from the ravages of extreme weather and sea level rise. If and when a global target is imposed on the absolute amount of GHG that can be emitted by the logistical activities, full account must be taken of the vital contribution they will make to human welfare in a climate-changed world. This is a topic that requires further investigation. As acknowledged by the IPCC in its fifth Assessment Report: 'Little research has so far been conducted on the inter-relationship between adaptation and mitigation strategies in the transport sector' (Sims et al, 2014: 622).

Optimizing the routeing of freight flows

If the amount of freight and the locations between which it is transported are fixed, it is still possible to reduce tonne-kms, vehicle-kms, fuel consumption and emissions by routeing the freight flows more efficiently. This can be done at different levels. At the upper levels of a supply chain, the pattern of bulk flow between raw material sources and different tiers of production can be optimized using various linear programming techniques. This involves solving the classic 'transportation problem', as originally defined by Hitchcock (1941), which tries to distribute product flows in a way that minimizes transport costs while staying within supply and demand constraints. Minimizing these costs will often simultaneously minimize distance travelled, fuel and emissions, though not necessarily so. If the minimization of CO_2 emissions is factored into the objective function, the pattern of flow may have to be re-optimized. This was confirmed by modelling undertaken by Elhedhli and Merrick (2012: 379) which showed 'the addition of carbon costs into the decision process for supply chain can change the optimal configuration of the network. The addition of carbon costs created a pull to reduce the amount of vehicle kilometres travelled.'

Optimizing the repositioning of empty containers within port hinterlands can also yield significant CO_2 savings. A survey of large UK shippers identified this as 'one of the largest potential sources of carbon savings in the maritime supply chain' (McKinnon, 2012b). After an inbound container has been 'de-stuffed' at an import location it is generally returned to a port. From there it will often be despatched to another inland location to collect an export load. Triangulation between the import and export locations, possibly via inland container holding points, can significantly rationalize empty container movements (Hjortnaes et al, 2017). Only 10 per cent of empty containers in the hinterlands of US ports are 'loaded with export cargo shortly after being unloaded of import cargo and without coming back empty first to the maritime terminal' (Rodrigue, Comtois and Slack, 2017). This is partly because most shipping lines are not prepared to exchange or pool their containers. If they were willing to commit some or all of their containers to a pool of non-liveried 'grey boxes' this organizational constraint would be eased (van Marle, 2012). This 'grey box' concept has been debated for many years and trialled on several occasions but proven not to be viable or scalable. To work effectively a pooling system also requires data sharing and a system for matching empty containers with export loads, both of which have been lacking until recently. Rodrigue, Comtois and Slack argue that what is needed is a 'virtual container yard... a "clearinghouse" where detailed information is made available to the involved actors.' Research by the Boston Consulting Group (Sanders et al, 2015: 3) also highlighted the need for 'a neutral, global clearing house for information about container demand and availability and the potential for interchanges at specific locations.' It estimated that helping to avoid 'all movements of empty containers for carrier-specific reasons would allow the industry to reduce carbon dioxide emissions by more than 6 million tonnes annually.' These carbon savings would accrue from reduced container movement globally onboard vessels as well as across port hinterlands. Having identified a major commercial opportunity in the rationalization of empty container repositioning, BCG set up and spun off an online container exchange called Xchange,[5] which is now partnering with many shipping lines and traders.

At lower levels in the supply chain, where much of the freight is distributed (or collected) in multiple-drop rounds, one encounters the vehicle routeing problem (VRP), a freight variant of the travelling salesman problem which has taxed the minds of mathematicians for several centuries (Dantzig and Ramser, 1959). The objective is to optimize the routeing of the vehicle(s) around a set of locations within various operational constraints, such as limited vehicle capacity, delivery time windows and driver hours restrictions. The inclusion of the time dimension into the optimization process creates a vehicle routeing and scheduling problem (VRSP). Optimization is usually defined in commercial terms primarily to reduce operating costs, delivery times and fleet size (Freight Best Practice Programme, 2005). Computers have been used to solve this problem for over 50 years, though it is only over the past 30 years that computerized vehicle routeing and scheduling (CVRS) has been extensively commercialized. NACFE (2016) reported that 95 per cent of the truck-kms run by 17 large North American fleets were subject to 'route optimization'. Companies replacing manual routeing with CVRS have typically reported transport cost savings of 5–10 per cent (Dekker, Bloemhof and Mallidis, 2013). In many cases, the distances their vehicle travelled would be reduced by a similar margin and with it the tonne-kms moved. As Eglise and Black (2015) point out, however, one cannot assume a linear relationship between the minimization of vehicle-kms and the minimization of emissions. After all, on some routes vehicles may be able to travel a higher proportion of kilometres at more fuel-efficient speeds. This raises the possibility that if CVRS software packages were modified to minimize CO_2 emissions rather than time, distance or cost, vehicle routeing could be used more effectively as a decarbonization tool. Several studies have tested this hypothesis over the past decade and broadly confirmed it (eg Palmer, 2007; Liu et al, 2014). Some have even suggested that emissions could be minimized at little or no cost penalty. Figliozzi (2010: 6), for example, found that 'in congested areas, it may be possible to reduce unhealthy or GHG emissions with a minimal or null increase in routeing costs.' Research in the expanding field of 'pollution routeing' has modelled the trade-offs that can be made between emission reductions and more traditional operational metrics, taking account of vehicle speed, traffic congestion and departure times (eg Bektaş and La Porte, 2011; Franceschetti et al, 2013). Although these remain largely theoretical exercises, they are preparing the ground mathematically for a time when minimizing CO_2 emissions becomes a transport management priority.

Giving emission reduction a stronger weighting in route optimization is only one way in which CVRS can strengthen its contribution to the decarbonization of freight transport. Although a relatively mature product, it is constantly evolving and taking advantage of developments in analytics, the use of Big Data, GPS and vehicle telematics. For example, in what has been called one of the 'largest commercial analytics projects ever undertaken', UPS has developed a vehicle routeing system called ORION (On-Road Integrated Optimization and Navigation) which uses 'prescriptive analytics' to achieve a step improvement in the efficiency with which its parcel delivery vans are routed (Davenport, 2016). For the company this has been a 'game-changer', with its 55,000 ORION-optimized routes saving annually 10 million gallons of fuel, 100,000 metric tons of CO_2 emissions and $300–$400 million in cost avoidance (UPS, 2016b). The use of geographical information systems (GIS) and real-time routeing data permits the dynamic re-routeing of vehicles while on the road to avoid congestion and this further helps to cut fuel consumption and emissions (International Energy Agency, 2017a). While in developed countries CVRS is becoming much more sophisticated, levels of adoption in much of the developing world remain low. Its diffusion across the large and rapidly expanding truck fleets of emerging markets such as China, India, Indonesia and Nigeria, where traffic congestion is rife could significantly cut the carbon intensity of trucking worldwide.

Summary

This chapter has reviewed a much wider range of freight transport demand management options than most previous work on the subject. It ended optimistically with a summary of current developments in the routeing of freight flows and transport equipment which can reduce both freight movement and carbon emissions at minimal or zero net cost. Attempting to reduce freight traffic levels by forcing a reversal of well-established spatio-economic processes like globalization and inventory centralization would have a much higher carbon mitigation cost and require radical public intervention. The chapter has also looked beyond what can be done internally within the logistics sector to suppress freight demand to external technological and socio-economic developments that will have a bearing on future freight traffic levels. Few of these developments are currently incorporated into long-term freight models, partly because they are very difficult to quantify but also because our understanding of their relationship with logistics is still rather sketchy. It is not possible, therefore, to say how all the various trends discussed in this chapter will collectively influence future freight traffic levels. In a low-carbon scenario, they could combine to dampen global demand for logistics, but alternatively they could reinforce the three-fold growth in freight tonne-kms predicted by the International Transport Forum (2017b), making deep decarbonization by 2050 an almost impossible task. Given these uncertainties, we must drive down the carbon intensity of logistics to minimize the CO_2 emitted by each additional tonne-km of freight movement. The next four chapters explore how this can be done.

Notes

1 The Paperless Project (2018) Facts about paper: the impact of consumption [online] http://www.thepaperlessproject.com/facts-about-paper-the-impact-of-consumption/ [accessed 6 February 2018].

2 TaxiBots are described as 'self-driving cars that can be shared simultaneously by several passengers' (International Transport Forum, 2015b).

3 Drewry (2018) Maritime research, consulting and financial advisory services [online] https://www.drewry.co.uk/ [accessed 6 February 2018].

4 This Index was calculated as follows: the annual value of manufactured goods imported into the United States from 14 Asian countries was divided by the US annual domestic gross output of manufactured goods to give a 'manufacturing import ratio'. The US Reshoring Index measures year-to-year changes in this ratio.

5 Container xChange (2018) [online] https://www.container-xchange.com/public/index.html [accessed 6 February 2018].

Shifting freight to lower-carbon transport modes

04

Transport modes[1] vary enormously in the amount of CO_2 they emit per unit of freight movement. An express air freight service can be 500 times more carbon-intensive than bulk delivery by electric train. An obvious decarbonization strategy is therefore to send as much freight as possible by those modes with the lowest carbon intensity. To use the jargon, this involves altering the 'freight modal split'. This option features prominently in the decarbonization frameworks outlined in Chapter 1 and in the policy statements of national governments. It was by far the most frequently mentioned means of cutting freight-related emissions in the Intended Nationally Determined Contributions (INDCs) made by governments to the COP21 Paris Accord. In the 13 per cent of 158 INDCs that explicitly referred to freight transport, freight modal shift accounted for just under half of all the policy initiatives (Gota, 2016) (Figure 4.1).

This is not surprising because for many decades, since long before climate change entered the political agenda, many politicians have seen modal shift as a panacea for the environmental ills of the transport sector. Government efforts to get more freight onto greener modes have had a rather disappointing history, however. This has not deterred the latest generation of policy makers from placing modal shift at the centre of decarbonization plans for the transport sector.

The potential for freight modal split nevertheless varies widely around the world. In most countries it involves shifting freight from road to rail and waterborne services. According to UN data, the road share of the freight market ranges from 96 per cent in North Africa to only 10 per cent in the Russian Federation, with other regional averages spread quite evenly between these extremes (Figure 4.2).

The possible contribution that modal shift can make to freight decarbonization will clearly be much higher in those regions with greater dependence on trucking.

Figure 4.1 Proportion of INDC documents mentioning freight-related decarbonization measures

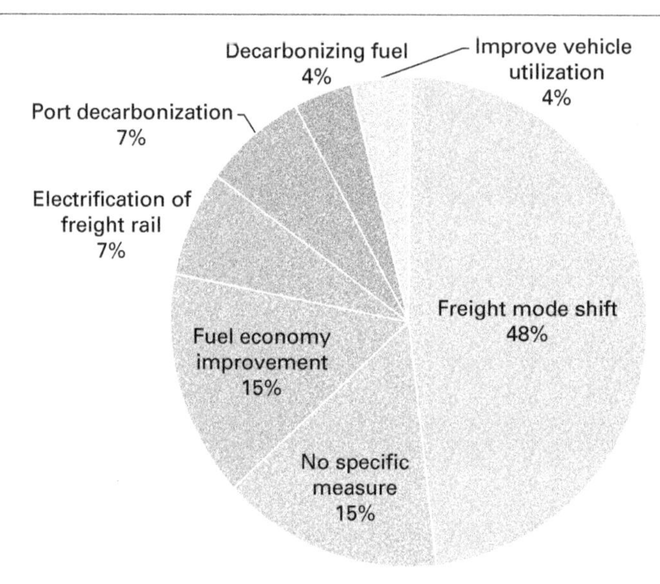

DATA SOURCE Gota (2016)

Figure 4.2 Inter-regional variations in the road share of combined road and rail tonne-kms (2015)

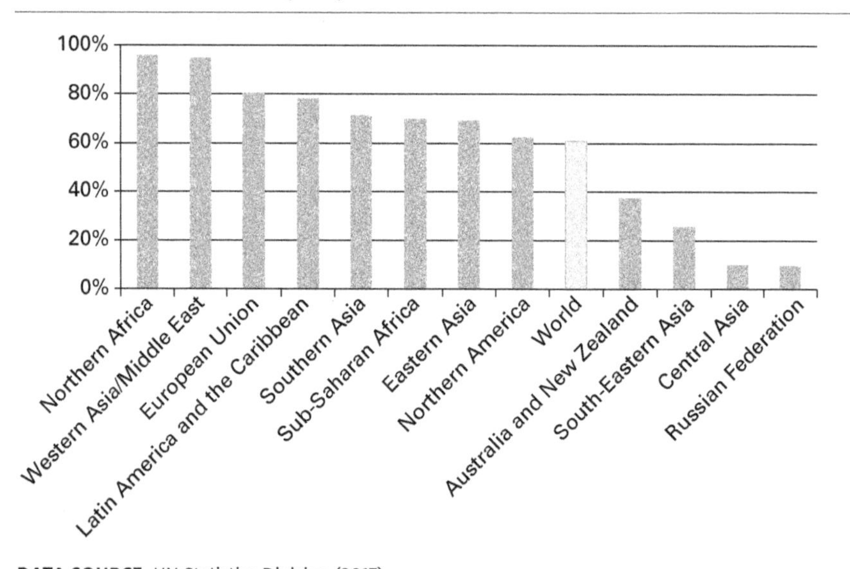

DATA SOURCE UN Statistics Division (2017)

The next section summarizes available data on the carbon intensity of different freight transport modes and offers advice on the interpretation of the figures. We then review trends in freight modal split in several parts of the world, showing how in most countries, lower-carbon modes, particularly rail, have actually been losing market share. The reasons for this erosion of rail's share are examined, as they indicate the scale of the challenge facing policy makers intent on reversing this trend. The concluding section considers how businesses might be encouraged to make use of lower-carbon modes.

This chapter will focus on the surface movement of freight within countries, the logistical arena in which the modal split issue is most pertinent. The prevailing view is that within the intercontinental freight market, the air cargo and maritime sectors are fairly discrete and opportunities for modal shift relatively limited. This situation may change as the development of new overland freight routes between the Far East and Europe create new modal options for current users of both air and shipping services. Some reference will therefore be made to the prospects for Asia-Europe modal shifts.

Quantifying the carbon savings from modal shift

Over the past decade governments, academic researchers, consultancy firms, trade bodies, NGOs and individual companies have published data sets that can be used to compare the carbon intensity of all the freight transport modes. Figure 4.3 has been compiled with data from one of the most widely respected and cited sources of carbon intensity values based on UK freight transport operations (DEFRA, 2017).

The global overview provided by Sims et al (2014) provides a range of values for particular modes. It suggests, for example, that the carbon

Figure 4.3 Average carbon intensity of freight transport modes: gCO_2/tonne-km

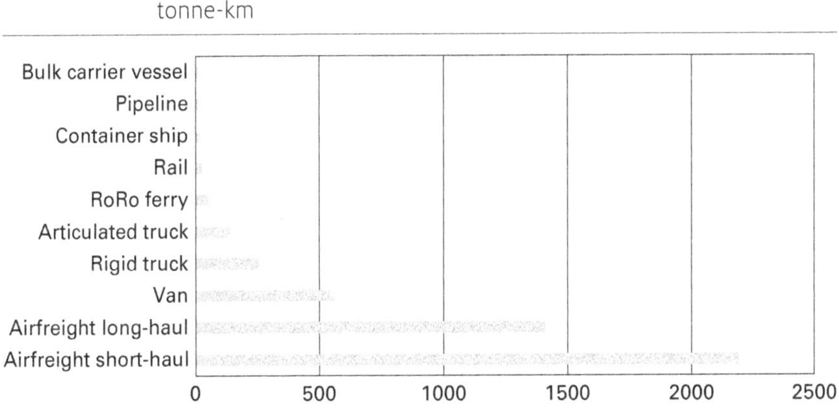

DATA SOURCE DEFRA (2017)

intensity of a large heavy-duty vehicle (HDV) lies between 70 and 190 gCO_2 per tonne-km and for a diesel-powered freight train between 25 and 60 gCO_2 per tonne-km. Combining low and high values within these ranges suggests that rail could at best have a 760 per cent carbon advantage over HDVs and at worst only a 14 per cent advantage. This variability makes it difficult to decide which emission factors one should use when analysing the potential carbon savings from modal shift at national, sectoral or company levels. The variations are usually the result of five factors:

1 *Sampling frame*: this can range from a single vehicle on a test track to fleets of thousands of vehicles or vessels in real-world operations. Clearly the larger and more cross-sectional the sample the more representative the average will be.

2 *Data sources*: an important distinction can be made between emission factors derived from lab-based testing and factors based on actual transport operations. A comparison of carbon intensity values for UK road freight revealed that the laboratory estimates were significantly higher (McKinnon and Piecyk, 2010). While this need not always be the case, it is preferable to use operational data where this can be accurately obtained for large and varied samples.

3 *Load factor assumptions*: the carbon intensity of a freight movement is highly sensitive to the loading of the vehicle. Table 4.1 shows how the average carbon intensity of moving freight in a 44-tonne truck can vary by a factor of four depending on the degree of empty running and consignment weight.

   Published carbon intensity data sets vary in the assumptions made about load factors. Sometimes where the organization releasing the data is keen to promote a particular mode it will assume higher utilization for its favoured mode. As little general data is collected and published on the loading of freight vehicles, it is difficult to know how average load factors vary between modes. As Gucwa and Schäfer (2013) explain, the load factor exerts a strong influence on the energy efficiency of freight transport modes. It is important, therefore, for tables of carbon intensity values to be accompanied by a list of the assumed modal load factors or, as in the case of emission factor tables compiled by DEFRA (2017), for separate emission factors to be provided for different degrees of loading.

4 *Country circumstances*: international variations in modal carbon intensity values can reflect differences in transport regulation, geography, infrastructure and market conditions. For example, in countries with

Table 4.1 Variations in the carbon intensity of a trucking operation with variable loading and empty running: gCO_2 per tonne-km for a UK truck with maximum weight of 44 tonnes and 350 bhp

Load	% of truck-kms run empty										
Tonnes	**0%**	**5%**	**10%**	**15%**	**20%**	**25%**	**30%**	**35%**	**40%**	**45%**	**50%**
10	81.0	84.7	88.8	93.4	98.5	104.4	111.1	118.8	127.8	138.4	151.1
11	74.8	78.2	81.9	86.1	90.8	96.1	102.1	109.1	117.3	127.0	138.6
12	69.7	72.8	76.2	80.0	84.3	89.2	94.7	101.1	108.6	117.5	128.1
13	65.4	68.2	71.4	74.9	78.9	83.4	88.5	94.4	101.3	109.5	119.3
14	61.7	64.4	67.3	70.6	74.2	78.4	83.2	88.7	95.1	102.7	111.8
15	58.6	61.0	63.8	66.8	70.3	74.2	78.6	83.7	89.7	96.8	105.3
16	55.9	58.2	60.7	63.6	66.8	70.5	74.6	79.5	85.1	91.7	99.7
17	53.5	55.7	58.1	60.8	63.8	67.2	71.2	75.7	81.0	87.2	94.7
18	51.4	53.5	55.8	58.3	61.2	64.4	68.1	72.4	77.4	83.3	90.4
19	49.6	51.5	53.7	56.1	58.8	61.9	65.4	69.5	74.2	79.8	86.5
20	48.0	49.8	51.9	54.2	56.8	59.7	63.0	66.9	71.4	76.7	83.0
21	46.6	48.3	50.3	52.5	54.9	57.7	60.9	64.5	68.8	73.9	80.0
22	45.3	47.0	48.8	50.9	53.3	55.9	59.0	62.5	66.5	71.4	77.2
23	44.2	45.8	47.6	49.6	51.8	54.3	57.2	60.6	64.5	69.1	74.7
24	43.2	44.7	46.4	48.3	50.5	52.9	55.7	58.9	62.7	67.1	72.4
25	42.3	43.8	45.4	47.3	49.3	51.7	54.3	57.4	61.0	65.2	70.3
26	41.5	42.9	44.5	46.3	48.3	50.5	53.1	56.0	59.5	63.6	68.5
27	40.8	42.2	43.7	45.4	47.3	49.5	52.0	54.8	58.1	62.1	66.8
28	40.2	41.5	43.0	44.6	46.5	48.6	51.0	53.7	56.9	60.7	65.3
29	39.7	41.0	42.4	44.0	45.7	47.8	50.1	52.7	55.8	59.5	63.9

SOURCE McKinnon and Piecyk (2010)

higher truck size and weight limits the carbon intensity of road freight operations is likely to be lower, while the intensity values for rail freight are lower in countries that can accommodate exceptionally long trains and the double-stacking of containers.

5 *Primary energy*: where a freight mode is partly powered by electricity, assumptions about the electricity/liquid fuel split and the carbon intensity of the electricity can be critical. This, for example, helps to explain wide international variations in CO_2 emissions per tonne-km for rail (Figure 4.4).

In the case of electrified freight services allowance must be made for the emission intensity of the primary energy source, whereas for freight operations powered directly by liquid fuel it is generally 'tank-to-wheel' (TTW) emission factors that are quoted. These factors exclude emissions from the energy supply chain extending back to the original source of the fuel in an oil well, field, forest etc. As transport modes switch to alternative fuels and electricity generation is decarbonized it will become increasingly important to compare the carbon intensity of modes on a well-to-wheel (WTW) basis. At present, well-to-tank (WTT) emissions for oil-powered vehicles, vessels and trains represent a similar proportion to WTW emissions (Table 4.2), but these differentials will widen through time as modal energy mixes change.

Figure 4.4 Average carbon intensity of rail freight operations in selected countries and regions in 2005 and 2015: gCO_2 per tonne-km

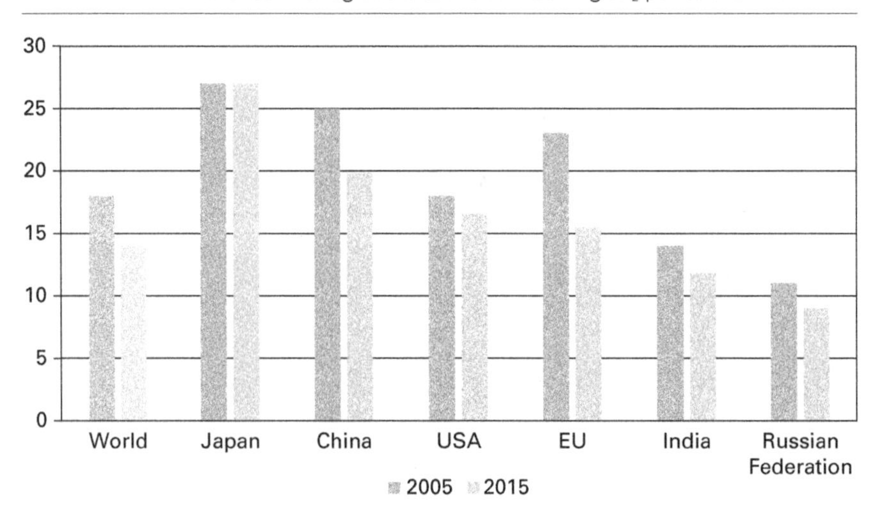

SOURCE International Energy Agency (2017b)

Table 4.2 UK Well-to-Tank and Well-to-Wheel emissions factors for differing freight modes: gCO_{2e} /tonne-km

	Well to tank (WTT)	Well to wheel (WTW)	WTT %
Diesel van	131	546	24%
Petrol van	233	854	27%
Battery van	64	398	16%
Rigid truck	47	197	24%
Articulated truck	19	79	24%
Freight train	8	34	24%
Container ship	3	16	19%
Bulk carrier	1	4	25%
Short-haul air cargo	219	1998	11%
Long-haul air cargo	160	1457	11%

SOURCE DEFRA (2017)

As discussed in Chapter 2, migrating from a TTW to a WTW basis enlarges the system boundary of the carbon calculation. As carbon auditing capabilities improve, this boundary is likely to be further extended to include emissions from the manufacture and maintenance of the vehicles, the construction and maintenance of infrastructure and related IT and administration. Only when these other activities are factored into the carbon intensity calculation is it possible to make a truly comprehensive assessment of the potential GHG savings from reallocating freight traffic between modes. Van Lier and Macharis (2014), for example, made a full life-cycle assessment (LCA) of the movement of freight by barge in Flanders, including CO_2 emissions from the construction and maintenance of waterway infrastructure and the manufacture and maintenance of barges. Nahlik et al (2015) conducted a cross-modal LCA of freight movements to, from and within California to compare GHG reduction options. Holistic assessments of this type are particularly important where the proposed modal shift is large enough to require an expansion of infrastructural capacity (Facanha and Horvath, 2006). Research by van Essen and Martino (2011), for example, suggested that fulfilment of the EU's ambitious 2050 modal split targets for long-distance rail freight and passenger traffic would require a 65–82 per cent increase in the capacity of European rail infrastructure. The resulting construction work can carry a large GHG penalty and even when spread over the planned life of the project may significantly narrow rail freight's carbon advantage over road haulage.

When taking a longer-term perspective on the contribution of modal shift to logistics decarbonization, one should also recognize that advances in technology are likely to decarbonize freight transport modes at differing rates. One cannot assume that the carbon intensity differentials between the various modes will remain static. So applying modal averages based on, say, 2010 data to the target modal split for 2030 or 2050 is likely to yield a misleading estimate of the potential carbon savings from modal shift. Forecasts exist of future improvements in the carbon efficiency of individual freight modes but they are seldom integrated into multi-modal calculations of the longer-term carbon impact of modal shifts. The interpretation of modal carbon intensity averages also requires several other qualifications, discussed more fully in McKinnon (2010a).

Freight density

The average density of the freight moved varies significantly between modes. Rail and waterborne modes carry a preponderance of heavy primary commodities, such as coal, aggregates, cement, steel, oil and chemicals, whereas road specializes more in the movement of lighter manufactured and packaged goods. This is significant because carbon intensity is conventionally measured in terms of consignment weight rather than volume, ie CO_2 per tonne-km rather than CO_2 per m^3-km. As very little volumetric data is collected and published, one has little choice but to use a weight-based metric. It should be recognized, however, that this intrinsically favours modes carrying denser products. For example, coal (anthracite) has an average density almost three times that of the typical mix of groceries delivered to a supermarket (0.3 tonnes per cubic metre as opposed to 0.86 tonnes per cubic metre). Partly because of this difference in density, moving 100 tonnes of groceries by road would require much more vehicle movement, energy use and emissions than moving 100 tonnes of coal by rail. Expressing carbon intensity in tonne-kms rather m^3-kms substantially inflates rail's environmental advantage, particularly as coal, a relatively dense product, has traditionally been the main commodity transported by rail worldwide. Coal accounted for 38 per cent of rail tonne-kms in 2007 and 49 per cent in China (Thompson, 2010).

Directness of the routeing

Measurement of the distance component in the tonne-km figure also tends to bias modal comparisons of the carbon intensity in favour of rail and waterborne modes. The rail and waterway networks have a much lower density and connectivity than the road network and as a result, freight consignments often follow more circuitous routes by train and barge. For example, in the case of a freight haul between Rotterdam and Prague, the rail distance would be 13 per cent greater than the most direct road distance (922 km as opposed to 1,041 km) (McKinnon, 2010a). Woodburn (2007) also quotes examples of rail distances for freight trains being increased by, respectively, 9 per cent and 15 per cent to release capacity on trunk lines for passenger trains and to route trains with larger deep-sea containers via lines with higher loading gauge[2] clearances. Rail and waterborne transport may have lower carbon intensity in terms of tonne-kms, but because freight travels more circuitously across their networks they often generate more tonne-kms than the equivalent door-to-door road journeys. Allowance should be made for this distance bias when calculating the carbon savings from a modal shift.

Treatment of intermodal traffic

An increasing proportion of freight moves on intermodal (or co-modal) services, usually combining a rail or waterborne trunk haul with road feeder movements at one or both ends. This is necessitated by the lower density of the rail and inland waterway networks and the alignment of most production and distribution operations with the road network. As discussed later, a fundamental rebalancing of the freight market towards rail and water will require a dramatic increase in intermodal services. Carbon intensity calculations are mainly mode-specific, however, and rarely are emission factors quoted for door-to-door movements comprising more than one mode. Several studies have compared the carbon intensity of rail–road intermodal services with trucking (eg Vanek and Morlok, 2000; IFEU and SGKV, 2002; Freight Best Practice Programme, 2010b; and Craig, Blanco and Sheffi, 2013) and found that on average the former offer carbon savings of between 30 and 60 per cent. The analysis by Craig, Blanco and Sheffi of 400,000 intermodal shipments in the United States, nevertheless, revealed that 'unlike trucking, which provides more consistent carbon intensity across lanes, the intensity of intermodal varies considerably.' They conclude therefore that 'for a shipper considering a prospective change to intermodal transportation, the average may be a poor estimate of the actual savings on a specific lane' (p.51). Much of the variability in carbon intensity is due to differing degrees of circuity in the routeing of consignments via intermodal terminals. Across a sample of eight intermodal journeys criss-crossing mainland Europe, the intermodal distance was on average 8 per cent greater,[3] with the intermodal

Table 4.3 Differences in road and intermodal distances between European cities

		Road (km)	Intermodal (km)	Extra intermodal distance (km)	%
Paris	Hamburg	887	914	27	3%
Amsterdam	Milan	1032	1125	93	9%
Copenhagen	Zurich	1337	1416	79	6%
Brussels	Prague	901	1059	158	18%
Rome	Warsaw	1847	1827	–20	–1%
Marseilles	Cologne	1020	1087	67	7%
Vienna	Madrid	2387	2635	248	10%
Budapest	Bordeaux	1933	2163	230	12%
	Total km	11344	12226	882	8%

SOURCE McKinnon (2010a)

distance increment ranging from -1 to 18 per cent (Table 4.3) (McKinnon, 2010a). In an innovative analysis of how patterns of internal trade in the United States would change if tight limits were imposed on GHG emissions, Taptich and Horvath (2015) highlight 'intermodal terminal availability' as a key determinant of the areas over which particular products could be traded within particular GHG budgets, what they call 'freightsheds'. They show that for certain commodities, such as paper and meat/seafood, 'increasing the availability of intermodal terminals will increase low-GHG accessibility nationwide' and 'also have beneficial impacts on GHG reduction potentials of mode-switching policies' (p.11326). The predicted growth of intermodal freight volumes around the world should justify the establishment of more modal interchange points, reducing the length of road feed movements and making intermodal routeing more direct and carbon efficient.

In summary, there is little dispute about the overall ranking of freight modes in terms of carbon intensity and the fact that modal shift to rail and waterborne services cuts CO_2. The debate is more around the magnitude of modal differences in carbon intensity and the net CO_2 savings achievable by altering the modal split. One must exercise caution in interpreting many of the intensity estimates in the public domain and ensure that the figures quoted for the different modes are strictly comparable. Much of the problem lies in the averaging of empirical data drawn from small and unrepresentative samples. There is a need for greater granularity in modal comparison using emission data disaggregated by vehicle/train type, service/duty-cycle type, geography and commodity to allow companies to assess the carbon benefits of switching mode for specific flows on particular corridors. It is, after all, at this level that most modal choice decisions are made. The Global Logistics Emissions Council and EU Learn project are currently co-ordinating efforts to refine carbon measurement along these lines. The availability of more disaggregated modal emissions data will also assist in the calibration of macro-level freight modal split models.

Freight modal split trends

It is very difficult to analyse global trends in freight modal split on the basis of currently available data. The best that one can do is review recent trends in several of the world's major freight markets that collectively account for a large share of total freight movement. In this section we will examine trends in the EU, the United States, China and India. Almost all the available freight modal split data relates to domestic transport within national borders. At this level of analysis, one encounters wide variations in the mix of transport modes used, largely reflecting differences in geography. For example, island nations and countries endowed with extensive river systems tend to be more heavily reliant on waterborne transport. In some of these countries, there is a three-way split between road, rail and waterborne transport, whereas in others the modal choice is basically between road and rail or road and waterway. While coastal shipping and inland waterways have the potential to significantly increase their share of domestic freight markets, it is generally felt that the dominant freight modal shift in the coming decades will be from road to rail as rail networks are more extensive and have the capability to distribute a broader range of commodities. Incidentally, air is generally excluded from modal split analyses based on physical freight flows because it represents such a small percentage of the total freight movement. In EU countries in 2014 it represented only 2 million tonne-kms out of 3.5 billion (0.06 per cent) (Eurostat, 2017c).

Between 1995 and 2014, the freight modal split trend in the EU was regressive in environmental terms (Figure 4.5). Road increased its share of tonne-kms from 45 to 49 per cent while rail, coastal shipping and pipelines, modes with markedly lower carbon intensities, each dropped 1–2 percentage points (Eurostat, 2017c). The inland waterway share of around 4 per cent remained fairly constant over the 20-year period. Although in absolute terms rail tonne-kms rose by 6 per cent, rail's share of the freight market

Figure 4.5 EU28 freight modal split trend 1995–2014: per cent of total tonne-kms

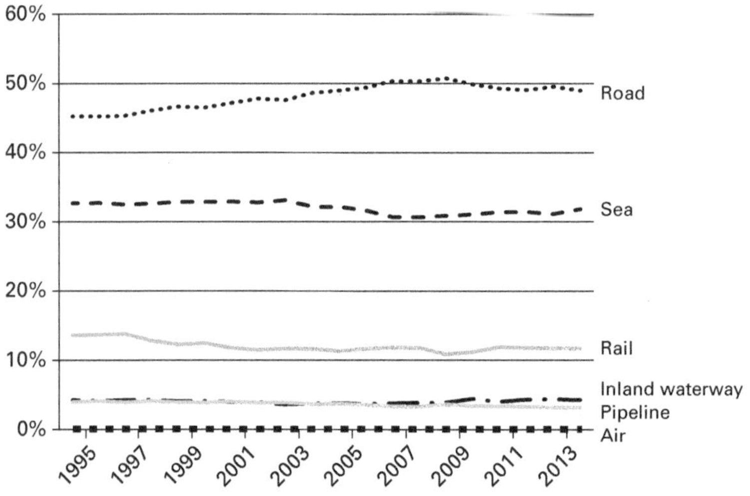

SOURCE Eurostat (2017c)

contracted from 13.6 to 11.7 per cent. EU average figures for rail also mask wide variations in rail's fortunes in individual member states (Figure 4.6).

Over the more recent period from 2006 to 2015, its share of the combined road–rail market fell by as much as 35 per cent in Estonia and increased by an impressive 10.5 per cent in Austria, with the percentage change in other countries spread across this range. Most of the countries experiencing an erosion of rail market share are in East European states. This is mainly due to the upgrading of road infrastructure in these countries over the past 25 years but also to some residual rebalancing of their freight markets from the days when central economic planning favoured the use of rail (International Energy Agency, 2009).

Having reviewed these trends, the European Court of Auditors (2016: 7) deemed the 'performance of rail freight transport in the EU... unsatisfactory in terms of volume transported and modal share', especially as, since 1992, the European Commission has 'set shifting of the balance between different modes of transport as a main objective.' The experience of EU countries on the right side of Figure 4.5, and of the UK between 1997 and 2013, shows that it is possible for rail to grow its market share, though these gains have been small relative to the modal shift target that the EU has set for 2030. This target is to have 30 per cent of freight tonnage travelling over distances greater than 300 km to move by rail or water by 2030, and 50 per cent by 2050. Achievement of its 2011 Transport White Paper target

Figure 4.6 Changes in rail share of the combined road–rail freight market in EU countries between 2007 and 2015

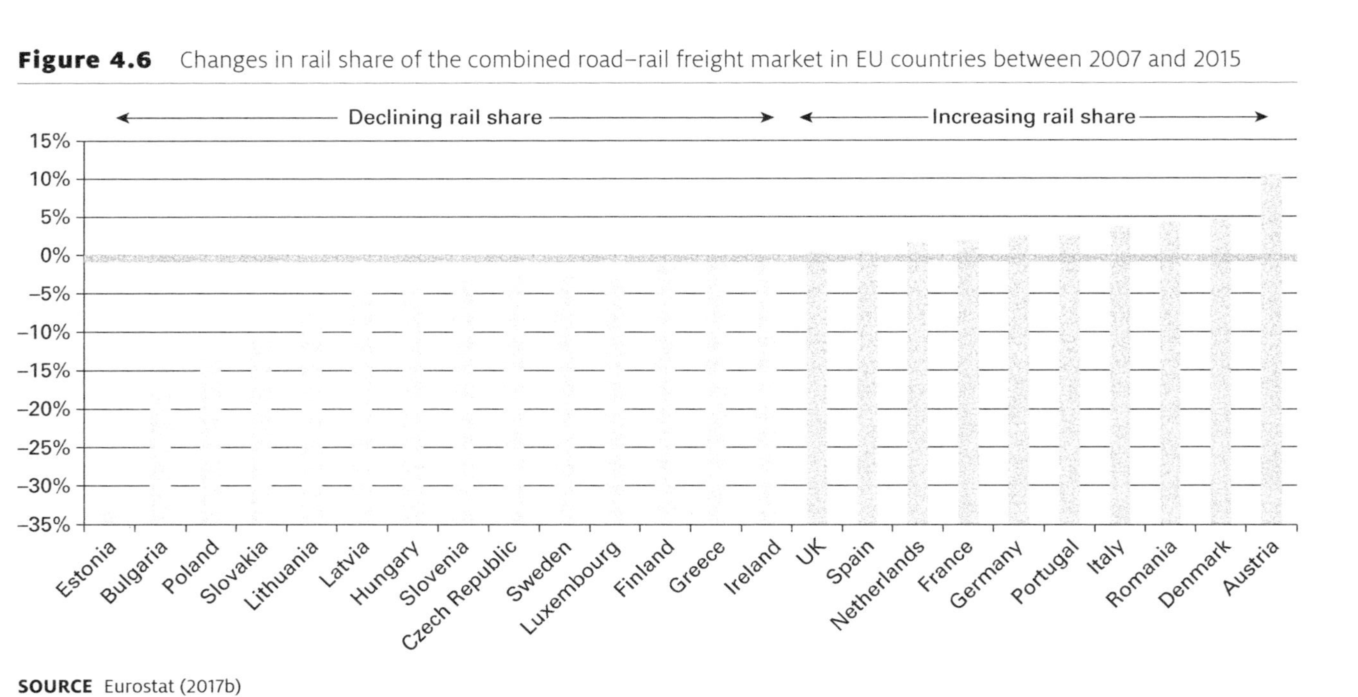

SOURCE Eurostat (2017b)

Figure 4.7 Projected change in EU freight modal split with and without EU 2011 Transport White Paper target

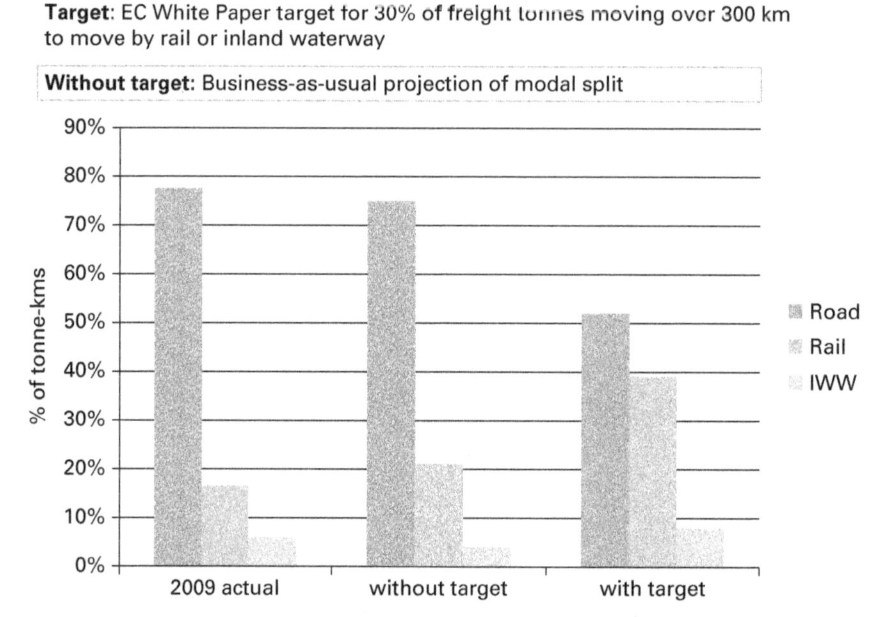

DATA SOURCE Tavasszy and van Meijeren (2011)

would result in rail's share of the European freight market growing from 11 per cent in 2009 to 39 per cent by 2030 (Tavasszy and van Miejeren, 2011) (Figure 4.7).

There seems very little prospect of this target being met. The EU aspires to have a land-based modal split more like that of the United States, though its fragmentation into 28 separate member states currently prevents this. The integration of the US economy at a continental scale creates market conditions conducive to the movement of freight by what are predominantly long-haul modes, namely rail, inland waterway and pipeline. Between 1995 and 2006, the US rail freight share rose from 27 to 34 per cent of tonne-kms, but over the next seven years it fell back to 29 per cent, partly as a result of waterborne transport, rebranded in 2007 as 'America Marine Highway', expanding its role. Over the period 1995–2013, the United States marginally increased its overall reliance on trucking as it captured just under half of all domestic tonne-kms (Figure 4.8). So, again, in the United States the modal shift trend has not been carbon-reducing.[4]

The EU and US trends, however, have been heading marginally in the wrong direction in comparison with India, where there has been a very

Figure 4.8 US freight modal split trend 1995–2013: per cent of total tonne-kms

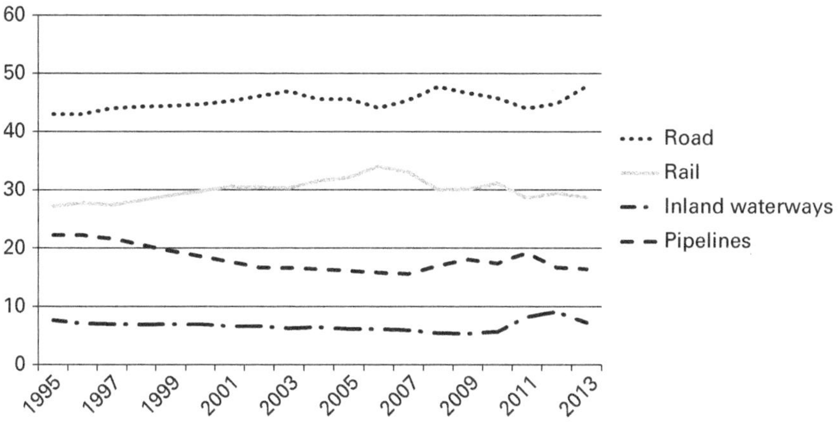

SOURCE US Bureau of Transportation Statistics (2017)

pronounced shift to road. In India, the rail network has been haemorrhaging freight traffic to road for many years. Between 1990 and 2015, rail's share of the total road–rail freight market in India roughly halved (from 63 to 31 per cent) mainly as a result of inefficiencies, under-investment, the subsidization of passenger services from freight revenues, and upgrading of road infrastructure and services. Plans have been devised to arrest and reverse the precipitous decline of India's rail freight market, but this will be difficult as, over the past decade, many companies have realigned their production and distribution systems with respect to the road network. Commercial property has gravitated to points of high accessibility on the road network, often distant from the nearest railway line, creating 'logistical lock-in' to a heavily road-dependent modal split.

In China, waterways are the dominant mode of domestic freight transport, accounting for 53 per cent of total tonne-kms in 2015 (National Bureau of Statistics of China, 2016), though there are wide regional variations in modal split across the country (Tian et al, 2014). Waterways captured 56 per cent of the huge growth in Chinese tonne-kms between 2008 and 2015 (up by 60 per cent) (Figure 4.9).[5]

Road secured 37 per cent of this freight traffic growth while rail actually lost about 2 per cent of its traffic over this period, partly because of a lack of capacity. The erosion of freight from rail to road is less of a concern in China than in India, partly because it has been much more modest, but also because so much Chinese freight moves by water. Although the energy intensity of these waterborne services is significantly more higher than rail

Figure 4.9 Freight modal split trend in China 1990–2015: trillion ton-kms

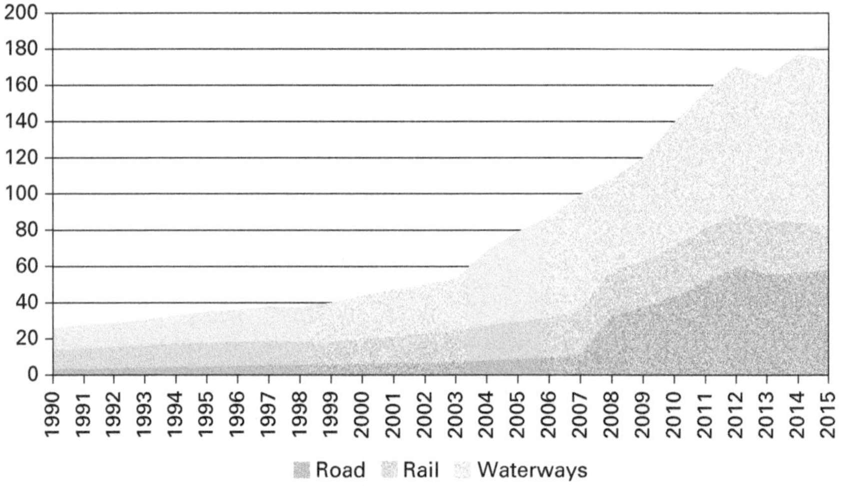

SOURCE National Bureau of Statistics of China (2016)

in China, it is well below that of trucking (Hao et al, 2015). As 'surface freight tonne-kms' are projected to grow six-fold in India and China by 2050 the future modal split trends in these countries will be fairly critical to the global decarbonization of logistics by then (International Transport Forum, 2015a). Hao et al (2015) see modal shift, mainly from road to rail, as a key element in a 2050 freight scenario for China which would result in freight-related GHG emissions falling 32 per cent below the business-as-usual trend.

BAU forecasts of rail's share of the total road–rail freight market suggest, however, that in much of the Asia Pacific region the trend is likely to continue downwards between now and 2050 (International Transport Forum, 2017b). It is also predicted to decline in Africa and, to a lesser extent, Latin America. Only in the developed regions is rail expected to maintain or expand its share of land-based tonne-kms. The US rail share is projected to remain fairly constant, while in Europe it is expected to show a modest rise. There are no similarly long-term and geographically extensive forecasts of the share likely to be held by waterborne transport. The ITF modelling of the future road–rail split combined with the statistical review of past modal split trends does not inspire confidence in government NDC claims that modal shift will deliver most of the required carbon reductions from freight transport over the next 30 years. This raises two questions. First, why has it proved so difficult to get companies to make more use of cleaner, lower-carbon transport modes, and second, what new policy initiatives will be needed to induce a major shift to trains, barges and ships? These questions are addressed later in this chapter.

This section has examined freight modal split trends primarily at a national level. Significant CO_2 reductions can also be made if transcontinental movements of international trade switch to lower-carbon modes. The next section reviews recent trends in this sector of the global freight market.

Modal realignment at a transcontinental level

Long-distance movement of international freight is overwhelmingly dominated by shipping and air cargo services. It would not be meaningful to express the traffic split between these modes in terms of the total amounts of freight moved by sea and air because air competes for only a very small proportion of the freight carried by sea, almost all of it containerized. In 2016, container shipping transported 65 times more freight tonnage than air cargo (Seabury Consulting, 2017). The air freight share of this combined market declined from 2 per cent in 2000 to 1.5 per cent in 2016. In carbon terms, this has been a favourable trend as the average carbon intensity of long-haul air freight services is, according to DEFRA (2017), 112 times greater than that of container shipping. This huge difference in carbon intensity extended over distances of several thousand kilometres translates even quite small modal shifts into relatively large savings in CO_2. For example, let us suppose that in 2016 10 per cent of air cargo tonnage had transferred to container shipping. This would have increased the tonnage carried by sea container services by only 0.15 per cent, but cut CO_2 emissions from the combined sea and air market by 7 per cent. This 15.7-million-tonne annual reduction in emissions would be roughly equivalent to half of all carbon emissions from freight movement in the UK in 2016.

It is often argued that the wide difference in transit times between air and ocean services effectively makes them separate markets with minimal opportunities for modal shift. On a door-to-door basis it typically takes eight times as long to ship a product from China to Europe by sea as by air. For many time-sensitive consignments, whose high value density and high inventory costs can justify air freight rates typically three to four times higher than container shipping costs, distribution by air is the sensible, and often only, option. There are, nevertheless, certain categories of air cargo that could switch. For example, quite small changes in the relative cost of air and shipping services could shift traffic for which the trade-off between higher cost and shorter transit time is marginal. Better planning can also reduce a company's dependence on air freight services. Use of these services is often an emergency response to production delays, poor forecasts or service disruptions, some of which could be avoided if supply chains were better managed.

A sharp increase in air freight rates would put shippers under pressure to use air cargo services more sparingly. As these rates are very sensitive to the price of kerosene, a hike in oil prices could tilt the modal balance for some shippers. A strong environmental case can also be made for taxing aviation fuel, something which, remarkably, does not happen at present. This taxation is prevented by the Warsaw Convention of 1944, to which most countries in the world are signatories, and creates a glaring environmental anomaly. Aviation is by far the most carbon-intensive transport mode, while the CO_2 emitted high in the atmosphere has a radiative forcing effect two to four times higher than that of emissions released on the ground surface. Aircraft contrails also have a global warming effect. Despite all this, the energy used in aviation is virtually tax-free. In contrast, trucking companies operating in Europe paid an average of €0.52 per litre for diesel fuel (excluding VAT), despite the fact that their operations have a carbon intensity 18 times lower than that of long-haul air cargo operators (DEFRA, 2017). The extension of the European Emissions Trading Scheme (ETS) in 2012 to include aviation was intended to go some way toward correcting this anomaly, but as it excludes flights to and from Europe it has had no effect on transcontinental air cargo services. There is also little prospect of aviation fuel being subject to taxation in the short to medium term.

The ability of container shipping to capture traffic currently moved by plane has been subject to conflicting pressures in recent years. On the one hand, wide adoption of slow steaming by container shipping lines has increased air freight's relative transit time advantage. This practice, which is discussed in more detail in Chapter 7, was phased in from 2008 onwards mainly to economize on fuel. It has lengthened maritime transit times by an average of 15–20 per cent. According to Cariou (2011), it is likely to have reduced CO_2 emissions from container shipping by around 11 per cent between 2008 and 2010. Smith et al (2014) estimated that from 2007 to 2012, daily consumption of fuel by container ships carrying more than 8,000 TEU fell by 71–73 per cent. After allowing for an increased number of days per voyage, carbon reductions per TEU-km were still substantial. These direct CO_2 savings will have dwarfed any indirect CO_2 increases from modal shifts from ocean to air resulting from the deceleration of vessels. Some of this decline in service quality has been offset by improved price competitiveness on many trade lanes where freight rates have been depressed mainly by over-capacity. Air cargo rates on some routes have also declined, but by a smaller margin. For some companies contemplating a switch of traffic from air to sea this recalibration of cost/transit time trade-offs may be appealing, but not enough to induce a major modal shift.

For most air freight users the transit time gap for end-to-end movements is just too large to seriously consider a modal shift to ocean transport. There are, however, two intermediate options that can give companies a more favourable balance of cost to transit time and yield significant carbon savings: sea–air services and transcontinental rail services (Table 4.4).

Sea–air services involve combining ocean and air freight legs, mostly on Far East–Europe journeys. Geodis, the French logistics conglomerate which claims to have 'pioneered the sea–air concept' claims that it can offer services that are 30–50 per cent quicker than sea freight at rates 30–50 per cent cheaper than air freight. On a long sea–air service on the Far East–Europe trade lane, a consignment is typically shipped by sea from, say, China to the Middle East, where at an intermodal hub such as Dubai, it transfers to an air freight service for final delivery to a European destination. On a short sea–air service, a hub in Malaysia or Singapore might be used as the modal transfer point. Table 4.4 compares the typical transit times, costs and CO_2 emissions of sea–air services with other modal options. With average carbon savings of 45–50 per cent relative to a full air freight service, sea–air services can represent an attractive decarbonization option for companies moving products in the mid-range of time sensitivity. They currently account, however, for only a small share of transcontinental freight movement.

Rail freight services between China and Europe are a more recent phenomenon and are currently enjoying an annual volume growth of around

Table 4.4 Relative CO_2 emissions, transit times and costs for trans-Asian freight services: indicative index values

	Rail	Air	Sea	Sea-air
CO_2 emissions	100	2100	150	1450
Transit time	100	20	200	110
Cost	100	475	50	230

SOURCES Based on DB Schenker (2015) and Reidy (2017)

65 per cent, admittedly from a very low base. The first commercial service began in 2011, carrying laptops and monitors for Hewlett-Packard between Chongqing and Duisburg. In 2016 a total of 370 freight trains left China bound for Europe and a total of 300,000 tonnes of freight was carried by rail on the two main Eurasian routes.[6] The quantity of freight moving by rail on this continental 'landbridge' currently constitutes a tiny fraction of total Far East–Europe freight tonnage, only around 0.06 per cent of maritime container traffic in 2016 (*Economist*, 2017c; UNCTAD, 2017). It is, however, expected to grow rapidly over the next few years with the support of the Chinese government's 'One Belt One Road' (OBOR) policy. With transit times of 15–20 days, Eurasian rail services offer shippers a modal option that is twice as quick as sea container services at rates well below those of air freight. To reduce carbon emissions these new services will have to divert freight from air rather than sea. The carbon intensity of the rail service per tonne-km is roughly twice that of sea, though the train follows a much more direct route to Europe (typically 11,000 km as opposed to 22,000 km (Reidy, 2017)) and is thus able to reduce total emissions per consignment by a third or more (D B Schenker, 2015). This CO_2 differential is tiny when compared with the twenty-to-one carbon advantage that trans-Asian freight train services have over air cargo links.

As shipper confidence in overland rail links between China and Europe builds, infrastructure is upgraded, rolling stock capacity expands and transit times shorten, increasing numbers of shippers are likely to follow the example set by Hewlett Packard, BMW and others and switch from planes to trains on the Silk Road routes.

Barriers to freight modal shift

A large literature has accumulated over the past 50 years on the factors influencing companies' choice of particular transport modes and the modal decision-making process. For many businesses this is a strategic decision made at the time when they reorganize their production and/or distribution system, and can be difficult to alter in the short to medium term. Where they have committed to road-based logistics, it can be very difficult to persuade them to switch to rail or waterborne transport (Steer Davis Gleave, 2015). This is mainly because these alternative, lower-carbon modes suffer from three sets of interrelated disadvantages which we can label geographical, temporal and operational.

Geographical

As discussed earlier, rail and waterway networks have much lower density and connectivity than road networks. Historically this was less of a problem when canals and railways were the dominant modes of freight transport and businesses had little choice but to locate their factories and warehouses adjacent to their networks. With the development of motorways and the rise of the road haulage industry there was a fundamental realignment of what geographers call the 'space economy' to the road network. In most countries, only a small fraction of industrial premises now have direct access to the rail or canal networks. For example, a survey of 2,809 warehouses in Germany revealed that only 180 (6.4 per cent) had direct rail access (Rolko and Friedrich, 2017). The road feeder movements that these premises require to access rail and waterborne services can add significantly to door-to-door delivery times and costs, making these alternative modes relatively uncompetitive. Moreover, the amount of traffic handled by these intermodal services correlates with the number of terminals at which freight can gain access to rail and waterway networks. Where there is little traffic, there are few access points serving wide catchment areas with long average feeder movements. This is a 'chicken and egg' situation in which a major growth in intermodal volumes could stimulate a virtuous cycle of more terminals, shorter feeder distances and greater competitiveness. The challenge lies in achieving the critical mass of intermodal business needed to initiate this cycle. In Europe the intermodal market has been growing much faster than the rail freight market as a whole, though from a relatively low base. Between 2005 and 2015, intermodal tonne-kms rose by 28 per cent by comparison with a 1 per cent increase in total rail tonne-kms (BSL, 2017).

It is in larger countries that rail and waterborne services are able to exploit their comparative advantage in long-haul movement. It is often said, for example, that rail is only competitive over distances greater than 300–500 km. Such glib references to break-even distances are misleading as they over-simplify the relationship between haul length and modal competitiveness. Nevertheless, other things being equal, the longer the haul, the stronger tends to be the economic case for rail. This particularly applies to intermodal services where the road feeder distances are short relative to the length of the rail line haul. Unfortunately, the opposite applies in most small countries where the average length of haul is simply too short for the railways to capture a large share of the freight market.

Temporal

There are primarily two senses in which time presents an obstacle to modal shift. First, rail and waterborne services are generally slower than their road-based equivalents, putting them at a disadvantage, particularly in competition for more time-sensitive, higher-value traffic. It is estimated that goods moved by rail across Europe travel at an average speed of only 18 km per hour (European Court of Auditors, 2016). Second, rail freight services must adhere to much tighter scheduling than trucking operations, making rail an intrinsically less flexible option. All movements on the rail network are timetabled, with passenger and freight trains allocated paths at specific times. This rigid scheduling contrasts with the freedom that companies have to despatch trucks at different times of the day, making it easier for them to synchronize freight movements with production and warehousing operations. As these operations have become more flexible and subject to just-in-time pressures, they have become less compatible with the fixed scheduling of freight trains. Having to share infrastructure with rail passenger services has, in some countries, exacerbated this scheduling constraint because the growth in passenger volumes has created bottlenecks on commuter lines around major cities. The acceleration of passenger services on mainlines has also widened the speed differential, with slower freight trains often being forced to weave slower paths across the network.

Operational

Modes vary enormously in their optimum consignment size: for rail this is the trainload and for inland waterways a full bargeload, both substantially greater than the typical truckload. To exploit their scale economies, rail and waterborne services need the freight to be tendered in large quantities either directly by the shipper or via a freight forwarder. This suited their traditional customer base in the coal, metals, chemicals, oil and aggregates sectors moving primary commodities in bulk. Rail and waterways already carry a large share of these staple flows, but two of them, coal and oil, will shrink over the next few decades as the world abandons fossil fuels. Some of this traffic loss might be substituted by biomass and lower-carbon fossil fuels such as natural gas, but there is a serious risk that, in many countries, the switch to alternative energy will reduce rail and water's share of the freight market. To increase their market share they will have to replace bulk flows of primary commodities with large quantities of manufactured goods and packaged products whose typical consignment size is much smaller.

In the UK, the railways have been successful in attracting traffic of this type from retailers and hinterland flows of international containers with a range of manufactured goods. So rail freight businesses can change their product mix, though this presents greater technical, operational and marketing challenges.

Another operational constraint is the lack of interoperability between the railway systems of different countries or rail freight companies. The track and loading gauge of the network, signalling systems, locomotive power ratings, wagon couplings etc can all differ, preventing the through movement of trains and the exploitation of rail's long-haul advantage. This has been a particular problem in Europe; international differences in technical standards and working practices have impeded the development of continental rail freight services comparable to those in the United States.

Summary

Because of one or more of these constraints, many businesses do not consider modal shift to be a serious option. They are essentially captive to a particular mode, usually road. In some countries, only a small part of the freight market is genuinely contestable in the sense that real competition exists between modes. In forecasting the potential contribution of modal shift to the decarbonization of logistics, one must therefore make a realistic assessment of the prospects of getting firms in particular sectors and locations to switch freight to a greener mode.

In addition to the fundamental constraints outlined above, lower-carbon modes can also be inhibited by a series of regulatory and commercial constraints. For example, there has traditionally been a high level of government intervention in the rail freight sector, which has not always been in its best interests. In many countries the rail freight system is state-owned and sometimes used in the pursuit of other policy goals such as employment creation and economic development. Often commercial discipline and customer focus are lacking and the quality of management poor. Whether nationalized or not, railway companies are often subject to common-carrier obligations, tariff restrictions and limits on their ability to market their freight services and partner with other businesses.

There is often a good deal of inertia on the shipper's side, as well as failure to adequately review the modal options even at critical times when the logistics strategy is being reframed. Logistics managers are often risk-averse and reluctant to entrust their freight traffic to an alternative mode. Some of this reluctance can be due to reports of bad experiences circulating in the logistics industry (Flanders and Smith, 2007), many of them relating to problems long corrected but still tarnishing the image of the mode. There is growing evidence, however, of companies reappraising their modal options for environmental reasons. While there are few examples of companies opting for a greener mode that is more expensive or offers inferior service, more account is now being taken of environmental criteria in the modal choice decision.

In summary, there are many legitimate reasons for companies continuing to rely heavily on trucking and resisting governmental efforts to get them to transfer more of their freight to lower-carbon modes. Achieving the ambitious targets that have been set for modal shift will require more radical initiatives, such as those discussed in the next section.

Future policies, trends and practices

The most radical means of redistributing freight among modes would be to restrict the capacity of higher-carbon modes and/or grossly inflate the cost of using them. The former strategy was widely applied to the road freight sectors of countries such as the UK, the United States, France and Australia during the interwar period. Quantitative licensing systems were then used to limit the number of trucks on the road, ostensibly to suppress 'destructive competition' in 'immature' road haulage industries but mainly to ease competitive pressures on the railways. These regulatory efforts to preserve rail's pre-eminence proved largely ineffective and were abandoned in the 1970s, '80s and '90s (McKinnon, 1998). There is currently no political appetite to reintroduce such a *dirigiste* approach to managing the division of freight traffic between modes. On the other hand, use of the price mechanism to favour greener modes commands wide political support, particularly where it involves internalization of the environmental costs of freight transport.

The imposition of taxes that reflect the marginal social costs of transport has spawned a large literature and generated much political debate, particularly within the EU. The European Commission, in its 2011 White Paper on transport, committed to 'proceed to the full and mandatory internalization of the external costs' of all transport modes by 2016–20 (European Commission, 2011: 29). Little progress has been made on the implementation of this policy, though preparatory research has been done to update the monetary values of transport externalities to be factored into a future environmental tax calculation. The latest EU Handbook compiled by Ricardo-AEA (2014) substantially raised its 'central value' for the cost of a tonne of CO_{2e} from €25 in the 2008 edition to €90, bringing it into line with several other studies. If a separate carbon tax were imposed at this level on a maximum-weight articulated truck in the UK running an annual distance of 100,000 km, it would incur an additional annual charge of £6,960. This would represent 37 per cent of the combined fuel tax and excise duty incurred by this class of vehicle annually and raise its overall operating costs by roughly 7 per cent. On its own, a cost increase of this magnitude would be unlikely to induce much modal shift. Not only climate change costs would be internalized, however. If taxes rose to absorb the full basket of externalities the percentage impact on total operating costs could be several times higher. Such a hike in truck taxation would be very strongly resisted and very unlikely to secure much political support. Hauliers could legitimately argue that the UK already taxes road freight operations very heavily by international standards and that much of the existing taxation should be seen to be recovering environmental costs. Indeed, the current taxes on heavy lorries in the UK would be equivalent to a carbon tax of €245 per tonne CO_{2e}, 2.7 times the recommended EU carbon price. Furthermore, differential fuel taxes are already used in some countries to promote the use of alternative modes. Differential levels of duty can be imposed on the fuel consumed by different modes. For example, in the UK the diesel fuel tax is over five times higher for trucks than for freight trains. In a country with high levels of transport taxation, like the UK, introducing an additional carbon tax at the recommended level would be unlikely to induce much modal shift. In low-tax countries, the inflationary effect on road freight rates and potential modal split impact would be more pronounced, but political opposition correspondingly greater.

A return to quantitative licensing and the full internalization of climate and other environmental costs may be considered too draconian in the current political arena, but may eventually have to be implemented not just to rebalance the modal split but to achieve the necessary degree of logistical decarbonization overall. In the meantime, it is easier and more common for governments to use 'carrots' rather than 'sticks' to entice businesses to make greater use of greener modes. Various forms of financial incentive have been and are being used for this purpose, including capital grants for rolling stock/vessels and terminal development, discounted infrastructure access payments and operating subsidies/revenue support grants. The EU's Marco Polo programme, which ran in two phases between 2003 and 2013, has been one of the world's largest freight modal shift initiatives, costing around €60 million per annum. It is estimated that between 2003 and 2012 it removed 3.56 billion truck-kms from Europe's roads (equivalent to 3.5 million journeys between Paris and Berlin), cutting CO_2 emissions over this period by 4.36 million tonnes (European Commission, 2013).

Many other developments are encouraging greater use of lower-carbon modes. They are helping to overcome the three categories of constraint discussed in the previous section.

Geographical

Over a 20–30-year time frame for the deep decarbonization of logistics, one can envisage more industrial and logistics property being drawn to rail-side and canal-side locations to facilitate the switch to these lower-carbon modes. Even if in the early years companies make little use of these modes, they will expand their modal options over the lifetime of the asset. Land use planning policies at national and local government levels could exert more pressure on companies to locate freight-generating premises adjacent to rail and waterway networks. They could, for example, make direct access to these modes a condition of planning permission or provide financial incentives to real-estate companies to prioritize the use of rail- and canal-side land for logistics-related property development. This assumes, of course, that rail and barge operators will service these properties.

For the foreseeable future, however, intermodality will be the main way of overcoming the relative inaccessibility of rail and waterway networks. In recognition of this fact, governments and international organizations are stepping up their support for intermodal services. Many are now doing this within a corridor framework. The EU, for example, has defined nine intermodal corridors criss-crossing the continent, each with a champion, an investment budget and strategic plan. They will receive through the EU's Connecting Europe Facility (CEF) a total of €24 billion to upgrade intermodal connections and thereby help the EU meet its 2030 target of getting 30 per cent of freight moving more than 300 km onto rail or water. India's 'Dedicated Freight Corridor' (DFC) project is promoting intermodal rail services across a quadrilateral corridor network linking Delhi, Mumbai, Chennai and Howrah (McKinsey & Co, 2010). By expanding rail's share of freight movement in these corridors and generally enhancing modal efficiency, the DFC plan aims to cut freight-related emissions by a factor of 4.5 over the next 30 years. More details of this ambitious plan to reverse the steep decline in rail's market share can be found in a report by the Indian National Transport Development Policy Committee (2014).

Temporal

The concept of synchromodality, now being heavily promoted in Europe, can help to relieve temporal constraints on the use of low-carbon modes. In its original formulation the concept aims to synchronize intermodal connections to allow consignments to switch modes during a transit with minimum delay and disruption (Verweij, 2011). Tavasszy, Behdani and Konings (2015) have called this 'synchronized intermodality' and argued that it represents a 'horizontal integration' of different transport modes to provide a seamless through-movement of consignments.[7] When carefully choreographed by a logistics service provider the total transit time can be competitive with direct road movement. The service can be sold to shippers as a mode-neutral solution in which differing combinations of modes are used at different times and on different routes at the logistics service provider's discretion. Zhang and Pel (2016) developed a model to compare the performance of conventional intermodal and synchromodal systems and applied it to container flows from the port of Rotterdam along the Rhine corridor. They found that synchromodal transport has 'clear benefits from a societal and environmental perspective, as it facilitates modal shift from road transport (reduced by 16 per cent) towards rail transport, yielding a reduction in CO_2 emissions (of 28 per cent)' (p.8).

The degree of modal shift and level of CO_2 savings can be augmented where the concept of synchromodality is recast in a supply chain context. Dong et al (2017: 5) argue that:

> The meaning of 'synchro' in synchromodality needs to be broadened from the synchronization (and scheduling) of the different transportation modes towards the synchronization (and scheduling) of transportation with other supply chain activities such as inventory management and the setting of service levels.

Not all categories of inventory need to be transported at the same speed. Cycle stocks that are routinely replenished to meet stable demand can often be moved by slower modes, leaving the faster ones to handle orders placed at short notice to meet more variable demand. Commodity flows can be stratified by time sensitivity and variability and allocated to different modes accordingly. As shown in Figure 4.10, this can involve reducing the amount of time less time-sensitive stock spends in storage at either or both ends of the journey and increasing the amount of time spent in transit. This 'mobilization' of some categories of inventory allows companies to use slower modes without increasing the total order lead time. Using real-world company data, Dong et al (2017) modelled several scenarios for

Figure 4.10 Switching slow-moving inventory from fixed locations to vehicles to permit modal shift

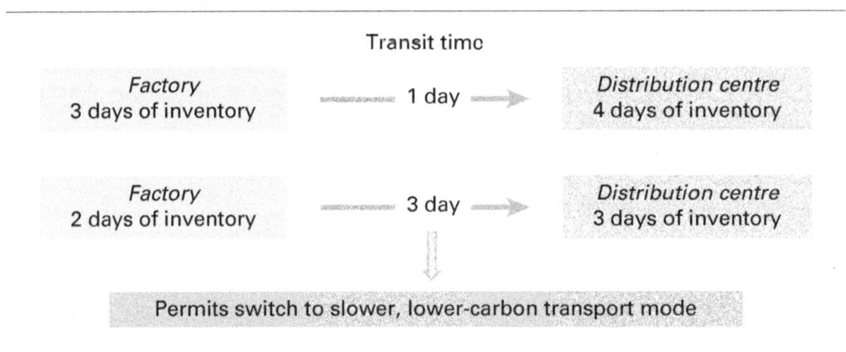

'synchromodality from a supply chain perspective' (SSCP), and found that it could increase the intermodal share from 30 to 70 per cent, cutting CO_2 emissions by 30% and logistics costs by 6%.

Rail freight services are also benefiting indirectly from the construction of high-speed passenger rail lines. This is releasing capacity for freight trains on existing lines and easing scheduling conflicts between passenger and goods services. In some countries, such as France, the high-speed lines also carry express freight services, allowing rail to diversify the range of services it offers. Rail and waterborne services in many countries are also benefiting directly from upgrades to track infrastructure, signalling systems, IT and intermodal terminals, which are accelerating services, improving reliability and increasing the flexibility of freight schedules. Meanwhile, worsening congestion on road networks, tightening restrictions on truck drivers' hours and reductions in truck speed limits are narrowing the gap.

Operational

The increasing willingness of companies to engage in supply chain collaboration and share logistics assets can help to consolidate freight in viably sized trainloads. Freight forwarders and intermodal operators have traditionally had the role of consolidating traffic from numerous shippers to fill freight capacity across all modes. Their efforts can be reinforced by shipper-led initiatives to work together to bundle freight into flows that are large enough to justify new rail or waterborne services. A good example of such collaboration is that between Procter and Gamble and Tupperware. The combination of P&G's heavy loads of detergent with Tupperware's light loads of plastic boxes along a corridor stretching from Aalst and Mechelen in Belgium to Thiva and Athens in Greece, strengthened the economic case for a rail trunk haul. P&G had already been using rail on this route, but as a result of the collaboration Tupperware switched its traffic from road to rail, raising the utilization of the containers from 55 to 85 per cent, cutting transport costs by 17 per cent and saving over 200 tonnes in CO_2 emissions annually (Muylaert and Stoffers, 2014).

The interoperability of freight train services in EU member states is steadily improving as a result of the gradual implementation since 2012 of the EC Directive 2008/57/EC, which aims to ensure common technical specifications across European railways. This not only relates to technical aspects of infrastructure, energy, rolling stock and telematics but also includes the management of train operations. The European Rail Traffic Management System (ERTMS) intends to remove 'technical barriers [that] are hampering the development of rail transport at the European level'. In its so-called 'Fourth Railway Package' presented in 2013, the European Commission set the objective of creating a 'Single European Railway Area' within which national railway undertakings conform both to common technical standards and market conditions. While reasonable progress has been made on the technical aspects of the package, the market-related proposals have proved controversial and are taking much longer to agree. This is reckoned to be preventing European railways as a whole from exploiting their full potential and significantly increasing their share of the European freight market.

Summary

Wide variations in the carbon intensity of freight transport modes make modal shift an obvious means of decarbonizing logistics. It is not surprising, therefore, that this policy option features prominently in government climate change plans for the transport sector. However, the potential carbon benefits of freight modal split are possibly being overestimated, and the practical problems of getting large numbers of companies to switch mode underestimated. Many modal split policies have fallen well short of expectation and have often done little more than slow the contraction of the rail and waterways modal share. A series of recent developments, including synchromodality, supply chain collaboration and improved interoperability, may help these lower-carbon modes win more traffic.

This chapter has focused on the modal split issue, but it is also important to examine its interrelationship with other decarbonization initiatives in the logistics sector. This reveals a conflict between modal shift aspirations and efforts to increase the carbon efficiency of road transport, which is, after all, the dominant freight mode. The more efficient trucking becomes, the harder it will be for alternative modes to capture a larger share of the freight market. This presents planners and policy makers with a dilemma and demands a more realistic appraisal of the net contribution that freight modal shift can make to decarbonization.

Notes

1. The term 'modality' is increasingly being substituted for 'mode' as part of the general trend to make words longer and sound more technical. In this book I prefer to use the simpler, shorter word 'mode'.
2. The loading gauge measures the maximum clearance around a vehicle travelling on the rail or road networks. In the case of rail it is determined by the height of tunnels and bridges and distance between station platforms.
3. These estimates were obtained from the EcoTransIT calculator.
4. The modal split data for Europe, the United States and China cited in this section come from the OECD/ITF website.
5. The compilation of Chinese tonne-km statistics was changed in 2007–08 and so figures before and after this date are not strictly comparable.
6. Estimate based on data from the EU–China Research Centre in the College of Europe.
7. For a discussion of the relationship between synchromodality and related concepts such as multimodality and co-modality consult SteadieSeifi et al (2014).

Improving asset utilization in logistics 05

Logistics, like most industries, underutilizes its assets. If companies used their logistics assets more efficiently, they could cut energy use and related GHG emissions. As 90 per cent of logistics emissions come from freight transport, attention naturally focuses on the use of vehicle capacity. Survey data suggests that much of this capacity is underutilized and as a result it takes more vehicle trips than strictly necessary to move a given amount of freight. There is an economic as well as an environmental cost associated with these additional trips, giving companies a financial incentive to minimize them. It is the close correlation between the economic and environmental benefits that makes raising vehicle load factors such an attractive decarbonization option. Few decarbonization measures in the logistics sector are so closely aligned with good business practice and offer such low-carbon mitigation costs. This seems not to have been appreciated by the authors of the COP21 NDC statements referring to freight transport, since only 4 per cent of them refer to improved vehicle utilization as a means of cutting CO_2 emissions in this sector (Figure 1.3). The measure nevertheless features prominently in the main transport decarbonization frameworks discussed in Chapter 1 and has been the subject of much research in recent years.

Bold claims have been made for the potential benefits of ensuring that freight vehicles run full. According to a much-quoted statistic, underused capacity in the European trucking system was worth €160 billion and represented 1.3 per cent of the EU27 CO_2 footprint (ALICE, 2014). There are many good reasons, however, for vehicles carrying much less than their maximum load, rendering underloading endemic and virtually impossible to eliminate. To obtain a more realistic estimate of the contribution that improved asset utilization can make to logistics decarbonization one needs to understand why so many freight vehicles run empty or only partly loaded. The measurement of utilization has also to be carefully scrutinized because load factor calculations are highly sensitive to the choice of metric. In an ideal world, the loading of trucks, trains, ships and aircraft carrying freight would be regularly monitored using a range of indices and the resulting data made available for research purposes. In practice, there is a dearth of utilization data across all the main freight modes, making it very difficult to conduct macro-level assessments. So a good starting point for our examination of this decarbonization option is to review the available utilization data and thereby establish the empirical foundations of research on this topic.

Utilization metrics in the freight transport sector

Choice of indicators

The measurement of vehicle utilization in the freight sector usually distinguishes empty running, typically expressed as the percentage of the total distance that is travelled without a load, and the load factor, defined as the percentage of available capacity utilized on a laden trip. Although they may seem straightforward, both measures are open to differing interpretations. In the case of empty running, it is important to determine if a vehicle returning with only handling equipment, such as wooden pallets or roll cages, is considered empty. The repositioning of an empty container by a trucking, rail or shipping company can be considered a loaded movement where it is earning them revenue, despite the fact that no freight is actually being carried. Surveys must therefore clearly define what is meant by empty.

Measuring the utilization of laden vehicles is more complicated, partly because the density and 'stackability' of freight varies so much but also because many freight journeys make multiple stops to collect and/or deliver product.

Density and stackability of freight

All utilization metrics are ratios of the actual load carried to the maximum carrying capacity of the vehicle (Caplice and Sheffi, 1994). Three different units can be used for this ratio: weight (tonnes), cube (m^3) (space occupied in the vehicle) and deck area (m^2) (vehicle floorspace covered by the load).[1] Almost all the available statistics express utilization in weight terms, because this is the easiest of the three units to quantify consistently. This is acceptable for dense loads that reach the maximum weight before cubic or floorspace limits are reached. In the case of low-density loads that do not

Figure 5.1 Impact of varying product density on utilization of vehicle space and weight capacity

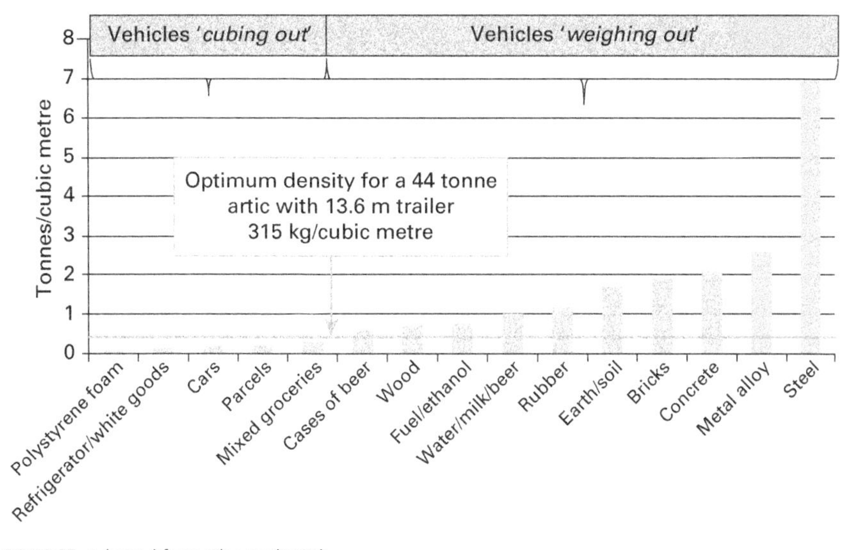

SOURCE Adapted from Glaeser (2010)

reach the weight limit, cube or deck area metrics give a much better indication of capacity utilization, the latter where the product cannot be stacked very high and is subject to a floorspace constraint. Where utilization of these low-density loads is expressed purely in weight terms the impression is given that capacity is being underutilized when in fact all the space may be completely filled. Figure 5.1 shows how the average density of a variety of products varies and indicates the optimum density for moving them by a 44 tonne articulated truck in the UK, where the optimum is defined as simultaneously meeting weight and volumetric limits.

The pricing system for freight transport services favours consignments that have a density close to the optimum for a given mode. For each mode it is possible to calculate a 'chargeable weight' using a volume/weight conversion factor related to the optimum density (Lovell, Saw and Stimson, 2005). For air cargo, trucking and shipping, each cubic metre of freight is multiplied by, respectively, 167 kg, 333 kg and 1,000 kg to calculate the chargeable weight for the consignment. The chargeable weight is then compared with the actual weight of the consignment and the higher value used to determine the freight rate. Pricing freight-carrying capacity in this way is designed to optimize its utilization in both weight and volume terms and in the process maximize revenue.

While commercial exploitation of vehicle capacity is multidimensional, research on the subject has tended to focus on only one of these dimensions – weight. This has been a major limitation because assessing average load factors solely on the basis of weight underestimates the true utilization of vehicles as it fails to recognize that a substantial proportion of loads 'cube out'[2] before they 'weigh out'. This underestimation of capacity utilization is becoming more prevalent as the average density of freight declines and more loads fall into the 'cubing out' category. The scale of the problem is illustrated by data from the UK Government's annual road haulage survey. This was one of the few national freight surveys to ask vehicle operators if their loads were constrained by weight, volume or both. It shows that there was a sharp reduction in the proportion of loads constrained only by truck weight limits (ie 'weighing out') between 2000 and 2010 (from 30 to 3 per cent), a modest increase in the proportion solely constrained by volume (from 24 to 30 per cent) and a huge increase in the share doubly constrained by weight and volume (from 8 to 37 per cent) (McKinnon, 2015). This trend was due in part to the 2001 increase in the maximum truck weight in the UK from 41 to 44 tonnes (on 6-axles), but has also reflected a change in the average density and stacking height of loads. If one only referred to the weight-based measure of utilization, one would conclude that the loading of trucks in the UK had drastically deteriorated over the decade as the proportion running at full weight had dropped to a tenth of the 2000 figure. In reality, the situation had greatly improved between 2000 and 2010 as the proportion of kilometres run by vehicles simultaneously meeting weight and volumetric constraints jumped almost five-fold. This highlights the danger of relying on weight-based utilization metrics, but unfortunately in the few countries that actually monitor vehicle loading, these are the only ones available.

Multiple drop/collection trips

Where a vehicle travels directly between origin and destination with the same load (or empty), its state of loading is obviously constant and easily measured. Where it is routed around several locations to deliver and/or collect consignments of varying size and weight, determining the mean utilization is much more complex, particularly where, as in the parcel sector, the number of stops can exceed 100 in a single trip. The problem is not simply one of recording and analysing all the delivery/collection data; it involves defining a maximum utilization for this type of trip which does not encourage perverse behaviour. If, for example, the objective were to maximize average utilization across the trip, carriers would have to deliver the heaviest/largest load last or collect it first. The vehicle would then run with the heaviest/largest load for the longest distance. This might yield impressive vehicle fill statistics, but it would also maximize fuel use and emissions. In an article about this 'load factor paradox' Arvidsson (2013) explains that 'increasing the load factor is usually regarded as a way of improving efficiency, but we observe that under certain conditions improving the load factor affects economic and environmental sustainability by increasing total costs and emissions' (p.56). In the case of multiple drop/collection rounds, vehicle utilization objectives must be subordinated to the greater goal of minimizing economic and environmental costs and vehicle routeing calculations calibrated accordingly.

Data availability

There are huge differences between freight transport modes in the amount of publicly available data on vehicle utilization. By far the most data is available for road freight operations, particularly by trucks with a gross weight in excess of 3.5 tonnes (McKinnon, 2010a). Some utilization statistics exist for rail freight, shipping and air cargo operations, though the metrics used are often inappropriate for an environmental analysis.

In almost all countries, the road freight sector is highly fragmented and very diverse, making it necessary to randomly survey large samples of operators to get reasonably accurate estimates of empty running and loading. Participating companies generally have few qualms about confidentiality as the data is anonymized and aggregated across sub-sectors and regions. In contrast, much of the world's rail freight is handled by state-owned monopolies that are often reluctant to divulge commercially, and often politically sensitive data about the utilization of equipment. In countries with privatized rail freight sectors, oligopolistic structures discourage companies from being open about key competitive variables like equipment utilization. In the case of shipping and airfreight, most of the available utilization data relates to only a part of the sector (such as deep-sea container services), making it very difficult to generalize for the transport mode as a whole.

Global assessments of the availability of utilization data for the main freight show that outside the EU and North America, data is very sparse (International Energy Agency, 2009; GLEC, 2016). GLEC (2016: 70) lists the main sources of utilization data by transport mode and region.

The next sub-section reviews available data on the empty running of freight vehicles and their average loading on laden trips.

Trucking

European Union EU member states are required by statistical directive to collect and report data on a range of road freight indicators. Only one utilization metric is included, percentage of empty running, which Eurostat (2017d), the EU's statistical agency, defines as where 'there are no goods or empty packaging in the lorry, the trailer or the semi-trailer, which are therefore completely empty'. Eurostat has by far the most comprehensive set of truck empty running statistics in the world. It estimates that in 2016, 20 per cent of the total distances travelled by trucks in the EU28 was run empty (Eurostat, 2017d). Empty running within national borders was much higher (23 per cent) than for international journeys (12 per cent). The lower level of international empty running is partly attributed to the 'liberalization of international road freight transport' and the greater economic incentive that operators have in finding return loads for longer, more costly cross-border trips. The average level of empty running across the continent dropped from 22 per cent to 20 per cent between 2010 and 2016, indicating a slight improvement in this key component in the carbon intensity of road freight operations. These average figures, however, conceal exceptionally wide variations between EU countries (Figure 5.2).

Figure 5.2 Percentage of truck-kms run empty in European countries 2016

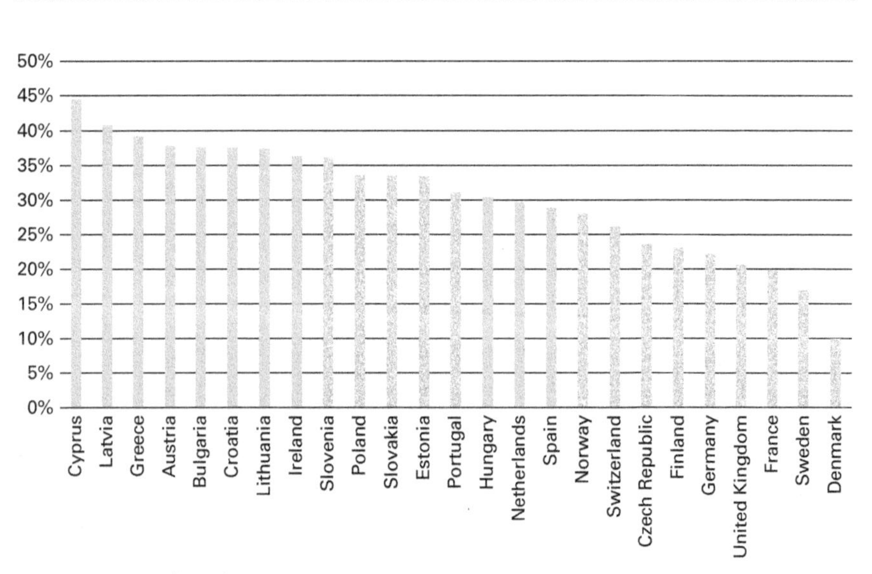

SOURCE Eurostat (2017a)

In 2016, this varied from 10 per cent in Denmark to almost 45% in Cyprus, with a fairly even spread of national values across this range. Some of this extreme variability may be attributable to national differences in the minimum weights that governments set for the inclusion of trucks in the survey.[3] Most of it is likely to be due to differences in economic geography, the nature of the freight market and regional imbalances in freight flows, though this is a subject that requires further research.

Eurostat does not require national governments to provide data on the utilization of vehicle capacity and most do not collect it. It has published data on the average weight of truck payloads in EU member states, but in the absence of data on the carrying capacity of the vehicles transporting these payloads it is not possible to determine the average vehicle utilization.

The European Environment Agency (2010) published weight-based truck utilization data for 10 EU countries for the period 1997–2007. It suggested that there were wide variations in average levels of utilization, ranging from over 80 per cent in Spain to around 30 per cent in Portugal. While some international variation is to be expected, variations of this magnitude suggest that there are inconsistencies in definitions and methodology. The EEA data also indicated that in every country the utilization trend was downward over the various periods for which data was available. Neither the EEA nor EU have published more recent data of this type and so it is not known if this decline in load factors has continued. Since the EEA data is weight-based, it is possible that the decline in utilization was partly due to a drop in the average density of road freight in these countries, as discussed earlier in the UK context.

United States A survey by ATRI (2016) of the operators of 107,000 trucks/tractor units in the United States found 20.9 per cent of their annual mileage was run empty, an almost identical figure to that of the EU for 2016. Truck operators belonging to the SmartWay Partnership[4] reported a significantly lower average empty running figure (16.7 per cent), though this may partly reflect the fact that membership of SmartWay signifies a strong commitment to efficiency and environmental sustainability (SmartWay, 2015). It could also be due to a differing mix of carrier types in the SmartWay sample. SmartWay has also been a source of published data on the volumetric utilization of US trucks. Data for 2012 indicated that on average, 80–85 per cent of the space in Class 8 trucks, the heaviest category and workhorse of the US trucking industry, was filled. This figure included vehicles reaching the maximum weight limit before all the space was occupied. It seems remarkably high (particularly as the maximum gross weight of US trucks is relatively low by international standards – only 36.3 tonnes). Again, some allowance must be made for the fact that the SmartWay sample is unlikely to be representative of the US truck fleet as a whole.

Latin America Data published by the Inter-American Development Bank (2015) suggests that the average level of truck empty running in Latin American countries (in 2012) was significantly higher than in the EU and United States, at 43.4 per cent. As in the EU, there were wide international differences ranging from 54 per cent in Argentina to 30 per cent in El Salvador.

Other countries Transport Research Support (2009) (a joint initiative of the World Bank and DFID) compiled a set of truck empty running estimates for 12 developing countries/emerging markets (six in Africa and three each in Latin America and Asia). In most cases a single average was given, though for some countries it was a range. If one excludes Malawi, for which the range was exceptionally wide, the mean empty running figure was 30 per cent, only slightly higher than the average figures recorded in Europe and the United States.

Summary Empirical data on truck utilization is confined to relatively few countries, much of it is out of date, and some of the figures appear suspect. Some of the international differences, particularly between EU countries, are so wide that they lack credibility and cast doubt on the consistency of the methodology. In most of the countries with published data, between a fifth and a third of truck-kms are run empty. In some countries, particularly in Latin America, the figure rises to around 50 per cent, suggesting that no return journeys find a backload. In the EU, where time-series data is available, the empty running proportion has dropped slightly in recent years. Data on average load factors in the trucking sector is particularly poor and does not provide an acceptable basis for modelling decarbonization potentials and trends, except possibly in two or three EU countries. This situation is particularly disappointing as road is by far the dominant mode of freight transport and is expected to remain so.

A new approach to measuring truck utilization was pioneered in the UK around 20 years ago and was applied in eight sectors until 2009 when the government withdrew funding for the programme. This involved 'synchronized audits' of a standard set of utilization and energy-related key performance indicators (KPIs) across numerous truck fleets carrying particular categories of product, such as food and drink, non-food retail supplies, parcels and building materials (McKinnon, 2010a). Vehicle utilization was measured by weight, floorspace and load height and benchmarked across the participating fleets. The trip data collected also permitted retrospective analysis of the potential for empty running (McKinnon and Ge, 2004). These so-called transport KPI surveys provided a deep insight into the utilization of truck capacity and its impact on carbon intensity.

Rail freight

In a rail freight system, empty running can take various forms:

- locomotive running light (ie without a train) – known as 'deadheading' in the United States, done to reposition locomotive capacity and sometimes drivers;
- a trainload of empty wagons being returned as a complete set, for example to a coal mine;
- individual empty wagons in a train also comprising loaded wagons.

It has not been possible to find any published data on empty running within the rail freight sector. In some countries, most notably the United States, the three forms of empty running are simply subsumed within more generalized measures of utilization and productivity.

Railway companies, both directly and via organizations such as the International Union of Railways (UIC), the European Community of Railways and Infrastructure Companies (CER) and the American Association of Railroads (AAR), release data on the extent of their networks, the amount of rolling stock they operate, train-kms and tonne-kms. This data can be used to calculate utilization and productivity indices. Between 1982 and 2000, for example, average annual tonne-kms per locomotive in the United States rose from 30 million to 70 million, mainly as a result of the 1980 Staggers Act which liberalized the US rail freight system. Similar indices can be calculated for rail freight operations in other parts of the world using data on gross and net tonne-kms, train-kms, numbers of wagons and locomotives, compiled by the UIC (2017).

The problem with the available metrics for rail is that they do not show how efficiently freight is being moved from an environmental standpoint. They give no indication of the proportion of the available carrying capacity actually being used. Average tonne-kms per km of track, per locomotive or per wagon are poor proxies for average load factor. In the absence of actual load factor data, one might infer from data in the AAR and UIC reports that utilization of the equipment has probably been improving, but it would be preferable to have the actual numbers to confirm this.

There have been a few examples of independent researchers trying to estimate the utilization of freight trains. Woodburn (2011), for example, assessed the utilization of container trains in the UK by filming their movements to and from the main deep-sea ports. This showed that the number of empty slots on the container trains varied significantly between train operators and in some cases were higher than the figures they quoted.

Shipping

There are also different types of empty running in the maritime sector, for example:

1. *Repositioning of empty bulk tankers*, for the oil or dry trades, to collect their next load. For example, on 14 November 2014, Lloyds List estimated that there were 479 empty oil tankers being repositioned worldwide. There is little potential to backload bulk tankers as they are designed to carry only one commodity and the flows of that commodity are usually unidirectional.
2. *Empty slots on container and roll-on roll-off vessels*. This can also be considered a form of underloading and expressed as a percentage utilization figure.
3. *Repositioning of empty containers*. In 2016, 26 per cent of the 700 million containers (TEUs) moving through ports were empty (UNCTAD, 2017). Boston Consulting Group (2015a) estimated that moving empty containers cost $15–20 billion and emitted 19 million tonnes of CO_2. These figures indicate the scale of this operation and the potential benefits of rationalizing the repositioning of 'empties'. Rodrigue, Comtois and Slack (2017) argue, however, that 'positioning of empty containers is... one of the most complex problems concerning global freight distribution' and subject to numerous constraints.

The Clean Cargo Working Group (2014) published data on the utilization of container ship capacity disaggregated by trade lane. Shipping lines belonging to the CCWG are responsible for 70–80 per cent of all deep-sea container movements and so this data set provides a global perspective. Utilization is measured with respect to two 'limiting factors': the maximum number of TEU slots on the container ship and its maximum 'deadweight'. It is averaged for round trips and only takes account of loaded containers.

The exclusion of empty containers from the calculation removes a significant proportion, probably around a quarter, though this is justified by the decision to allocate CO_2 emissions from container shipping only to loaded containers. It is not known to what extent the observed underutilization is due to slots on the vessels being unoccupied or filled with empty containers. Overall utilization averages around 70 per cent.

Weight-based utilization in the maritime sector differs in one important respect from that of other freight modes – the need for vessels to take on ballast water to maintain stability and ensure that they are level at the bow and stern (what is known as 'trim'). In essence, when the vessel is empty or lightly loaded, ballast water[5] is substituted for cargo to keep the vessel properly balanced. Vast quantities of ballast water are transported by ships. It was estimated in 2010 that 3–5 billion tons of ballast water was being moved annually by international shipping (Godey et al, 2010 – quoted by Lindstad et al, 2015). When trucks or air freighters are empty their gross weight and fuel consumption are much reduced. In the case of ships, there is a much smaller reduction in gross weight, fuel and CO_2 when they are underloaded because of the need to top up with ballast water.

Air cargo

There is very little 'empty' operation in the air freight sector, for several reasons:

1. In 2015, 52 per cent of air cargo was moved in the bellyholds of passenger aircraft (Airbus, 2016). By definition, freight represents only a part-load on these aircraft. Even if the cargo compartment was empty, the plane would still be carrying a load of passengers and luggage.
2. Express air cargo is channelled through the hub-spoke networks of so-called 'integrators', such as FedEx, DHL and UPS. Their network services handle sufficient volumes of traffic to generate a load on virtually all radial routes to and from the hubs.
3. The high cost of air transport ensures that any empty flights by air freighters are kept to an absolute minimum. The exceptions would be planes flown empty to specific locations to pick up a dedicated load, such as a large piece of equipment.

According to IATA (2017b), load factors of air freighters have been averaging around 45 per cent in recent years, though they have been rising since early 2016. This statistic requires some qualification, however. First, as it is a weight-based measure, it does not take account of the cubic capacity in the aircraft or the density of the load. Second, this utilization data relates only to all-cargo aircraft and excludes bellyhold capacity on passenger planes. It therefore provides only a partial view. Bellyhold has been increasing its share of total airfreight traffic. Between 2007 and 2017, roughly two-thirds of the growth in global air freight capacity (measured in tonnes) was in the bellyholds of passenger aircraft (IATA, 2017b). This has been diverting freight that would otherwise have been carried in freighters, constraining their load factors, but improving the relative carbon efficiency of passenger aircraft with bellyhold capacity. Airbus (2016) predicts that the bellyhold share of global airfreight will rise from 52 to 62 per cent by 2035, more closely linking future trends in the carbon intensity of air cargo to those of passenger aviation.

Overcoming constraints on vehicle loading

Freight transport operators are often accused of making prodigal use of transport capacity and clogging the roads with truck movements that could have been avoided. Monbiot (2017) articulates a widely held sentiment when he claims that 'Lorries shifting identical goods in opposite directions pass each other on 2,000-mile journeys. Competing parcel companies ply the same routes, in largely empty vans. We could, perhaps, reduce our current vehicle movements by 90 per cent with no loss of utility.' This suggests that companies are not just environmentally reckless but also financially irresponsible because every 'unnecessary' vehicle-km costs money, reducing competitiveness and profitability. They therefore have a strong monetary incentive to use transport capacity as efficiently as possible. Why, then, according to the available statistics, is there so much underutilization of this capacity?

I have answered this question at some length in previous publications (eg McKinnon, 2014a and 2015), pointing out that there are often perfectly good reasons for freight vehicles, across all modes, carrying a lot of empty space that could be filled with freight. These reasons can be distilled into seven basic factors:[6]

1. logistical trade-offs;
2. lack of information;
3. scheduling;
4. dimensional incompatibility;
5. business silos and lack of collaboration;

6 traffic imbalances;

7 regulation.

The remainder of this chapter will examine each of these factors and consider what can be done to ease their constraining influence on capacity utilization.

Logistical trade-offs

The optimization of cost and service trade-offs is fundamental to the management of logistics. Companies make a cost trade-off when they pay more for one activity to make a greater saving on another, thereby cutting costs overall. They make a cost–service trade-off when they provide a higher-quality, more expensive service to win enough additional revenue to increase profitability. Both forms of trade-off can involve sacrificing vehicle fill for higher business goals. Transport, after all, is just one activity in the so-called 'logistics mix', which also comprises inventory management, storage, materials handling and associated information processing. Company surveys in Europe (A T Kearney, 2008) and the United States (Establish Inc, 2010) have suggested that, on average, freight transport represents 45–50 per cent of total logistics costs, slightly more than the combined cost of financing inventory and physically storing/handling it (Figure 5.3).

Figure 5.3 Breakdown of logistics costs in the United States (2010) and Europe (2008)

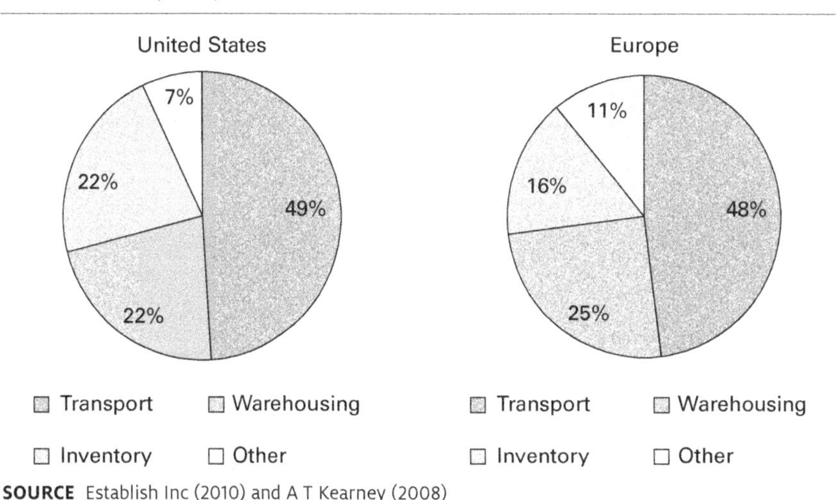

SOURCE Establish Inc (2010) and A T Kearney (2008)

These proportions vary significantly around the world, with the transport share often being significantly larger in less developed countries (Rantasila and Ojala, 2012).

It is common for companies to trade off higher transport costs for lower inventory and materials handling costs. As discussed in more detail later in this chapter, delivering goods on a just-in-time (JIT) basis to cut inventory often entails running vehicles less full, but as the inventory savings exceed the transport cost increment, total costs are reduced. Similarly, distributing goods such as groceries in roll cages can minimize handling costs at reception bays but at the expense of poorer vehicle utilization. Faster and more frequent delivery can also generate more sales and increase market share, though to the detriment of transport efficiency. Companies can therefore behave perfectly rationally in a cost-minimization or profit-maximizing sense when they underload their vehicles. The 'companies' here are the shippers, ie the users rather than the providers of the transport service. As Santén (2016: 65) explains: 'Transport providers view shippers' logistics activities as constraining efficient transport and attainment of high load factor, for example, by demanding shorter lead times and delivering smaller shipments with a higher frequency.' She suggests that shippers may be 'unaware of their negative influence on load factor' and may 'need to be informed about how they can take action to improve the load factor performance.'

Logistical trade-offs are not always made on a rational basis, however. Often they have not been subjected to thorough analysis, particularly in the case of service–cost trade-offs, where it is difficult to model the relationship between service level and revenue. The trade-offs can also be distorted by the uneven distribution of power and status across management functions (McKinnon, 2003). Production, marketing and sales departments usually exert a stronger influence on corporate strategy than logistics. In such circumstances, maximizing vehicle fill can be accorded much lower priority than meeting production and sales targets. As has been discussed, many green logistics initiatives are confined to the operational and functional levels and have their environmental benefits offset by higher-level strategic and commercial decisions. Nevertheless, over the past few decades logistics has been strengthening its position in the corporate hierarchy and gaining greater clout in asset-related decision making. Advances in IT and materials handling technology are also helping companies to recalibrate logistical trade-offs, often in ways that enhance transport efficiency. Future increases in the relative cost of transport, resulting, for example, from greater internalization of environmental costs, congestion charging and higher energy costs, will also force companies to give capacity utilization greater precedence.

Lack of information

Efficient utilization of logistical assets requires the collection, processing and sharing of large amounts of data. Managers need real-time visibility of these assets to determine how much unused capacity is available and where it is located. In the case of vehicle fleets they must be able to find matching loads that can fill the spare capacity while satisfying a range of operational and commercial constraints. For example, a potential backload for a returning vehicle must follow a similar route and schedule and be economically attractive to the operator. In the pre-internet age, load matching was fairly primitive, relying mostly on informal personal networks and capturing only a small percentage of the available opportunities. The development of online freight exchanges over the past 20 years has revolutionized this load-matching process, greatly facilitating and extending the search for loads and establishing new online trading mechanisms for the buying and selling of freight capacity. In the developed world, the online trading of logistics capacity is now a mature market offering a range of procurement options from the one-off backload to the auctioning of a company's national or multinational transport demands over a three- to six-month period. These demands can also be optimally structured before entering the online tendering process to further enhance the matching of freight flows with vehicle capacity. In emerging markets, online trading of freight capacity is at an earlier stage in its development. In some developing countries, most notably India, there has recently been a proliferation of online freight exchanges as there was in the UK and United States prior to the bursting of the 'dot-com bubble' of the early 2000s. This 'shake-out' consolidated business in the hands of a few large sites within which the probability of a successful vehicle–consignment match was much enhanced. Online markets may evolve similarly in emerging markets.

In both developed and developing markets, the use of mobile devices is decentralizing and accelerating the load-matching process, particularly among the large population of owner–drivers that dominate the trucking industries of most countries. A smartphone and an app can give drivers real-time access to details of available loads, permitting wider and more rapid uptake of e-freight services. Uber's diversification into the freight sector, what has been called the 'Uberization of freight' (LEK Consulting, 2017), may also reinforce this global trend in electronic load matching.

A lack of information also inhibits asset utilization in other ways. Uncertainty about the demand for transport a few days, or even a few hours, ahead makes it hard for carriers to manage vehicle capacity effectively (Sanchez-Rodrigues, Potter and Naim, 2010). Getting advanced information from their clients can give them more time to 'right-size' the vehicle to the job, consolidate loads and optimize routeing. A practice known as Collaborative Transportation Management (CTM), first promoted in the United States in the late 1990s, allows carriers to extend their planning horizon by getting more closely involved with the planning of shippers' logistics operations. According to several studies (eg Esper and Williams, 2003), CTM has significantly raised vehicle load factors, particularly where order lead times are short (Chan and Zhang, 2011). The pooling of transport data more generally across supply chains offers high potential for improving vehicle fill. Until recently we lacked the technical capability and standards to exploit this potential. With the advent of cloud computing and widespread adoption of business data standards this is no longer the case. GT Nexus, market leader in the provision of cloud computing services to the logistics sector, 'allows companies to manage their international and domestic transportation, tackling supply chain complexity with integration to partner systems and a single control layer for monitoring the lifecycle of transportation' (GT Nexus, 2017). GS1 (2017), with two million member companies in 150 countries, describes its data standards as the 'digital DNA' of commerce, 'enabling enhanced efficiency, safety and sustainability for a wide range of businesses and their customers.' The main obstacle to the sharing of operational data is now corporate reluctance to relax rules on data confidentiality, though this is easing as trust between supply chain partners builds and companies recognize the need for greater logistical collaboration. This trend is elaborated upon later in this chapter.

Tight scheduling

Over the past few decades delivery operations have become more tightly synchronized with the internal operations of factories, warehouses and shops. This close coupling of transport with the various business processes that it serves has reduced the time available for full loads to accumulate. Vehicles are then despatched only part-loaded to meet production or retail schedules. Rigid adherence to the JIT principle in manufacturing and 'quick response' in retailing are often blamed for trucks running around half-empty. The companies applying these principles are then accused of sacrificing transport efficiency, and by implication the environment, for less inventory, lower handling costs, higher sales etc. Put simply, JIT may not be compatible with low-carbon logistics.

However, dismissing JIT as being fundamentally damaging to the environment would be premature, for two reasons. First, JIT is not just a stock control system, it is:

> A business philosophy that cuts waste and raises productivity across production and distribution processes. Although JIT may raise the carbon intensity of delivery operations, the carbon efficiency gains within factories, warehouses and shops may offset the additional emissions [from transport] (McKinnon, 2016c).

If one conducted a full life-cycle analysis of JIT, spanning all the production, storage and materials-handling activities that it affects there could well be a net reduction in GHG emissions. Second, many companies have been able to mitigate the adverse effects of JIT on vehicle utilization by redesigning their inbound logistics and often channelling supplies through consolidation hubs. The industrial geography of sectors closely wedded to the JIT principle, such as automotive and aerospace, has also adapted in ways that minimize transport excesses. For example, suppliers subject to the tightest JIT pressures often locate branches in the vicinity of major car assembly plants, thereby minimizing delivery distances to the production line. McCann (1998: 3) examines how JIT creates a 'rationale' for 'increased concentration of... spatially diffuse inter-firm purchasing linkages.'

Despite these qualifications, it may become necessary in the longer term to decelerate the movement of freight to save energy and cut carbon emissions (McKinnon, 2016c). Many of the CO_2 savings from this admittedly radical option could accrue from improved utilization of transport capacity within time frames that are adjusted more to environmental than to commercial imperatives. Of the 13 ways of decarbonizing supply chains identified by the World Economic Forum/Accenture (2009: 17), the second most highly rated is 'despeeding' which, among other things, would yield a 'potential loadfill improvement within increased time windows'. The wider contribution of supply chain deceleration to decarbonization is discussed more fully in the next chapter.

In the short term, vehicle loading would benefit from improvements to the reliability of logistics schedules. When managers lack confidence in these schedules they are less likely to backload their vehicles for fear of delaying the next outbound delivery. Research has shown that prioritization of the outbound flow of goods from factories and warehouses is one of the main reasons that managers give for not backloading returning vehicles (McKinnon, 1996). The merging of consignments into fuller loads also hinges on their arrival at a consolidation point within predefined time windows. At present, delays are rife across the freight transport system, particularly in the developing world where transport infrastructure, administrative processes and working practices are often poor. For example, TCI and IIM (2012) estimated that delays to long-haul trucking movements on major inter-regional corridors in India could represent 15–25 per cent of total transit time. A series of surveys in the UK over the period 1998–2009 found that approximately a quarter of road freight journeys were delayed, but only by 30 minutes or less (McKinnon et al, 2009). A third of these delays were attributed to traffic congestion and the remainder to other factors such as queueing at distribution centres, vehicle breakdowns and staff absenteeism. Worsening congestion is exacerbating the reliability problem in many countries and probably as a consequence depressing vehicle load factors, though as yet there is no empirical evidence to confirm this link. As, worldwide, traffic is growing faster than infrastructural capacity, congestion-related delays are likely to increase in frequency and duration. There are, nevertheless, actions that companies can take to minimize their impact on logistical efficiency (McKinnon et al, 2009). For example, where distribution is by articulated vehicle increasing the ratio of trailers to tractor units (known as the 'articulation ratio') can decouple loading/unloading operations from the vehicle trip, reducing the impact of traffic delays on load factors. China planned to increase the proportion of tonne-kms moved by 'drop-and-hook' (ie articulated) vehicles to 12 per cent by 2015 and 15 per cent by 2020 (relative to a 2005 baseline), thereby reducing energy consumption in the road freight sector by 1.8 per cent by 2020 (Clean Air Asia, 2013). Track-and-trace systems, which are now widespread across the developed world but are yet to be rolled out in many developing countries, give transport managers advance warning of vehicles deviating from schedule and allow them to take corrective actions, some of which can improve vehicle loading.

Dimensional incompatibility

In an ideal world, products would fit snugly into the packaging, the packaging would fit the handling unit (such as pallets and roll cages) and the handling units would neatly interlock to fill the available space in the vehicle or container. There would then be minimal wastage of space at each level in the so-called 'unit load hierarchy'. Samuelsson and Tilanus (1997) adopted this broad definition of 'space efficiency' in a survey of carriers moving 'less than truckload' freight and, by multiplying a series of 'partial efficiencies' at each level, found that the actual products often occupied only a small proportion of the available space in vehicles. They estimated that in less-than-truckload distribution only 24 per cent of the space within a pallet load was actually occupied by products, similar to the 27 per cent share of the vehicle cube filled by pallet loads, revealing that the typical vehicle was transporting a lot of air. While primary and secondary packaging at, respectively, the individual product and carton levels is usually outside the control of logistics managers, the building of pallet loads (at the tertiary level) and loading of the vehicle is their responsibility. Their efforts to maximize space efficiency at these levels, however, are often constrained by differences in the dimensions of boxes, handling units and vehicle/container interiors. Twenty years ago, A T Kearney (1997) researched this problem and recommended standardization on a new size of 'efficient unit load'. The European Commission (2004) proposed a new 'European intermodal loading unit' (EILU) to make the dimensions of containers and swap-bodies compatible with those of the euro pallet, the handling unit used for most European palletized loads. The sizes of standard ISO containers, used in the maritime sector, are defined in imperial units, whereas euro pallets are measured in centimetres. If the 40-foot containers are four inches wider it is possible to carry 30 as opposed to 25 euro pallets, increasing carrying capacity by 20 per cent. An increasing number of containers are now 'pallet wide', though dimensional incompatibility between vehicles/containers and the modules they carry remains a significant constraint on load factors and, by implication, the carbon efficiency of logistics.

Business silos and lack of collaboration

Logistics has long been described as a 'boundary spanning' activity (Morash, Dröge and Vickery, 1996). It certainly merits this title because within companies it interfaces with other management functions like production, procurement, marketing and sales, while externally it physically connects all the businesses in a supply chain. Logistics works best when there is good cross-functional and cross-supply chain co-ordination. Unfortunately, much of the business world is 'siloed' and run by managers who optimize at a functional or company level and fail to exploit opportunities for collaboration.

This is often not their fault. Reporting systems and incentive structures typically discourage them from acting co-operatively. This widely prevalent and much-debated business phenomenon has an impact on the loading of vehicles. As discussed previously in this chapter, transport efficiency is often compromised because greater priority is given to the goals of other functions. Inter-functional relationships, however, do not always have to work to logistics' disadvantage. For example, closer dialogue between logistics and procurement managers can help to identify opportunities for backloading a company's vehicles with inbound loads from suppliers. These opportunities can be systematically reviewed during contract negotiations. Sometimes when there is a large potential for backloading it can be beneficial for supplies to be purchased on an 'ex-works' basis, transferring responsibility for delivery from the vendor to the buyer. For example, some supermarket chains use 'factory gate pricing' as a means of improving the utilization of their own vehicle fleets (Potter, Mason and Lalwani, 2007).

While better management of cross-functional relationships can help to raise load factors, many firms are likely to find external collaboration with supply chain partners a more effective means of filling their vehicles. Increasing numbers of firms are realising that most of the internal 'easy wins' have been exhausted and they are now struggling to find new ways of raising the productivity of their transport operations by themselves. To achieve a step change in vehicle utilization they have now to consider sharing logistics assets with other companies. This can involve collaboration with companies at the same level in the supply chain (horizontal collaboration) or with suppliers upstream and distributors/customers downstream (vertical collaboration). Vertical collaboration, which can be traced back to the 1990s, was initially motivated by a desire to cut inventory but is now also concerned with 'transport optimization'. Horizontal collaboration is a much more recent phenomenon. According to the EU CO3 Project (2013) it requires that 'multiple independent shippers pro-actively work together in clusters or communities to "bundle" their overlapping freight flows. Bundling in this context means that the compatible freight flows of the shippers are consolidated in space, as well as synchronized in time.'

For manufacturers and retailers, horizontal collaboration goes beyond simply outsourcing their distribution to the same logistics service provider (LSP). The traditional role of the LSP, after all, has been to group the traffic of several clients into more viably sized loads. In collaborative relationships, companies work together in other ways to exploit logistical synergies. This, for example, might involve adjusting schedules to synchronize flows and thereby combine loads in the same vehicles. For example, analysis of a logistical partnership between Nestlé and Pepsico in the Benelux countries revealed that this form of 'collaborative synchronization' could cut CO_2 emissions per tonne of product delivered by 54 per cent relative to each company running a separate fleet and by 26 per cent when compared with an LSP-managed groupage operation (Jacobs et al, 2014). This collaborative case study is one of several publicized in recent years to highlight the economic and environmental benefits and encourage other companies to do the same. Table 5.1 provides details of supply chain collaborations in Europe over the past decade, many of which have been in the fast-moving consumer goods (FMCG) sector.

This case-study evidence has been supported by computer modelling of collaborative scenarios based on operational data collected in company surveys. For example, Pan, Ballot and Fontane (2013) estimated that pooling the transport operations of two large French supermarket chains could cut CO_2 emissions by 14 per cent. Such pooling does not have to be confined to pairs of companies in bilateral arrangements. Two so-called Starfish studies in the UK have shown that the potential benefits of multilateral collaboration in the FMCG sector can be significantly greater. In the first Starfish project, analysis of transport data from 27 large FMCG

Table 5.1 Examples of supply chain collaborations in European countries

Collaborating shippers	Sector	Geography
Pepsico and Nestlé	FMCG	Benelux
Unilever and Kimberly Clark	FMCG	Netherlands
Nestlé and United Biscuits	FMCG	UK
Baxter, Colruyt, Eternit and Ontex	Healthcare, construction, wines & beverages, FMCG	Belgium-Spain
P&G and Tupperware	FMCG & household products	Belgium-Italy
Mars, United Biscuits, Saupiquet and Wrigley	FMCG	France
Tetley, Kellogg and Kimberly-Clark	FMCG	UK
JSP and Hammerwerk	Industrial equipment	Czech Republic
Spar and inbound suppliers	Retail chain, FMCG	Belgium

companies (equivalent to 8 per cent of all road freight tonne-kms in the UK) found that the potential savings of CO_2 from multilateral collaboration ranged between 4 and 20 per cent depending on the choice of scenario (Palmer and McKinnon, 2011). A second survey undertaken three years later with 10 FMCG companies suggested that the possible carbon reductions could be even greater (Palmer, Dadhich and Greening, 2016).

In Europe, organizations such as ECR, ELUPEG and Lean and Green have been extolling the virtues of logistical collaboration, while the EU has funded two major research projects on the subject (CO3 and NEXTRUST). These are not new initiatives, however. Supply chain collaboration has been heavily promoted for many years by trade bodies and consultancy companies, and yet still only a relatively small proportion of companies share their logistical assets. This form of collaboration has been much discussed, but little implemented. This has left many commentators on the logistics scene sceptical that supply chain collaboration will ever become truly transformational. It can be argued, however, that earlier exhortations were premature, that conditions are now changing and that we are on the eve of a new era when partnership will become the dominant paradigm. The changing circumstances can be grouped into five categories, what I call the 5Ms of logistical collaboration:

Motives. Financial, operational and environmental pressures on businesses are mounting, forcing them to find new sources of cost and emission savings. As the incremental benefits of transport rationalization at a company level diminish, managerial attention will shift to collaborative opportunities across the supply chain. There is also growing realization that only supply chain-level initiatives will deliver the deep reductions in logistics-related carbon emissions required over the next few decades.

Mindsets. A new generation of managers is taking over who are keener than their predecessors on collaborative working. Growing commitment to the 'sharing economy' at a personal level may spill over into corporate decision making and company reward systems adjusted to incentivize managers to seek out and progress joint initiatives across the supply chain.

Models. A combination of experience and research has increased understanding of what makes collaboration successful (Sanchez-Rodrigues, Harris and Mason, 2015). Collaborative business models have been refined and attuned to the needs of logistics (Palmer et al, 2012).

Mathematical techniques, based on game theory, have been devised to optimize the division of costs and benefits between collaborating companies. For example, during the course of the EU CO3 project, a software tool was developed which used the so-called Shapley Value[7] to optimize the 'gain sharing'. A new type of consultancy company has emerged, sometimes called a supply chain 'orchestrator', to act as a neutral third-party adviser, analyst, 'trustee' and general facilitator of collaborative relationships – essentially a logistical marriage broker. It can provide the necessary technical support to companies at various stages in the collaboration process.

Markets. Some collaborative initiatives have been shipper-led and excluded logistics service providers. This has naturally made some LSPs and their trade associations suspicious that shippers were forming cartels to bully them into lowering rates. On the contrary, it is now generally accepted that LSPs have a key role to play in logistical collaborations and can act as a catalyst in bringing shippers together. Collaboration can also reinforce other freight market trends such as the shift to lower-carbon transport modes. As discussed, supply chain collaboration can help to aggregate freight flows into train or barge loads, making modal shift more commercially and environmentally attractive.

Ministries. Governments have also been wary of industry collaborations, viewing them as potentially anti-competitive. Companies often give the risk that they will breach anti-trust legislation as a reason for not collaborating. Many European countries and the EU now accept that logistical collaboration yields wider societal benefit and, if properly implemented, does not infringe competition rules. Indeed, public bodies, such as ALICE,[8] the EU's technology platform for logistics, now actively encourage collaboration to increase the utilization of logistics assets.

Traffic imbalances

In an ideal world all freight flows would be geographically balanced by weight and volume at all spatial scales, from local to global. Vehicles might then be fully laden in both directions. Reality falls well short of this ideal. Freight movements are unidirectional, unlike personal trips which generally return to the home. Even where the value of physical trade between regions or countries is similar, differences in the value density of the products they exchange can result in very unbalanced material flows. These traffic imbalances are not only responsible for the empty running of freight vehicles; they also cause the underloading of vehicles in the direction with lighter flows. They are, nevertheless, the main reason for what can be considered 'structural' empty running. This is an inevitable consequence of the pattern of trade and is very difficult to eliminate. In most countries and regions, the current amount of empty running is significantly above this structural minimum and inflated by the factors outlined earlier, such as lack of information, tight scheduling, unreliability etc. Several studies have investigated the relative importance of these other constraints on return loading but have not attempted to estimate the extent to which empty running could be reduced if they were relaxed. In earlier research, we used data from an extensive survey of food deliveries in the UK over a 48-hour period to calculate retrospectively the potential for reducing empty running (McKinnon and Ge, 2004). Assuming that empty trips would have to be longer than 100 km in the UK to justify backloading, we estimated that approximately 43 per cent could have found a return load. Structural imbalances in freight flows over the two-day period would therefore have been responsible for roughly 57 per cent of the empty running. Once allowance was made for a realistic range of operational constraints, it would have been practical to backload only 2.4 per cent of the empty trips. The results of this analysis are not necessarily generalizable as they relate to one sector in one country in which the average length of haul was relatively short (92 km at the time of the study). The study did, nevertheless, indicate the extent to which traffic imbalances can restrict backloading.

For generations, carriers have used a tactic known as 'triangulation' to get around this traffic imbalance problem. As I explain elsewhere (McKinnon, 2015: 250):

> Instead of running vehicles on A-to-B-to-A bilateral routes, they send them on more complex inter-regional trips (eg A-to-B-to-C-to-A), which can allow them to exploit traffic imbalances in opposite directions along the route and thus raise the average load factor across the journey as a whole.

Improvements in data transparency, vehicle telematics and routeing algorithms are helping carriers apply the triangulation principle with much greater sophistication.

In the longer term, the development of a 'physical internet' (PI) would elevate this process to a much higher level (Mervis, 2014). Montreuil (2011) argues that just as emails find pathways through a single, open global internet and do so in a way that makes efficient use of server and telecommunication capacity, so one day freight consignments may have access to an analogous physical network. Ballot, Russell and Montreuil (2014) envisage the PI (or π) being a 'global logistics system based on the interconnection of logistics networks by a standardized set of collaboration protocols, modular containers and smart interfaces for increased efficiency and sustainability.' Although even its staunchest supporters accept that the PI is several decades away, bold claims are being made for the environmental benefits it will bring. Through its impact on freight vehicle utilization, the PI could make a significant contribution to logistics decarbonization in Europe. The development of the PI will be contigent on a series of interconnected revolutions in logistics information systems, materials handling and business practice occurring over the next few decades. It will, for example, require migration to a new system of interlocking, smart, reusable modular containers (Sallez et al, 2016). Arguably, the greatest challenge will lie in securing the necessary degree of logistical collaboration. As Saenz (2016) notes, 'achieving the level of collaboration needed to evolve towards a fully functional Physical Internet will require a profound change in business and managerial mindsets.' In a sobering review of barriers to the evolution of a PI, Sternberg and Norrman (2017: 750) advise 'caution when interpreting studies of the coming positive effects, as the models studied in [their] review promote extensive simplifications.' Even if the ideal vision of a PI proves unattainable, the results of recent PI 'roadmap' studies commissioned by ALICE suggest that by heading in this direction the European freight transport system can achieve a quantum improvement in capacity utilization.

Regulation

Although national governments are committed, explicitly or implicitly, to reducing carbon emissions from the freight transport sector, some of the regulations they impose inhibit efforts to improve vehicle fill. In particular, two sets of regulations in the trucking sector are worth highlighting: quantitative licensing and vehicle 'construction and use' regulations.

Quantitative licensing

As discussed in Chapter 4, this is a legacy from a time when governments felt it necessary to limit the capacity of the road haulage industry to preserve the railways and/or control what was deemed to be excessive competition. In most developed countries it has been largely abolished over the past 50 years, though some vestiges remain. Across the developing world the reform of freight market regulation began later, and many quantitative restrictions have yet to be relaxed. Even where they remain in force, however, the levels of enforcement are often low and so their practical impact is limited. Quantitative regulations typically take the form of limits on carriers' freedom to pick up and transport any load to any location. This denies them the right to collect a backload or top up an underloaded vehicle, even when it is economically and environmentally desirable to do so. Licensing can be used to confine an operator's activities to particular zones, routes or commodities. In both developed and developing countries it is common to restrict 'own account'[9] operators to carrying only their own freight in their own vehicles and to prohibit them from 'carrying for others'. In EU countries, for example, an average of 14 per cent of road freight is moved in 'own account' vehicles (DG Move, 2015). Debarring operators from picking up other company's freight can represent a significant regulatory barrier to backloading.

Another form of quantitative regulation which impairs average load factors across all modes is the restriction of cabotage, ie the right of a carrier to transport loads moving within the borders of a foreign country. Many countries ban or limit cabotage to reduce foreign competition in their domestic haulage markets. This means, however, that a foreign vehicle delivering an imported load cannot pick up any domestic backloads in the countries through which it must pass to return to its country of origin. Cabotage was substantially liberalized in the EU in 1998, though some controls remain. A High-Level Group on the Development of the EU Road Haulage Market (2012: 32) recommended further liberalization of cabotage rules, which it believed 'would permit increased load factors and improve economic efficiency whilst respecting working conditions and providing for environmental gains through lower CO_2 and pollutant emissions.' If implemented, however, the net load factor increase and CO_2 reduction would be relatively small, as cabotage represents only 3 per cent of road tonne-kms moved within EU member states and this domestic haulage accounts for two-thirds of all trucking activity in the EU (DG Move, 2015). Elsewhere in the world, the carbon benefits of cabotage liberalization would be larger, as, for example, in the maritime sectors of Indonesia, Malaysia and the Philippines (World Bank, 2014).

Construction and use regulations

Regulations impose limits on the size and weight of vehicles and hence their carrying capacity. Relaxing these limits allows companies to consolidate loads, thereby reducing vehicle kilometres, fuel consumption and CO_2

emissions. It can also improve the average load factor, though this depends on how this metric is defined, the density of the loads and the relative extent to which weight and size limits are raised. For example, increasing the maximum weight of a truck that is transporting a dense product and which 'weighs out before it cubes out' allows more of the available space in the vehicle to be filled. This was well illustrated by the changing pattern of truck utilization in the UK following the government's decision in 2001 to increase maximum gross weight from 41 tonnes to 44 tonnes. Over the following nine years, the proportion of loads simultaneously reaching weight and volume constraints in articulated lorries rose from 7 to 37 per cent. Partly as a consequence of this improved loading, around 136,000 tonnes of CO_2 were saved during the first three years of the higher weight limit (McKinnon, 2005). It is, nevertheless, argued by opponents of truck size and weight increases that since transport operators underuse capacity in the current generation of vehicles they do not deserve any extra capacity. Such reasoning is rather simplistic, however, as it fails to take account of all the causes of underutilization discussed earlier in this chapter.

Vehicle utilization is only one of many points of contention that have made increases in truck size and weight limits one of the most controversial issues in the field of logistics. In a more detailed review of this issue (McKinnon, 2014b) I describe it as a 'lethal cocktail of empirical evidence, sectional interests, political lobbying and emotion'. In assessing the relationship between truck size/weight limits and CO_2 emissions we must look beyond vehicle utilization to a host of interrelated factors including modal split, demand elasticity and energy use. Discussion of the case of raising truck size and weight issues therefore spans Chapters 4, 5 and 6 and cannot be easily pigeonholed within the book's conceptual framework. Since it is an issue primarily related to the provision and use of vehicle carrying capacity, we will examine the relationship here. Before doing so, it is important to clarify what is meant by increasing truck size and weight. The increase must be judged relative to some baseline. In an influential report, the International Transport Forum (2010) identified the baseline as 'workhorse' vehicles being responsible for the bulk of road freight movement in the countries they studied. Rather than refer to vehicles above the baseline being longer and/or heavier (LHV), the term 'high-capacity transport' (HCT) is increasingly being used to embrace both weight and volumetric increases. This is a generic term and does not imply any particular percentage increase in either dimension. Therein lies the problem of generalizing the results of research on the links between the move to HCT and CO_2 emissions.

Some of this research takes a vehicle fleet perspective, while other studies have attempted to make macro-level assessments at country or continental levels. Figure 5.4 summarizes the findings of six studies in the former

Figure 5.4 Estimated per cent reduction in carbon intensity of longer and heavier vehicles against baseline vehicle

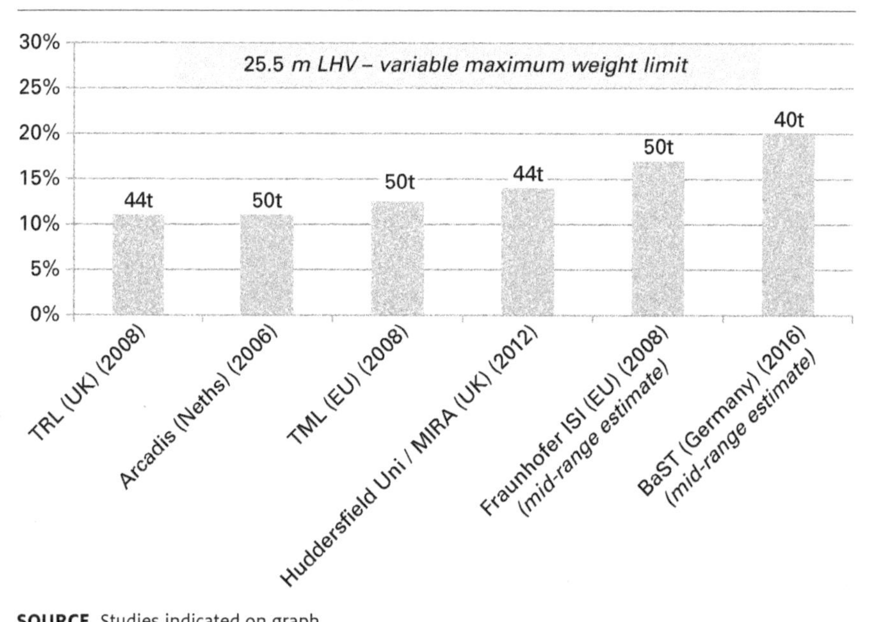

SOURCE Studies indicated on graph

Figure 5.5 Relationship between truck capacity and carbon intensity: 36 vehicle classes in nine countries and the EU: gCO_2 per metre³ ton-km

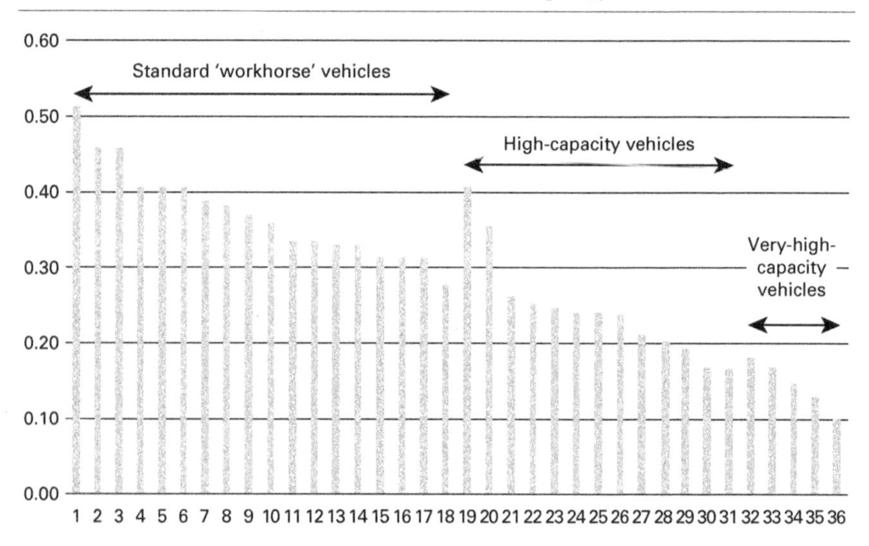

SOURCE International Transport Forum (2010)

category that have estimated the percentage reductions in CO_2 emissions from an increase in the maximum length of trucks to 25 metres with maximum gross weight in the 40–50-tonne range.

In most cases, this permits the substitution of two vehicles for three. Once allowance is made for the greater weight and size of the HCT configurations, net CO_2 savings were found to be between 10 and 20 per cent. Figure 5.5 uses data compiled by the International Transport Forum (2010) on a large sample of vehicles operating in different parts of the world to show how the carbon intensity of trucks declines as their capacity increases. Carbon intensity is measured here against a composite capacity index of weight, volume and distance travelled (metre^3–tonne-km) and assumes that the vehicle is optimally loaded with respect to this index. If the carrying capacity is well utilized, HCT can deliver significant decarbonization at a vehicle fleet level.

One cannot simply extrapolate these fleet-based carbon reductions to a country's freight transport system. At a national level, allowance must be made for three other factors:

1. *Greater circuity in the routeing of HCT vehicles.* Because of their greater size, in particular length, they are typically confined to a more restricted primary road network and denied access to inner-urban areas. Other things being equal, this increases average journey length relative to that of the baseline vehicle. Efforts have been made to improve the manoeuvrability of longer articulated vehicles by redesigning their coupling and installing 'steer axles' (Odhams et al, 2011). The adoption of a so-called 'performance-based standards' (PBS) approach to HCT, pioneered in Australia, has helped to tailor the vehicle to its route network, while a GPS-based 'intelligent access programme' (IAP) ensures that bigger trucks do not deviate from their specified networks. These initiatives help to minimize the distance penalty associated with HCT.
2. *Induced demand for freight movement.* Critics of HCT have long argued that making road transport cheaper encourages companies to move more freight. It might, for example, promote greater centralization of inventory or wider sourcing of supplies, both trends generating more tonne-kms and offsetting some of the carbon savings from load consolidation. Little *ex post* evidence of such a 'rebound effect' was found after the 2001 increase in maximum truck weight in the UK (McKinnon, 2005). In making *ex ante* estimates of the magnitude of induced traffic effects, researchers must choose or derive demand elasticity values. These measure the extent to which a reduction in the cost of road transport will

generate additional freight movement. The range of published elasticity values is quite large (De Jong et al, 2010), giving protagonists in the debate over HCT some latitude in their choice of value and weakening the scientific basis of the analysis.

3 *Modal shift.* HCT-related reductions in the unit cost of road transport can also induce a modal shift from lower-carbon rail or waterway services, as discussed in the previous chapter. Such a modal shift erodes the carbon savings calculated at a truck fleet level. Macro-level HCT studies have factored 'cross-modal' demand elasticity values into their calculations to estimate the strength of this rebound effect. The results are again highly sensitive to the choice of elasticity value, as illustrated by two studies published in 2008. One, undertaken by Transport and Mobility Leuven et al (2008) for the European Commission, modelled the effects of allowing LHVs on EU motorways and found that 'rail and waterway volumes would respectively decrease by 3.8 per cent and 2.9 per cent' (p.72). The other by Doll et al (2008), commissioned by the rail sector (CER), estimated that in a similar scenario EU rail tonne-kms would decline by between 11.6 and 12.9 per cent. In a review of eight European studies on the overall impact of 'megatrucks' for the European Parliament, Steer Davies Gleave (2013) observed that empirical studies, based on real-world evidence, 'show better outcomes than those predicted by desk studies, with lower modal shift observed' (p.76). A majority of the eight studies supported the view that relaxing truck size and weight yields a net reduction in GHG emissions overall.

The 'real-world evidence' that HCT significantly reduces the carbon intensity of trucking has been steadily accumulating. Following extensive trials in the Netherlands, NEA (2011), in a study for the Dutch government, concluded that 'LHVs definitely play a role in the reduction of CO_2 emissions, both regarding road transport and in connecting to intermodal transport... For road transport this is an important innovation in the effort to reduce CO_2 emissions' (p.34). After five years of HCT trials in five German states, it was concluded that '... the deployment of longer trucks demonstrated a positive impact on transport demand in terms of a reduction in the number of vehicle kilometres driven and accordingly also a reduction in levels of climate change gases...' (BASt, 2016). Looking to the future, Transport and Mobility Leuven (2017) see deployment of HCT as cutting CO_2 emissions from long-haul road freight by 3.5 per cent by 2030 and 7.5 per cent by 2050, while Greening et al (2015) anticipate that

HCT will have to account for a large share of 'demand-side' CO_2 reductions for the UK road freight sector to meet its 2035 carbon reduction targets.

Overloading of vehicles

In much of the developing world the overloading of vehicles can be as serious a problem as underloading. The engines of overloaded trucks have to labour, using excessive fuel and increasing the carbon intensity of the haulage operation. Research in Indonesia, where it is estimated that overloaded vehicles carry payloads on average 45 per cent above the legal limit, found that there was a linear relationship between the degree of overloading and the amount of CO_2 emitted (Wahyudi et al, 2013). Overloading causes excessive engine wear, degrading the vehicle's fuel efficiency even when carrying loads within legal limits, especially in countries where truck maintenance standards are low. The overloading of truck axles also damages the road, particularly in countries with a large proportion of unpaved roads. It was estimated that in South Africa the 10–15 per cent of trucks that were overloaded caused 60 per cent of all road damage (CSIR, 1997). When combined with under-maintenance of road infrastructure, truck overloading causes the road surface to become uneven and pot-holed, forcing all categories of traffic to run below their most fuel-efficient speed and causing frequent delays.

In many developing countries the 'workhorse' vehicle is a two- or three-axle open-top truck which, as it lacks a roof, can have freight stacked high and covered with a tarpaulin. Intensively competitive conditions in local haulage markets and under-investment in vehicle fleets make overloading endemic and very difficult to control. Enforcement of weight limits is often very lax, corruption rife and financial penalties small relative to the economic benefit of exceeding the maximum weight. In some countries, however, serious efforts are now being made to curb the level of overloading. In India and Indonesia, for example, weighing devices are being installed at toll plazas and operators given punitive fines for infringement of the weight regulations. This tighter weight enforcement regime is reckoned to be one of the reasons for the steep increase in demand for new heavy-duty trucks in India. Wider deployment of 'weigh-in-motion' devices across the road network also helps to discourage overloading (Pinard, 2010).

Improving vehicle utilization in urban areas

There have been over 40 years of research on ways of improving the loading of delivery vehicles in urban areas. Until recently, this rationalization of urban distribution was seen primarily as a means of easing traffic congestion, saving fuel and cutting emissions of noxious gases. Today these goals are being supplemented by the need to decarbonize city logistics. The European Commission (2011) in its transport white paper set an objective of achieving 'near-zero-emission urban logistics' by 2030, including CO_2 among the emissions to be eliminated. Meanwhile, the rapid expansion of online shopping has transferred logistical responsibilities from the consumer to the retailer or a carrier working on its behalf. Much of the freight that used to be moved in cars, buses and metros, where it was statistically 'invisible', is now transported in commercial vehicles whose carbon emissions are separately identified and subject to emission-reduction targets. Containing the growth of these emissions on the 'last mile' will require, among other things, greater emphasis on vehicle loading, particularly as online customers are being offered ever shorter delivery times.[10]

The conventional approach to increasing the average load factor of urban freight vehicles is to channel orders through a consolidation centre. At this centre, orders destined for a particular location, such as a shop or building site, are aggregated into loads that are large enough to be delivered directly. Multiple drop rounds are then replaced by a system of radial distribution with vehicles well filled in the outbound direction, but usually empty on the way back. Overall, this form of urban freight consolidation cuts vehicle kilometres, fuel consumption and emissions. In a detailed review of 24 urban consolidation centre (UCC) schemes between 1970 and 2010, Allen et al (2012: 481) found 'improvements in vehicle load factors ranged from 15 to 100 per cent, reductions in vehicle trips and vehicle kilometres travelled were typically between 60 and 80 per cent and reductions in greenhouse gas emissions from these transport operations ranged from 25 to 80 per cent.'

Many of these were pilot schemes which proved not to be financially sustainable in the longer term. In recent years, however, an increasing number of UCCs have become economically viable, saving their users money as well as offering a range of environmental co-benefits. For example, the Binnenstad service, launched in 2008, now consolidates retail supplies in the inner urban areas of 14 Dutch towns and cities (van Rooijen and Quak, 2010). Organized mainly by the receivers of the goods, the service is estimated to reduce operating costs by 10 per cent and CO_2 emissions by 40 per cent. While the main application of urban freight consolidation is in the retail sector, the concept has also worked successfully in the construction industry in London and Stockholm. The first London Construction Consolidation Centre reduced the number of vehicles delivering to four construction sites by 60–70 per cent and cut CO_2 emissions by 70–80 per cent (Transport for London, 2008). Transport for London has also pioneered a form of consolidation that does not require the establishment of a UCC. By encouraging the management of buildings like offices, hospitals and football stadiums to devise 'delivery and servicing plans' (DSPs) it is possible to consolidate the inbound flow of supplies into fewer, better-filled vehicles, possibly by channelling the orders through the premises of a 'lead supplier'. Analysis of five DSP projects in London found that both operating costs and externalities (including CO_2 emissions) were substantially reduced (Leonardi et al, 2014).

Summary

Companies running trucks are regularly criticized for underloading them and needlessly exacerbating congestion and pollution problems. Empty running and poor load factor statistics are often cited as evidence of mismanagement and a wasteful use of transport resources. This chapter has tried to correct common misconceptions about the underutilization of these logistics assets. It has questioned the validity of much of the available macro-level data on vehicle loading, which in most countries and for modes other than road is very limited. It has also shown that there are numerous interrelated reasons for vehicles not running full on every trip. Only by making a proper diagnosis of the problems of underloading, and in lower-income countries overloading, can we realistically assess the contribution that optimized loading could make to decarbonization. Many companies are quite prepared to sacrifice loading efficiency in the pursuit of other logistical and business goals. As this chapter has shown, however, it is possible to increase loading in ways that do not require a trade-off between cost and carbon. On the contrary, improving vehicle fill typically has a negative carbon mitigation cost, simultaneously improving the economic as well as carbon efficiency of road freight operations. Making these operations more competitive can frustrate efforts to shift freight from road to rail and waterborne services. As was discussed in the previous chapter, however, this is only one of many impediments to these low-carbon transport modes capturing a much larger share of the freight market.

Notes

1 Utilization of deck area is sometimes expressed in linear terms as 'load length'.
2 The term 'cubing out' generally applies to both the space and deck-area utilization.
3 EU rules give them discretion to set this minimum value within the range 3.5 tonnes–6 tonnes.
4 The SmartWay Partnership was established by the US Environmental Protection Agency in 2004 to help shippers and carriers to 'reduce their transportation footprint', primarily in environmental terms.
5 The movement of ballast water is a major source of environmental concern, not because of its impact on the carbon intensity of shipping operations, but because the uptake and discharge of ballast water can seriously disturb local marine ecosystems (NOAA, 2017).
6 This simplifies the list of 10 loading constraints presented in McKinnon (2015).
7 Lloyd Shapley won the Nobel Prize for economics in 2012 for his work on game theory.
8 ALICE stands for 'Alliance for Logistics Innovation through Collaboration in Europe'. It is partly funded by the EU and has over 100 member companies. http://www.etp-logistics.eu.
9 An 'own account' operator is a company, such as a manufacturer or retailer, whose main activity is not transport and which runs an 'in-house' fleet to move its own products.
10 The implications of the switch from conventional to online retailing for logistics-related CO_2 emissions is discussed in Chapter 8 as much of the research on this subject has been conducted in the UK.

Transforming energy use in the road freight sector 06

Logistics and energy

Logistics is a very energy-intensive business activity. The International Energy Agency (2016) estimated that in 2013 transport as a whole accounted for 26 per cent of final energy demand worldwide and that freight transport had a 40 per cent share of this total. This means that freight transport by all modes consumes around 10 per cent of the world's energy. No comparable data exists for freight terminals and warehouses, though on the basis of emissions calculations by the World Economic Forum/Accenture (2009), one can conservatively estimate that logistics as a whole is responsible for 11–12 per cent of global energy use. All but a small fraction of the energy used comes from the burning of fossil fuels, predominantly diesel. The combination of high energy intensity and heavy dependence on fossil fuel makes logistics a particularly difficult sector to decarbonize, as acknowledged by Guérin, Mas and Waisman (2014). Much of the effort to decarbonize logistics must therefore be directed at energy use, supplementing the carbon savings that can be made by applying the initiatives outlined in the previous three chapters. Nothing less than a transformation of logistics energy use will be needed to achieve the deep carbon reductions required over the next few decades. This transformation will entail both a step change in energy efficiency and a dramatic shift from fossil to renewable energy.

This chapter reviews the options for achieving both in the road freight sector. The next chapter does the same for shipping, air cargo and rail freight. Trucking gets a chapter all to itself for two reasons; partly because it accounts for 75 per cent of final energy use in the freight sector (International Energy Agency, 2017a), but also because energy consumption by trucks has generated much more research and attracted more attention from public policy makers than the other three modes. This and the following chapter are split into two parts. The first examines the contribution that new technology and operational improvements can make to increasing energy efficiency. The second assesses the GHG impacts of switching from fossil fuel to alternative low-carbon energy sources.

It is important to start with a few general observations about energy use in logistics.

a *Generalization.* While new technology will deliver much of the necessary reduction in energy use and emissions in logistics, there is no single technical fix but instead an array of technologies that can collectively cut the energy use per unit of freight movement, storage and handling. In assessing the contribution that technology will make, it is important to recognize international differences in both the current technological baseline and the rate at which technical innovations will be adopted. This makes it hard to generalize about global trends and decarbonization potentials in logistics energy use.

b *Measurement.* Various ratios are used to measure the energy efficiency of logistical activity. They generally have energy as the numerator; the units of measurement may vary (eg quantities of fuel, joules or megawatt-hours (MWh)) but it is always energy that is being measured. Logistical activity, on the other hand, can be measured in quite different ways. In the case of freight transport operations, the denominator can be vehicle-km, tonne-km, container-km, pallet-km or m^3 of product per km. Some researchers have even devised a composite tonne-km-m^3 metric to capture weight, volume and distance in a single variable (International Transport Forum, 2010). Where energy use is related to both the amount of freight carried and the distance it is moved, the term 'energy intensity' is normally used. This distinguishes it from 'energy efficiency' which typically relates energy use only to the distance a vehicle travels (eg litres of fuel per 100 km). This is an important distinction. Energy intensity is a combined measure of how well the vehicle is loaded and how fuel-efficiently it is operated. As the opportunities for improving vehicle loading were discussed in the previous chapter, attention here will focus on energy efficiency measured on a vehicle-km basis. It should be noted, however, that loading and energy use are inversely related. The fuel consumption of a heavy truck making the same journey with varying payloads was monitored by Coyle (2007). He found that on average, fuel consumption per km increased by 1.3 per cent for each additional tonne of payload weight. Some of the carbon benefits from improved loading are therefore offset by higher energy use per vehicle-km. However, since load consolidation reduces total vehicle-kms, overall energy consumption can be reduced. The heterogeneity of warehousing and freight terminal operations has long frustrated efforts to measure their energy efficiency on a consistent basis (Rudiger, Schön and Dobers, 2016). Warehouses vary enormously in their size, design, functionality and degree of mechanization and this is reflected in their energy use per unit of throughput. Benchmarking energy use per $metre^2$ of floorspace, $metre^3$ of cubic capacity or shelf space is therefore only meaningful for narrowly defined categories of building. Energy efficiency benchmarking can be applied more effectively at a disaggregated level to particular warehousing activities or equipment types such lighting, heating, cooling, materials handling etc, using a bottom-up approach to the analysis of potential energy savings.

c *Cost-effectiveness*. A distinction is sometimes made between the relative cost-effectiveness of cutting GHG emissions by efficiency improvement or switching to renewable energy. It is often easier to make the case for the former because it simultaneously cuts costs and GHG emissions and often provides a rapid financial payback on any investment that is required. The cost-effectiveness of repowering with greener forms of energy is much harder to determine because of uncertainties surrounding the future availability and cost of low-carbon energy, the development of new infrastructures to supply it and the actual GHG savings that alternative energy sources offer on a life-cycle basis. These issues are discussed later in this chapter.

Improving the energy efficiency of trucks: the role of technology

In the 1960s, 1970s and 1980s, advances in technology drove 0.8–1 per cent improvements per annum in the fuel efficiency of new trucks (International Energy Agency, 2007). This favourable trend was interrupted in the 1990s by the introduction and steady tightening of controls on NOx emissions from vehicle exhausts. To depress NOx levels it is necessary to raise combustion temperatures in the engine and this requires more fuel. So, in an effort to cut NOx emissions, potential fuel and CO_2 savings were lost, possibly by as much as 7–10 per cent in Europe up to 2007 (International Energy

Agency, 2007). Using real-world vehicle test data compiled by the German trucking magazine *Lastauto Omnibus*, Muncrief and Sharp (2015: 6) conclude that 'the fuel consumption of tractor trailers in the EU has not changed significantly over the past 13 years.' This claim has been disputed by Daimler, one of the largest EU truck manufacturers. Its independently audited data suggests that its new articulated vehicles in 2016 were 14.7 per cent more fuel efficient than their comparable predecessors of 2003, despite being three Euro emission classes higher (Euro VI as opposed Euro III) (Schuckert, 2016).

While there is dissent over the past trend in truck fuel efficiency, there is general agreement that the future trend can be sharply upwards. This has been demonstrated in the United States where the SuperTruck programme initiated by President Obama has managed to increase the brake thermal efficiency of a Class 8 truck, the workhorse of US long-haul trucking moving 70 per cent of US freight tonnage, from 42 per cent to 50 per cent (The White House, 2014). This has involved the co-ordinated deployment of several technologies in a prototype vehicle whose fuel efficiency is 50 per cent higher than the current average (3.5 km per litre as opposed to 2.3 km per litre). The White House report declared that 'if all Class 8 vehicles in the United States were SuperTrucks, the country would consume nearly 300 million fewer barrels of oil and spend nearly $30 billion less on fuel each year' (p.7). This would cut annual CO_2 emissions by roughly 150 million tonnes per annum. A series of European-based studies (eg AEA Technology, 2011; Schroten, Warringa and Bles, 2012; Delgado, Rodriguez and Muncrief, 2017; T&E, 2017) have suggested that refinements to existing technologies can significantly improve the fuel efficiency of new trucks in the short to medium term, with relatively short financial payback periods. A marginal abatement cost analysis by Schroten, Warringa and Bles (2012: 33) showed that most of the technologies were 'cost-effective under almost all assumptions', including having a payback period of less than three years. It estimated that collectively these truck technologies could cut long-haul fuel consumption by 33–36 per cent. The fuel efficiency of European trucks has traditionally been higher than that of their North American counterparts, as borne out in a literature review by Sharpe and Muncrief (2015). This is partly because much higher fuel taxes and prices in EU countries have incentivized manufacturers to build more fuel-efficient vehicles and because fuel consumption has been a more influential vehicle purchase criterion. So further gains in fuel efficiency in Europe will be from a higher baseline.

Given this substantial body of research confirming the future role of technology in decarbonizing trucking, we must find ways of incentivizing its uptake. In countries which tax diesel fuel heavily, such as the UK, high fuel prices provide a strong economic incentive. It is doubtful, however, whether in most countries tax policy and market forces on their own will provide sufficient impetus. Government efforts in Finland to get truck operators to sign up to a voluntary energy efficiency agreement, which included a commitment to technical upgrades, had only limited success (Liimatainen et al, 2012). T&E (2017: 14) explain that many fuel efficiency technologies are offered as optional 'eco-packages' by truck manufacturers and that carriers operating on slim profit margins cannot afford to acquire them. It is argued, therefore, that the adoption of these technologies has to be reinforced by the introduction of fuel economy standards for trucks, requiring all new vehicles to achieve at least a certain minimum level of fuel efficiency. Such standards are now in place in Japan, the United States and China and are being prepared for several other countries, including India, Mexico and Korea. The first phase of the US standards, introduced in 2014, applies only to tractor units, while Phase 2, starting in 2018, will also cover trailers. These Phase 2 regulations are described by ICCT (2016: 1) as 'ambitious, far-sighted and achievable' and likely to 'reduce carbon dioxide (CO_2) emissions from the vehicles affected by 1 billion metric tons over their lifetimes.'

Widespread adoption of fuel economy standards for trucks will be necessary to meet the so-called '35 by 35' target set by partners belonging to the Global Fuel Economy Initiative (2017). This aims to achieve a 35 per cent improvement in the average fuel efficiency of new trucks worldwide by 2035 (against a 2015 baseline). It is estimated that this would 'avoid 1–2 billion tons of CO_2 emissions per year by 2035 – an improvement of more than 20 per cent compared to a business-as-usual scenario' (p.2).

The EU has adopted a different approach to truck fuel economy. It has argued that because of the heterogeneity of truck configurations, it is necessary to establish a consistent system of energy and emission certification, measurement and reporting (CMR) before imposing fuel economy standards. Having rejected the physical testing of all truck variants as 'very burdensome', the European Commission has opted instead for the computer simulation of new trucks using a software tool called VECTO (Vehicle Energy Consumption Calculation Tool). This 'can be used cost-efficiently and reliably to measure the CO_2 emissions and fuel consumption of HDVs for specific loads, fuels and mission profiles (eg long-haul, regional delivery, urban delivery, etc.), based on input data from relevant vehicle components' (DG CLIMA, 2017).

Once this system is operating effectively, the Commission will specify CO_2 standards for particular truck configurations, probably in 2018. In the meantime, VECTO is introducing greater energy and CO_2 transparency into the European truck market, though this may 'only curb truck CO_2 emissions to a very small extent' (T&E, 2017: 14). Delgado, Rodriguez and Muncrief (2017: 2) regard 'waiting for official fleetwide CO_2 values' as an 'unnecessary delay' and believe that a simpler, quicker method of determining the 'regulatory baseline' could have been used. Their analysis suggests that, because of this delay, truck fuel efficiency is now increasing at a significantly faster rate in North America. When the EU joins the United States, Japan, China, India, Korea and Mexico in imposing truck fuel economy standards, approximately 70 per cent of global truck sales will be subject to such standards (International Energy Agency, 2017a).

In much of the developing world, hauliers are heavily dependent on imports of second-hand vehicles from Europe and North America. Only when these vehicles are internationally traded, typically after four to six years, do rising fuel economy standards in the developed world improve the carbon efficiency of trucking in lower-income countries. By then the fuel/ CO_2 performance of the vehicles may have significantly degraded. In the foreign market, the rate of degradation generally accelerates as a result of more intensive use, poor infrastructure, under-maintenance and a lack of qualified mechanics and spare parts. The dissemination of low-carbon truck technology to the developing countries, where much of the future growth in road freight will occur, is therefore relatively slow and ineffective. The governments of these countries can facilitate the technology transfer by controlling the maximum age and emission standards of imported trucks and operating a scrappage scheme for old vehicles, as happens in countries such as Mexico, Colombia and Egypt (International Energy Agency, 2017a). These policies can be difficult to enforce and fund, however.

Technological improvements to trucks can be classified into five categories: aerodynamics, engine/powertrain (transmission), tyres, lightweighting and control systems/platooning/automation.

Aerodynamics

Modern trucks are much more streamlined than their predecessors of 20–30 years ago whose 'brick' shape created a great deal of wind resistance, particularly at higher speeds. At slower speeds, most of the energy is required to overcome tyre rolling resistance on the ground and internal friction within the engine, gearbox and transmission. As the vehicle accelerates, aerodynamic drag becomes the main force impeding vehicle movement. The International Energy Agency (2017a: 75) notes that 'the drag force increases at the square of the speed; at typical highway speeds (90km/hr–120km/hr), it accounts for most of the tractive energy requirements.' In the early days of aerodynamic profiling this was reduced by installing so-called 'spoilers' over the cab, closing gaps at the sides of the vehicles (with 'fairings' or 'side skirts') and enclosing wheel arches (with 'spats'), which taken together was shown to cut fuel consumption by UK lorries by between 6 and 20 per cent (Freight Best Practice Programme, 2006). Since then a 360-degree approach has been applied to truck profiling which not only streamlines the front and sides of the vehicle but also the top, rear and underside. 'Teardrop' trailers with curved tops were introduced in the UK in 2007 and into France and Germany in 2014. They help to reduce drag caused by turbulence at the back of a trailer. This turbulence accounts for as much as a third of total air resistance and halving it can save around 4–5 per cent of fuel. In the United States, retractable flaps called 'boat tails' are becoming more widely installed to improve rear air flow. Until 2016, legal limits on truck length in the EU discouraged the use of boat tails, but these limits were relaxed by an EU Directive which granted 'derogations on the maximal lengths to make heavy goods vehicles greener by improving their aerodynamic performance.'[1] Meanwhile, it was claimed in 2014 that over 30,000 US trucks had 'undertrays' retrofitted to smooth airflow under the vehicle, cutting fuel consumption by up to 6 per cent (Morgan, 2014).

Among truck operators in developed countries, aerodynamic profiling has become one of the most popular means of improving energy efficiency. Across 17 large US truck fleets, the proportion of tractor units with aerodynamic profiling increased from 36 per cent to 64 per cent between 2003 and 2015, while the corresponding figure for trailer aerodynamics rose from 1 per cent to 21 per cent (NACFE, 2016). Aerodynamic kits will be increasingly fitted as standard to new tractors and trailers. Several of the large truck manufacturers have also designed concept vehicles with radically reshaped chasses which can drastically reduce drag, though within current vehicle length limits, at the expense of cubic carrying capacity. Transport and Mobility Leuven (2017: 6) suggests that the emission reductions from the streamlining of trucks in Europe are so great that 'legislators should consider introducing incentives to accelerate the market uptake of new technologies that improve vehicles' aerodynamic performances.'

As AEA (2011: 102) explain, however, the fuel benefit of aerodynamic profiling is 'not universal and is related to the vehicle mission profile'. Vehicles undertaking local collection/delivery work at relatively low speeds on a stop–start duty cycle obtain relatively small fuel savings from streamlining. The same applies to much of the road haulage in developing countries where a combination of poor infrastructure, congestion and vehicle age constrain vehicle speed even on long-haul operations. At around 35 kph, the average speed of an Indian truck is roughly half that of a European one. In China, where trucks travel more slowly and have a higher weight limit than their US counterparts, improvements to the rolling resistance of tyres can yield greater fuel savings than improved aerodynamics (Curry et al, 2012). As road infrastructure and vehicle performance are upgraded across the developing world, the fuel and CO_2 benefits of aerodynamic profiling will become more widely dispersed.

To date, attention has focused on the aerodynamic properties of the individual vehicle. Platooning offers the potential to streamline convoys of vehicles by linking them electronically and thereby reducing the gap between them. This is discussed below as a 'control' technology.

Engines/powertrains

In a review of low-carbon truck technologies, Atkins (2010) groups engine/powertrain developments into four categories:

1. *Waste heat recovery*. It is estimated that when a large truck engine converts energy to motion about 45 per cent of the energy is dissipated as waste heat. The recovery and reuse of this wasted energy, by means of a process known as 'turbo charging', can cut a truck's CO_2 emissions by 5 per cent (International Energy Agency, 2017a).
2. *Higher combustion efficiency*. This can be achieved, for example, by increasing the pressure at which fuel is injected into the engine, offering 2–3 per cent CO_2 savings
3. *Ancillary equipment*. Use of separate energy sources (such as batteries) to power ancillary equipment (such as air conditioning) rather than the main engine can cut carbon emissions by a few percentage points depending on the alternative source.
4. *Automated manual transmission*. This allows the vehicle to be driven more fuel-efficiently, particularly in urban drive cycles with frequent gear changes. Depending on the driving conditions, shifting from manual to automatic transmission can cut fuel consumption by 1–8 per cent.

Overall, the International Energy Agency (2017a) estimates that 'in the near term' engine improvement can raise the fuel efficiency of local delivery vehicles by 4 per cent and long-haul trucks by 18 per cent.

Hybridization also falls under the powertrain heading as the combination of battery power with an internal combustion engine (ICE) significantly reduces CO_2 emissions. It represents a transitional phase in the electrification of freight transport fleets. Its use in the road freight sector is still comparatively rare and confined to short-haul collection and delivery work where fuel and CO_2 savings of up to 30 per cent are reported (eg LowCVP, 2015; International Energy Agency, 2017a). Hybrids represented only 3 per cent and 1.4 per cent of new van sales in, respectively, the United States and EU in 2014 (Rodriguez et al, 2017) and are virtually absent from the heavy-duty truck sector. Vehicle numbers are growing, though from a low base. For example, UPS decided in 2016 to expand its fleet of hybrid electric delivery vans from 125 to 325. It is difficult to generalize about CO_2 reductions from hybridization as it comprises several different technologies, such as the standard diesel-electric, hydraulic, parallel hydraulic and even a flywheel variant, whose impact on carbon emissions vary with the vehicle body type and delivery duty cycle. Prototype hybrid long-haul trucks have been developed to demonstrate that hybridization has wider applicability in the trucking sector. For example, in combination with several other energy-saving technologies, Volvo's hybrid concept vehicle can achieve CO_2 savings of 30 per cent on a long-haul duty cycle. Overall, the cost-effectiveness of hybridization is constrained by the relatively high cost of the vehicles and the limiting effect of battery weight on carrying capacity.

Control systems, platooning and automation

Advances in vehicle hardware are being accompanied by major IT-related improvements, transforming the software side of trucking. Onboard computer monitoring and management of vehicle systems has been steadily upgrading fuel efficiency for many years. Some of the efficiency gains have been intrinsic to the powertrain, while others have supported vehicle maintenance activities and improved driving performance. For example, the installation of anti-idling devices has curbed the wasteful practice of allowing truck engines to idle while the vehicle is stationary. In the United States, idling accounts for around 8 per cent of truck fuel consumption, emitting around 20.3 million tonnes of CO_2 annually (equivalent to the total CO_2 emitted from the electricity consumed by 2.8 million American homes) (NACFE, 2014). Around 2.5 per cent of this fuel is burned to support heating, cooling and in-cab services during the drivers' rest periods (Vernon and Meier, 2012) but this still leaves much unnecessary idling than can be reduced by anti-idling devices installed in the vehicle or by engine 'stop–start' systems.

Fuel efficiency can also be enhanced while the vehicle is in motion by the use of 'adaptive cruise control' (ACC) where the vehicle speed is automatically adjusted to that of other vehicles in the traffic flow. It is claimed that this can cut fuel use by up to 10 per cent (Transport and Mobility Leuven, 2017) depending on the driver and traffic conditions. 'Predictive cruise control' (PCC) takes this technology to the next level and uses GPS data to adjust the vehicle speed in anticipation of the topography of the road ahead. Fuel efficiency gains of 2–5 per cent have been reported for PCC (Transport and Mobility Leuven, 2017; Rodriguez et al, 2017). These so-called 'advanced driver assistance system' (ADAS) technologies are well-established and extensively commercialized in Europe, but less so in the United States and hardly at all in less developed countries. Respectively, 50 per cent and 20 per cent of the tractor units in new articulated vehicles sold in the EU have built-in ACC and PCC, while the equivalent figures for the United States are 10 per cent and 2 per cent (Rodriguez et al, 2017). These systems are soon to be complemented by two more radical technologies which are still at the prototype stage: platooning and automation.

As mentioned earlier, platooning cuts fuel consumption by 'bunching' vehicles into closely coupled convoys to reduce air resistance. The coupling is electronic and gives control of the convoy to the driver of the lead vehicle. After much experimentation on test tracks, platooning has now been trialled on the road networks of several countries and shown to work effectively. The European Truck Platooning Challenge in April 2016 successfully demonstrated a cross-border application of the concept (Ministry of Infrastructure and Environment, 2016). On the basis of numerous trials, many figures are now quoted for the potential fuel savings from platooning (eg Ricardo et al, 2014; American Trucking Association, 2015; Cornelisson and Janssen, 2017). These average between 5–10 per cent. Most of the fuel savings accrue to the following vehicles, though the fuel efficiency of the front vehicle also increases because its rear wind turbulence is reduced. In trials reported by Lammert et al (2014), the fuel efficiency of the 'lead' vehicle rose by 5.3 per cent while that of the 'trailing' vehicle went up by 9.7 per cent. Estimates are clearly sensitive to a range of variables including the aerodynamic profiling of the individual vehicles, the traffic conditions, the speed of the convoy and the degree of separation between the vehicles. One study of a convoy of three trucks travelling at 80 kph found that a gap of 20 metres yielded a fuel saving of 5 per cent while one of four metres cut fuel use by 15 per cent. Having reviewed available research on the subject, Transport and Mobility Leuven (2017) estimates that by 2030 around 10 per cent of long-haul road freight in Europe will move in platoons, rising to 40 per cent by 2050. Other things being equal, this uptake of platooning would cut CO_2 emissions from long-haul European trucking by 1 per cent by 2030 and 4 per cent by 2050.

The trailing vehicles in platoons effectively become driverless, as the operator of the lead vehicle controls the whole convoy. The next step in this technological pathway will see individual trucks become driverless. Prototype autonomous trucks have already been trialled. In a much-publicized trial in 2016, a US Class 8 truck carrying beer travelled autonomously for 120 miles on the interstate freeway network in Colorado (Davies, 2016). Some commentators believe that automation will be implemented more rapidly in truck fleets than in car fleets partly because of their centralized control, but also because of greater commercial benefits in the freight sector. The International Transport Forum (2017a) has constructed scenarios of the possible roll-out of driverless trucks in long-haul and urban delivery operations. Its 'disruptive scenario' for long-haul road freight envisages rapid adoption from 2020 onwards, reaching a 90 per cent market penetration by 2040. In its 'conservative scenario' the adoption process does not start until 2030 and over the following decade reaches 25 per cent market adoption. At an urban scale, the penetration level varies between 20 and 56 per cent by 2040. How would such automation of trucking operations impact on fuel consumption and CO_2 emissions?

There is general agreement that computers can drive trucks more fuel-efficiently than humans. The International Energy Agency (2017a) suggests that automation has the potential to improve the fuel efficiency of a truck by 15–25 per cent. Roland Berger (2016) predict that by 2040 automation will cut the fuel consumption of US trucks by 10 per cent below what it would otherwise be, allowing for other efficiency improvements in the meantime. It is clearly difficult at this stage to accurately estimate the impact on fuel consumption of such a disruptive technology as truck automation at an aggregate level, particularly as over the predicted timescale for its implementation it is likely to be accompanied by platooning and electrification. There is little doubt, however, that at the adoption levels forecast by the International Transport Forum (2017a), vehicle automation could become one of the main ways of decarbonizing road haulage.

Tyres

In developed countries the main tyre innovations likely to cut carbon emissions are low rolling resistance, tyre pressure monitoring systems and automatic inflation. It is estimated that the current generation of low-rolling-resistance tyres can cut fuel consumption by up to 5 per cent depending on the vehicle speed. According to Meszler, Lutsey and Delgado (2015) there is the longer-term prospect of a second generation of low-resistance tyres offering 12 per cent improvements in fuel efficiency relative to today's standard tyres. Underinflation of tyres carries a fuel penalty: as a rule of thumb, 20 per cent underinflation reduces fuel efficiency by around 2 per cent (Department for Transport, 2006a). Although pressure monitoring systems exist, their current market penetration in the United States, EU and China is very low (Rodriguez et al, 2017). Devices which automatically inflate truck tyres to the correct pressure are also commercially available but are very rarely used. Transport and Mobility Leuven (2017) is, nevertheless, optimistic about the adoption of tyre technologies in Europe over the next few decades. It expects them to cut CO_2 emissions in the road freight sector by 7.5 per cent by 2030 and 12.5 per cent by 2050.

In some developing countries, most notably India, tyres offer quicker wins. It was estimated in 2013 that only 20 per cent of trucks in India ran on radial tyres, as opposed to a global average of 68 per cent and levels of tyre 'radialization' of 96 per cent in the United States and 100 per cent in Western Europe (Balachandran, 2015). Although slightly more expensive, radial tyres can cut fuel consumption by around 3–4 per cent because of their lower rolling resistance, offering a rapid payback on the additional investment. Their less rigid sides, by comparison with the previous generation of cross-ply tyres, require higher-quality infrastructure, but as road infrastructure is upgraded this becomes less of a constraint.

Lightweighting

Reducing the empty (or tare) weight of a vehicle can cut fuel consumption in two ways. In the case of weight-constrained loads, it allows more freight to be carried on each trip, thereby reducing vehicle-kms, fuel use and CO_2. An analysis of the lightweighting of two types of 'mass-constrained' trailer found that a 30 per cent reduction in trailer weight translated into reductions in energy intensity[2] of 18 per cent and 11 per cent (Galos et al, 2015). For low-density loads that 'cube out', the fuel savings accrue from the overall lightening of the vehicle. As discussed in the previous chapter, in some countries, such as the UK, an increasing proportion of trips are volume-constrained and so subject to this latter fuel-saving effect. Greszler (2009) considers this effect less significant. Overall, however, research suggests that the potential carbon savings from the lightweighting of trucks will ramp up from 1 per cent by 2020 to 2–3 per cent by 2030 and 2.7–5 per cent by 2050 (Ricardo-AEA, 2015). This will mainly involve the substitution of lighter materials such as aluminium, plastics and carbon fibre for steel (Helms, Lambrecht and Hopfner, 2003). Incorporating 3D printing into the truck manufacturing process may also make a contribution by hollowing out components. To achieve a net reduction in vehicle weight these light-weighting initiatives will have to offset the extra weight being added by aerodynamic profiling and other fuel-saving devices.

Operating trucks more energy-efficiently

Most of the discussion of technology and fuel economy standards relates to new vehicles. During their working life the energy efficiency of these new vehicles can seriously deteriorate if they are not properly maintained and operated. Numerous guides (eg Freight Best Practice Programme, 2010a) give operators maintenance checklists and alert them to the fuel losses associated with under-maintenance. A poorly tuned engine, a misaligned axle, a leaking fuel pipe and an underinflated tyre can each increase fuel consumption, and CO_2 emissions, by 1–3 per cent. Dealing with these problems often involves little more than effective fleet management and a committed workforce. Good working practice gains technical support from onboard diagnostic systems which electronically monitor the condition and performance of the vehicle and can detect a range of defects likely to impair fuel efficiency. Many of these systems now come as standard on new vehicles. Research by NACFE (2015a) has found that trucking companies introducing a 'rigorous programme of preventative maintenance' can cut fuel consumption by 5–10 per cent as well as enjoying a range of co-benefits including improved delivery reliability, greater safety and higher vehicle resale value.

It is, nevertheless, the operation of the vehicle that principally determines its lifetime CO_2 emissions. This puts the onus on the driver to ensure that it is driven fuel-efficiently and on the company to equip him or her with the necessary eco-driving skills. Driver training is widely regarded as one of the most cost-effective ways of cutting CO_2 emissions from road freight with among the shortest investment payback periods. The economic benefits do not simply accrue from fuel cost savings, but also from improved

Table 6.1 Impact of truck driver training programmes on fuel efficiency

Country	Type of training	Fuel efficiency improvement	Date
Canada (*SmartDriver*)	Online, classroom, on the road	up to 35%	current
US	Coaching, in-vehicle, real-time feedback	13.7%	2011
Switzerland	Various feedback technologies	5–15%	2011
Japan	Classroom	8.7%	2013
UK (*SAFED*)	Classroom, on the road	7%	2009
US	Coaching, in-vehicle, real-time feedback	2.6–5.2%	2014

SOURCE Partly based on Boriboonsomsin (2015) and AECOM (2016)

safety and lower insurance costs. Table 6.1 lists the percentage fuel reductions achieved by a sample of truck driver training programmes around the world (Boriboonsomsin, 2015: 15; AECOM, 2016). Wide variations in the reported values can be attributed to differences in baseline driving standards, the training methods and the nature and timing of the driver assessment. Fuel savings tend to be at their maximum immediately after the training session and diminish thereafter as the driver reverts back to their previous driving behaviour. This deterioration in post-training performance can be reduced or eliminated where the driver is monitored and given feedback on their driving style. Such monitoring and advisory systems are now widespread in Europe and North America, especially among larger carriers, though still comparatively rare in the developing world.

Drivers can also be given financial incentives to drive more fuel-efficiently. Boriboonsomsin (2015: 9) reports that in a study of 46 US truck drivers, 'providing financial incentives on top of individualized coaching and using an in-vehicle, real-time feedback system approximately doubled the fuel economy improvements.'

Fuel consumption and CO_2 emissions are highly sensitive to vehicle speed. Like all road vehicles, those carrying freight have an optimal speed range within which fuel consumption is minimized, a so-called 'sweet spot'.

In countries with good road infrastructure, much truck mileage is run above this speed range, even while staying within official speed limits. Reducing average speed can therefore cut fuel consumption and CO_2 emissions. NESCCAF et al (2009) estimated that reducing the speed of heavy US trucks by 1 mph within the 70–55 mph range saves between 0.7 and 1.0 per cent of fuel depending on the drive cycle. The American Trucking Association, which has been campaigning for a national truck speed limit of 65 mph (105 kph), argues that, other things being equal, a US Class 8 truck travelling at 65 rather than 75 mph uses 27 per cent less fuel (Garthwaite, 2012). To maximize the fuel savings, 'downspeeding' of the engine is required, which involves reducing engine revolutions per minute (NACFE, 2015b).

Many trucking companies in North America and Europe have voluntarily been cutting the maximum speed at which their vehicles are allowed to operate. This is easily implemented where a speed governor has been installed in the truck. This has been mandatory in the EU since 1992 and other countries such as Japan and Australia have similar regulations. The United States was going to make the fitting of speed limiters on heavy-duty trucks compulsory in 2017, but this has now been delayed. According to Barton Associates (2013), however, around three-quarters of US trucks already have these devices fitted. Schneider, the largest privately owned US trucking company, reduced the maximum speed for its fleet of 10,600 trucks from 101 kph (63 mph) to 97 kph (60 mph), cutting annual fuel consumption and emissions by, respectively, 17 million litres and 40,000 tonnes. In Europe, the French carrier, Geodis, limits the speed of its trucks to 82 kph (8 kph lower than the highway speed limit of 90 kph for trucks in France) as part of its carbon reduction programme.

Reductions in the maximum speed of trucks by these margins do not appear to have an adverse effect on business. NACFE (2011: 1) reports that 'four major US fleets have reported no significant loss of productivity on 98 per cent of their freight shipments after limiting their trucks' speeds. Of the shipments that were affected, the difference was measured in minutes per day.' Another study found that by cutting maximum truck speed to 65 mph from 70 mph and 75 mph, North American motor carriers could save between 2 per cent and 9.5 per cent on total operating costs (Barton Associates, 2013). Any loss of customer service is usually very marginal and increases in transit time can be more than offset by an acceleration of internal processes within industrial and commercial premises. This trade-off between time and CO_2 emissions in a supply chain merits further discussion.

Figure 6.1 Relationship between time and CO_2 emissions for supply chain activities

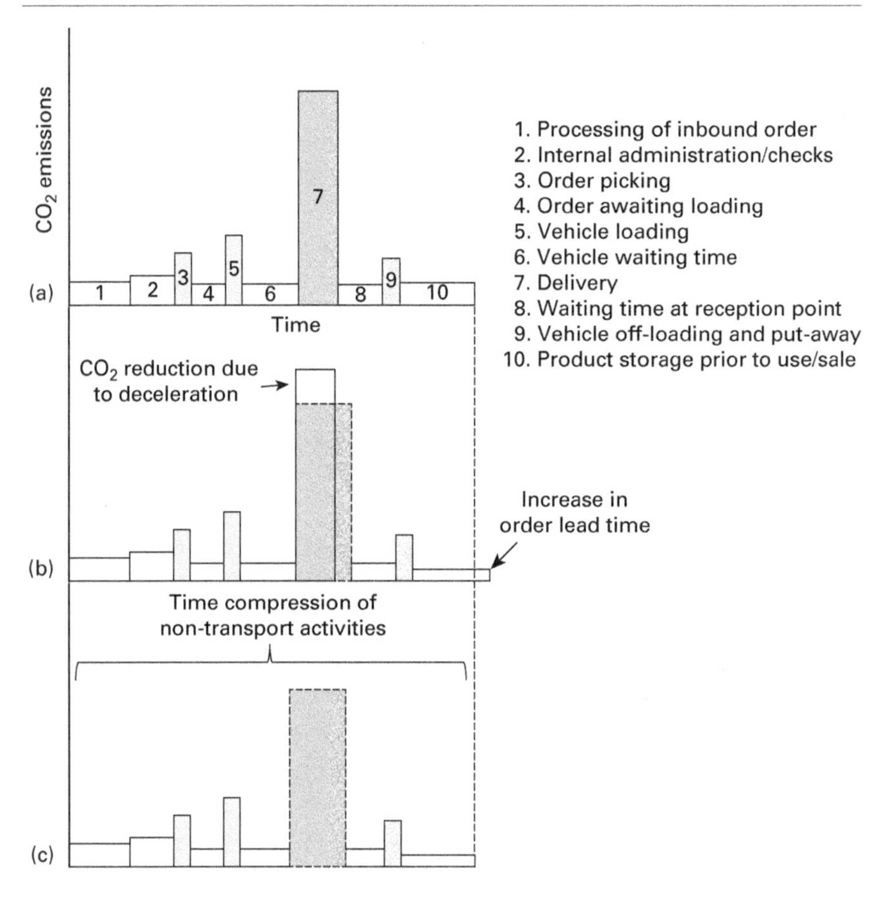

SOURCE McKinnon (2016a)

Figure 6.1 decomposes the fulfilment of a customer order into a typical set of 10 activities. These activities vary in the amounts of energy they consume and amounts of CO_2 they emit per hour. This 'time-based carbon intensity' (TBCI) is generally higher for those activities that involve physical movement and handling using mechanical equipment. Activities 3, 5, 7 and 9 fall into this category. Of these, activity 7, the movement of goods between industrial and commercial premises, is likely to have by far the highest TBCI. This was confirmed by a 'sustainable value-stream mapping exercise' undertaken Simon and Mason (2003) on an orange juice supply chain. The delivery operation accounted for 20 per cent of the total supply chain time but emitted two-thirds of total CO_2 emissions, giving transport a TBCI value 7.8 times greater than that of the other supply chain activities combined. By redistributing time from those activities with low TBCI values, mainly administrative functions within premises, to transport it is possible to reduce both its TBCI and that of the supply chain as a whole. In other words, we speed up the internal processes and slow down the transport. Figure 6.1 illustrates how this can be done in a way that keeps the order lead time constant, thus avoiding a deterioration in customer service. In graph (b) the delivery operation is decelerated and the transit time lengthened, with no compensating changes in other activities. In graph (c) other non-transport activities are time-compressed to offset the longer transit time, in most cases with little or no increase in CO_2 emissions. A proposal to slow down the movement of freight will no doubt be anathema to a logistics profession long accustomed to ever-faster delivery, but it is worth remembering that, of the supply chain decarbonization measures examined by the World Economic Forum/Accenture (2009), deceleration was ranked second in terms of carbon abatement and feasibility (Table 2.4). In preparing logistics for a low-carbon world, managers will have to seriously consider such a radical option. The wider use of deceleration as a means of decarbonizing logistics is discussed in greater detail in McKinnon (2016a).

Planned speed reductions must be distinguished from the effects of traffic congestion on the movement of freight by road. Congestion causes vehicles to run below their most fuel-efficient speeds and to stop and start more frequently. Research in Germany by Verband der Automobilindustrie has found that where a 40-tonne articulated truck travelling at an average speed of 50 kph has to stop once every kilometre because of congestion its average fuel consumption goes up by 86 per cent and for two stops rises by a factor of three. A macro-level modelling exercise by Kellner (2016: 13) using real-world data from online navigation services for the distribution of fast-moving consumer goods (FMCG) in Germany found that 'regular traffic congestion' increased total CO_2 emissions from delivery operations by 2.5 per cent. He calculated that for every 10 per cent increase in traffic delays, CO_2 emissions would rise by 0.25 per cent. The American Trucking Association estimated in 2008 that traffic congestion on 437 key bottlenecks in the US Federal Highway network would, over 10 years, cause trucks to consume 18.6 billion more litres of fuel and emit 45.2 million more tonnes of CO_2 (Moskowitz, 2008). There is a danger, therefore, that worsening traffic congestion will erode many of the fuel and CO_2 savings accruing from the technological and operational improvements outlined above. This is a particular problem in developing countries where the rate of traffic growth is far exceeding the provision of additional road space.

The conventional response to congestion is to expand infrastructural capacity, primarily for economic reasons. The environmental case for upgrading road infrastructure is more questionable and some would argue indefensible (Doherty, 1999; Campaign for Better Transport, 2017). Road building emits large quantities of GHG which, in terms of long-term carbon budgeting, must be set against reductions in energy-related CO_2 emissions from traffic using the new infrastructure during its design life (Facanha and Horvath, 2006). In a report for the World Bank, Egis (2010) found that the average carbon intensity of road construction varied from 90 tonnes CO_{2e} for a gravel road to 3,234 tonnes for an expressway. It also outlined various ways in which this carbon intensity could be reduced, particularly in developing countries. Construction company Skanska (2011), for example, used various mitigation measures to reduce by 27 per cent CO_{2e} emissions from the widening of a section of the M25 motorway around London. More effective traffic management can ease congestion levels within the existing infrastructure, reducing the need for new construction. A review of environmentally oriented road traffic management schemes, based on 'intelligent transport systems' (ITS) in Europe and North America revealed that they can cut GHG emissions by between 5 and 15 per cent (Barth, Wu and Boriboonsomsin, 2015).

Companies can also minimize the impact of traffic congestion on truck fuel efficiency by rescheduling deliveries into off-peak periods (referred to as 'off-hour' deliveries (OHD) in the United States). Using UK highway data, Palmer and Piecyk (2010) showed how by varying the start time of 56 long-haul freight deliveries over the period 5am–midnight it was possible to achieve significant savings in time, cost, fuel and CO_2 emissions. Emissions from the quickest journeys were actually 6 per cent lower than those with an average transit time.

The main interest in delivery rescheduling has, however, been in urban areas, where daytime traffic congestion is most acute. Sanchez-Diaz, Georén and Brolinson (2017) reviewed the impact of 'off-peak-hour delivery' schemes in New York, London, Paris and Stockholm and found that they can reduce CO_2 emissions by between 20 and 75 per cent, with an average figure of around 40 per cent. In an analysis of the effects of OHD schemes in New York, Bogota and Sao Paulo on freight emissions, Holguín-Veras et al (2016) estimated CO_2 reductions to be in the range of 45–67 per cent. They then extrapolated the results of this analysis to all metropolitan areas with a population of over 2 million, assuming that all freight deliveries would be made between 7pm and 6am. This would save just under 300 million tonnes of CO_2 annually; 62 per cent of this carbon reduction would occur in developing countries and 36 per cent of it in mega-cities with populations in excess of 15 million.

In the light of this evidence, the re-timing of deliveries to avoid congestion is clearly one of most promising means of decarbonizing road freight operations, particularly in cities. It is, however, subject to several constraints. First, as discussed in the previous chapter, application of the just-in-time principle has tightly coupled freight deliveries to production, materials handling and retailing operations, most of which still occur during the working day. Delivery rescheduling often requires some relaxation of JIT regimes which can be difficult to negotiate both internally and externally with supply chain partners. On the other hand, the widening adoption of 24/7 working across much of the economy and the increasing proportion of business premises capable of receiving 'unattended delivery' are facilitating the shift to night-time delivery. Second, night-time delivery exacerbates the urban noise problem and can disrupt the sleep of those living nearby. For this reason, many cities have imposed night curfews on freight deliveries. The noise problem has, however, been mitigated by vehicles getting quieter as well as local noise abatement schemes, reducing local environmental objections to OHD (Forkert and Eichhorn, 2012). The UK government has also issued good-practice guidelines for night-time delivery (Department for Transport, 2015c). Third, a shortage of truck drivers in many countries, such as the United States, UK and Germany, (McKinnon et al, 2017) has made it difficult for some companies to find enough drivers to work during the night at affordable rates. Finally, as discussed by Holguín-Veras et al (2014), the objectives of the shippers, carriers and receivers of the goods are often misaligned and the costs and benefits of OHDs unevenly distributed. They see the 'central question' of OHD as being 'how to engage all involved in a constructive, multi-stakeholder problem-solving approach', what they see as 'the Holy Grail of urban freight policy' (p.47).

In summary, the energy efficiency of trucks can be improved in many, mutually reinforcing ways. Taking a longer-term, global perspective, the International Energy Agency (2017a: 126) expects improvements in energy efficiency to be 'the largest contributor to emissions reductions' in the trucking sector by 2050, accounting for '30 per cent of cumulative GHG savings'.

Switching trucks to lower-carbon fuels

All but a tiny proportion of road freight movement is powered by liquid fossil fuel, mainly diesel but with some of the global small van fleet running on petrol/gasoline. This overwhelming dependence on fossil fuel can be explained by its exceptionally high energy density, its near universal availability and relatively low cost. In many countries, particularly in the developing world, the use of fossil fuels is subsidized by the government, either through direct payments or by under-charging for all the externalities they cause. The IMF valued these subsidies at US $5.3 trillion in 2015, equivalent to 6.5 per cent of global GDP (Coady et al, 2015). Given these circumstances, getting companies to switch to alternative low-carbon energy was never going to be easy, and so it is proving.

In this section we will look at six alternatives to the use of diesel and gasoline in the road freight sector:

- compressed and liquid natural gas (CNG and LNG) derived from fossil sources but potentially having a lower carbon content than diesel and gasoline;
- liquid biofuel, in particular biodiesel derived mainly from soya, rape seed, palm oil and recycled vegetable oil;
- biogas, in particular biomethane produced by the anaerobic digestion of food and animal waste;
- hydrogen, mainly electrolysed from water and used as an 'energy carrier' for transferring low-carbon electricity into vehicles via a fuel cell;
- batteries, recharged with low-carbon electricity;
- electrified highways directly transmitting low-carbon electricity into vehicles from overhead cables or under-road transmission systems.

Figure 6.2 Comparison of WTW emissions from different sources of freight transport energy: gCO_{2e} per vehicle-km

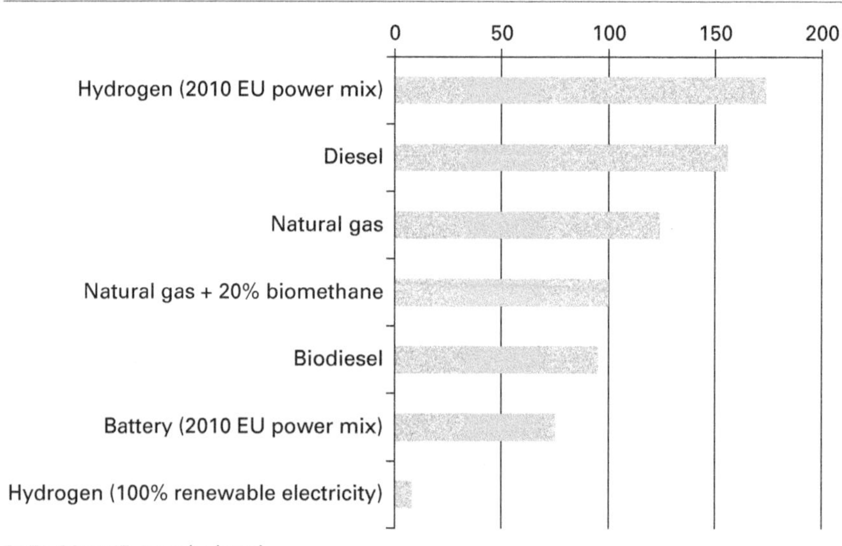

DATA SOURCE Rumpke (2010)

The last three options require the electrification of road freight transport and their potential to cut GHG emissions depends on the rate at which electricity generation will be decarbonized. Figure 6.2 provides an initial comparison of the average GHG intensity of alternative energy sources on a WTW relative to diesel basis, (Rumpke, 2010).

CNG and LNG

Natural gas is widely regarded as a 'transitional fuel' capable of easing the move from carbon-intensive fuels, like coal and diesel, to the low/zero-carbon non-fossil energy sources that will ultimately replace them. It has a lower carbon content than other fossil fuels and is relatively cheap and plentiful, particularly since the introduction of fracking has allowed countries to exploit extensive new gas reserves. Natural gas may play a transitional role in national energy policy, but is it likely to fulfil a similar role in the migration of the freight sector to low-carbon energy? Currently gas plays a very minor role in the movement of road freight. For example, industry estimates suggest that in 2015 only 0.07 per cent of European heavy-duty trucks were gas-powered.

The main environmental reason for switching trucks to natural gas is that it emits much less nitrogen oxide (NOx) and particulate matter (PM) per unit of energy than diesel and can therefore improve local air quality. Recent research suggests that on a life-cycle basis the use of fossil-sourced natural gas in trucks actually increases GHG emissions. A study of US Class 8 trucks running on CNG and 'centrally produced LNG' found that they increased life-cycle GHG emissions by, respectively, 0–3 per cent and 2–13 per cent when compared with diesel-powered vehicles (Tong, Jaramillo and Azevedo, 2015). The problem lies in the leakage of methane into the atmosphere both from the gas supply chain and the vehicle itself. As discussed in Chapter 1, methane has a global warming potential 21 times that of CO_2 and so even small quantities of these so-called 'fugitive' emissions can more than offset CO_2 savings from the switch from diesel to gas. Dominguez-Faus (2016: 11) estimates that in the United States, 'upstream methane leakage contributes between 7 and 11 per cent and vehicle methane leakage (ie methane slip) contributes between 5 and 9 per cent to the total carbon intensity of natural gas in long-haul trucks.'

Upstream leakage of methane is much greater in the CNG supply chain than in the LNG chain as the liquefaction of natural gas in centralized plants and its movement by truck or pipeline as a liquid can be more tightly sealed. Estimates of the level of methane leakage from natural gas networks vary widely, however (Heath et al, 2015). In the United States, the EPA suggests that it is below 1.5 per cent, while other studies claim that it could be as high as 3–3.5 per cent (Dominguez-Faus, 2017). European research by Ricardo (2016) revealed that there are substantial variations in the level of well-to-wheel (WTW) emissions for natural gases sourced from different locations and transported, stored and processed in different ways. For example, against an EU average of 19.2 kg CO_{2e}/joule, the GHG intensity of CNG varies between 12.3 in Northern Europe and 28.6 in South East Europe (Exergia, E3M and COWI, 2015). It is very difficult therefore to generalize about the GHG rating of natural gas up to the vehicle fuelling point. In recognition of the variability, Tong, Jaramillo and Azevedo (2015) adopt a probabilistic approach when comparing gas-powered and diesel vehicles.

The emission performance of natural gas trucks depends partly on methane leakage (or 'slip') but also on the efficiency with which it is burned. The engines of most gas-powered trucks have 'spark ignition', which is around 10 per cent less energy efficient than the 'compression ignition' used in vehicles. This erodes some of the CO_2 savings from the switch from diesel to gas. Dominguez-Faus (2016), nevertheless, found that in the absence of methane leakage from the vehicle, CNG and LNG trucks with spark ignition had a slightly lower GHG intensity on a WTW basis than their diesel counterparts. When methane slip at the vehicle level was factored into the calculation the CNG and LNG trucks had, respectively, 6 per cent and 4 per cent higher GHG emissions. On the other hand, an LNG vehicle with compression ignition, and lower methane leakage both upstream and from the vehicle, had a life-cycle GHG intensity 9 per cent below that of diesel. This demonstrates that particular engine, vehicle and fuel combinations can yield net GHG reductions, though they are still comparatively rare and financially unattractive.

The methane slip problem is particularly acute for so-called 'dual fuel' vehicles that combine gas with diesel in a retrofitted engine. Significant amounts of uncombusted methane leaks into the atmosphere when the engine switches between fuel types. Stettler et al (2016) measured GHG emissions from two types of truck with five different dual-fuel configurations on a TTW basis. CO_2 emissions dropped by up to 9 per cent, but the release of methane raised overall GHG emissions (expressed as CO_{2e}) to levels 50–127 per cent above those of equivalent diesel-powered vehicles. Companies investing in dual-fuel vehicles were attracted by their lower tailpipe CO_2 emissions and the refuelling flexibility they offered at a time when the density of gas refuelling points was low. Now that the methane slip problem in these trucks has been more fully investigated, the uptake of dual-fuel technology appears to have been premature and counterproductive in carbon terms. This highlights the importance of taking account of all GHGs and not just CO_2 when considering a switch to alternative fuels.

The need for dual-fuel capability will reduce as the density of gas filling stations increases and vehicles running solely on gas can undertake a broader range of delivery operations. Both the United States and the EU have set ambitious, long-term targets to increase this density, reducing the maximum distance between gas refuelling points to, respectively, 241 km and 150 km for CNG (International Energy Agency, 2017a).

Although upstream and vehicle methane leakages can be reduced through time, and gas refuelling infrastructures are developing, the International Energy Agency (2017a) decided to 'rule out' natural gas as a contributor to road freight decarbonization in the scenarios it has constructed for 2030 and 2050. This decision appears well-founded in the light of research by Camuzeaux et al (2015) which concluded that 'converting heavy-duty trucks (to natural gas) leads to damages to the climate for several decades' – around 70–90 years in the case of spark ignition engines and 50 years for more efficient engines with a compression injection system known as HPDI. The switch to CNG/LNG is likely to continue for air quality reasons but most of the resulting reductions in NOx and PM will, for the foreseeable future, probably be achieved at the expense of higher GHG emissions.

Liquid biofuel

When governments and companies began to look seriously for ways of decarbonizing transport there was great enthusiasm for fuels derived from biomass. The concept of plants absorbing CO_2 from the atmosphere during their growth cycle and then releasing it again when burned as a fuel had an appealing symmetry. It suggested that a major shift from fossil to biofuel would create a new ecological balance and, in the process, drastically cut transport-related GHG emissions. This encouraged national governments and the EU to mandate the blending of biofuels with fossil fuels at levels which vehicle engines could accommodate with minimal loss of performance. The EU Biofuels Directive of 2003 set a target biofuel share of 5.75 per cent for transport fuels by 2010. This, however, proved very difficult to achieve and has still not been reached in most EU countries.

In the meantime, more research has been done on GHG emissions from the production and distribution of biofuels as well as their use, much of which suggests that on a WTW basis they can actually have a higher GHG intensity than conventional fossil fuel. In the freight sector, by far the most widely used biofuel is biodiesel, which comes mainly from three feedstocks: used cooking oil (UCO) collected from restaurants and factories; rape seed and soya beans, grown mainly in Europe and North America; and palm oil sourced from countries in the tropics, mainly Indonesia and Malaysia, which account 80 per cent of the global total (Searle, 2017). On a life-cycle basis, biodiesel from the first of these sources yields significant GHG savings, though supplies of UCO are very limited. The cultivation of biofuel plants is a much more plentiful source of feedstock but releases large quantities of GHG, particularly N_2O which has a global warming potential 310 times that of CO_2 (over 100 years). Where this cultivation entails land use change, particularly the destruction of tropical rainforest to clear land for palm oil plantations, WTW (or, more appropriately, 'field to wheel') emissions of biodiesel can substantially exceed those of diesel fuel. Merely displacing food production onto previously uncultivated land also carries a GHG penalty. These are examples of what are called 'indirect land use changes' (ILUC) caused by the re-orientation of agriculture to energy production. The inclusion of ILUC significantly complicates the life-cycle calculation of biofuel GHG emissions and often reveals total emissions higher those of the fossil fuels they are supposed to replace. Biofuel production is also criticized for distorting food markets and inflating the price of basic foodstuffs.

Mounting evidence that some biofuels exacerbate rather than mitigate climate change has caused a reappraisal of official policy on renewable fuels. It has, for example, prompted the European Commission to include 'rigorous sustainability criteria' in its latest Renewable Energy Directive. These specify that 'to be considered sustainable, biofuels must achieve greenhouse gas savings of at least 35 per cent in comparison to fossil fuels', with this figure rising to 50 per cent in 2017 and 60 per cent in 2018 for new production plants (European Commission, 2017a). These savings are calculated on a full life-cycle basis and 'biofuels cannot be grown in areas converted from land with previously high carbon stock such as wetlands or forests' or 'produced from raw materials obtained from land with high biodiversity.' Both the EU and California have introduced low-carbon fuel standards for biofuels and the United States as a whole plans to do so in 2018. This should help to ensure that the use of these renewable fuels genuinely reduces road freight emissions on a WTW basis. Assuming that there is an adequate supply of biodiesel meeting these environmental criteria, it is likely that regulators will steadily increase the percentage of biodiesel blended into diesel fuel. Most trucks can currently run on blends of between 7 and 20 per cent biofuel (B7–B20), with the higher percentage blends more common in US truck fleets.

Biomethane

Chemically, biomethane resembles CNG and LNG, though unlike these other forms of natural gas it is not a fossil fuel. It is derived from fresh organic matter processed today rather than from vegetation deposited deep in the earth millions of years ago. Using a chemical process called anaerobic digestion, various forms of organic waste can be converted into a biogas which, with further processing, becomes biomethane. It has a similar range of applications to natural gas derived from fossil sources and can be distributed in either compressed or liquid form, but, on a life-cycle basis, emits much less GHG than its fossil equivalents. This is partly because of the nature of its production and distribution, but also because the conversion of organic waste into fuel prevents it from rotting outdoors or in landfill sites where it would release methane and nitrous oxide, both powerful global warming gases, into the atmosphere. Within the vehicle, biomethane is subject to the same 'methane slip' problem as other forms of natural gas, but on a WTW basis still offers a substantial GHG reduction relative to diesel (Figure 6.2). Indeed, it is considered the most 'climate friendly' of all the biofuels. UK supermarket chain Waitrose reports that running its articulated trucks on biomethane achieves an 83 per cent reduction on WTW CO_2 emissions, while saving 35 per cent in fuel costs (Laney, 2017).

There are, however, several constraints on the fuelling of trucks with biomethane. First, although biomethane production capacity is steadily expanding, it is still very limited and the transport sector must compete for the available supplies. A large proportion of biomethane currently goes into the gas grid network to supplement other forms of natural gas and to make it available to a range of users. Ironically, efforts to reduce the high level of food wastage in the supply chain, discussed in Chapter 3, will scale down one of the main feedstocks for biomethane production. Second, biomethane use, like that of CNG and LNG, is constrained by a lack of refuelling points. Combining biomethane and diesel in dual-fuel vehicles can extend their range. Although these vehicles are afflicted with the methane slip problem discussed earlier, their life-cycle emissions are much lower than those of dual-fuel trucks combining diesel with CNG or LNG. Third, in most countries biomethane is more expensive than diesel at the current level of world oil prices and needs preferential tax rates to incentivize adoption. Despite these limitations, it is predicted that biomethane will significantly increase its share of truck fuel over the next few decades, but from a very low base.

Electrification of road freight transport

The decarbonization of electricity generation lies at the heart of all climate change mitigation policies. To meet the COP21 target of keeping the increase in average global temperature 'well below 2°C' by 2100, most of the world's electricity will need to be zero carbon by 2050. This will be achieved by switching from fossil fuels to renewable and nuclear energy and installing carbon capture and storage (CCS) systems on any remaining coal-, oil- or gas-fired power plants. The International Energy Agency (2016) discusses in detail how this transformation of the electricity supply industry will occur, the obstacles to be overcome, the policy interventions it will require and the likelihood of it happening. Globally, the average carbon content of electricity is declining but not currently at a fast enough rate to reach the 2050 target. The steep decline in the cost of solar and wind power in recent years is likely to accelerate the switch to renewables. At present the average carbon intensity of electricity and its rate of decline also vary widely from country to country (Figure 3.4). This makes it difficult to generalize, both geographically and temporally, about the future carbon intensity of activities powered by electricity. For the purposes of this discussion, we will assume, admittedly rather optimistically, that the electricity supply industry worldwide will soon get onto a decarbonization pathway that will deliver low-carbon electricity[3] by 2050. The challenge then for the logistics sector will be to maximize the decarbonization leverage it can gain from the use of this electricity. For warehouses, freight terminals and electrified railway lines already connected to electrical grids the transition to low/zero-carbon electricity should be seamless as the energy mix gradually shifts away from fossil fuel. For the vast majority of freight transport operations that currently rely on in-vehicle combustion of fossil fuel this transition will be much more difficult and costly. The long-haul road freight sector has traditionally been seen as 'hard to electrify' (Energy Transitions Commission, 2017), though in the light of recent research and development this view may prove overly pessimistic.

Broadly speaking, there are three ways of electrifying trucks to give them access to low/zero-carbon electricity.

Hydrogen

This gas can serve as an 'energy carrier'. It is almost always found in compounds with other elements from which it must be separated. This separation can be done mainly in three ways (Alternative Fuels Data Center, 2017). Roughly half of the hydrogen produced today comes from the 'reforming' of methane by mixing steam with natural gas at high temperature, and about a third is extracted from the oil-refining process.[4] Most of the remainder comes from the electrolysis of hydrogen-rich compounds, of which the most abundant is water. It is when this electrolysis uses low-carbon electricity that hydrogen offers a means of indirectly distributing that electricity to vehicles. More electricity is used to compress the hydrogen into high-pressure storage units in fuelling stations and onboard the vehicles. The hydrogen is then pumped into a fuel cell in the vehicle that converts the embodied energy back into electricity. As the International Energy Agency (2017a: 98) explains: 'Trucks using fuel cells and hydrogen are essentially electric trucks using hydrogen stored in a pressured tank and equipped with a fuel cell for onboard power generation.'

The use of hydrogen as an indirect means of electrifying road freight vehicles is a controversial subject. Its critics contend that too much energy is wasted in the process to make this a viable option. Bossel (2004) estimated that in the case of compressed hydrogen, only 23 per cent of the original energy used in the electrolysis process is converted into vehicle motion.[5] This compares very unfavourably with the use of batteries for energy storage and distribution, which transfer approximately 69 per cent of the original energy to the wheels. Indeed, Nicolaides, Cebon and Miles (2017) argue that hydrogen trucks would be only half as energy efficient as diesel vehicles. It is argued that zero-carbon electricity will never be so cheap and abundant that it can be squandered in this way. On the contrary, Bossel (2004: 58) argues that 'renewable energy will be precious and therefore should be distributed and used intelligently. Wasteful electrochemical conversion processes such as electrolysis and fuel cells will be avoided wherever possible.' From this he concluded that 'renewable energy is better distributed by electrons than by hydrogen' (p.59). Over the past decade, technical improvements in electrolysis and fuel cells have reduced energy losses in the hydrogen supply chain, though they are still substantial. A recent study for the German government

Table 6.2 Relative energy efficiency and cost of low-carbon energy carriers (assuming transmission of 100 kWh of electricity at six euro cents per kWh)

Low-carbon electricity pathway	WTW energy efficiency	Euro cent per truck-km
Electrified road systems	77%	19
Battery	62%	20
Hydrogen	29%	55
Power to gas	20%	70

SOURCE German Ministry of the Environment (quoted by Akerman 2016)

showed hydrogen continuing to lag well behind other low-carbon electricity pathways in terms of energy efficiency and cost per truck-km (Table 6.2).

In its energy-transfer role, hydrogen also has several other disadvantages:

1 *Need to create a network of hydrogen filling stations*. Currently these stations are supplied by gas tanker, or where the volumes are large enough, a pipeline. Switching the processing of hydrogen from 'steam methane reforming' and oil refineries to the electrolysis of water with low-carbon electricity makes it possible to localize the production and distribution.

2 *Very high cost of hydrogen-powered vehicles*. These vehicles are predicted to be much more expensive than their diesel fuel equivalents. Shabani and Andrews (2015: 482) report that the initial hydrogen vehicles used by Californian ports have a capital cost '2.5 times higher than a same-sized diesel truck' though the running cost is 30–40 per cent cheaper. They also expect that the purchase price 'will fall considerably with higher-volume production'. It is predicted, for example, that the unit cost of fuel cells and hydrogen storage tanks will drop significantly from currently high levels. The International Energy Agency (2017a:102) on the other hand, anticipates that even 'where fuel cell technologies achieve very wide market deployment' hydrogen fuel cell trucks would still be 25–90 per cent more expensive.

3 *Vehicle space requirements*. Even when compressed at high pressure (70Mpa), hydrogen requires four times as much storage space as the equivalent amount of diesel fuel on a conventional truck. This can

encroach on the carrying capacity of vehicles transporting low-density loads in countries where there are tight limits on vehicle dimensions.

Advocates of the use of hydrogen in the transport sector stress several advantages:

1. Hydrogen, compressed at 70MPa, has six times the energy density of a battery in volumetric terms and 300 times in terms of weight (International Energy Agency, 2017a). This means that hydrogen fuel cell propulsion systems can be used in heavy, long-haul trucks, something that until recently was not considered possible with batteries. Hydrogen was therefore seen as a way of distributing low-carbon electricity to sectors of the road freight market that it could not access by other means. As discussed below, the recent launch of battery-powered long-haul trucks by Daimler and Tesla and the trialling of highway electrification now casts doubt on this traditional view.
2. Trucks running on hydrogen have lower running and maintenance costs than diesel-powered vehicles. The comparison, however, should be with low-carbon alternatives to diesel trucks and, on this basis, hydrogen propulsion has been shown to be a relatively expensive way of decarbonizing long-haul road freight (Mottschall, 2016).
3. Hydrogen is considered a 'versatile' fuel that can be produced in various ways from different feedstocks, either in centralized facilities or locally. Low-carbon methods of producing hydrogen are much less versatile, however, and exclude the feedstocks currently responsible for over 80 per cent of hydrogen supplies, namely natural gas.

One of the main deployments of hydrogen trucks has been in the hinterland of Californian ports where the main environmental justification has been to improve air quality rather than lower carbon emissions. In this respect, hydrogen fuel cell vehicles resemble electrical vehicles in having zero exhaust emissions. Several truck manufacturers, including Kenworthy in the United States and Scania in Europe, are developing prototype long-haul trucks running on hydrogen, while the parcel carrier, UPS, is keen to trial hydrogen-powered Class 6 trucks in areas where battery-operated vehicles lack sufficient range (Shumaker and Serfass, 2017). As hydrogen fuel cell technology is still at an early stage in its development, it is hard to predict its long-term impact on road freight decarbonization. As Transport and Mobility Leuven (2017: 4) has acknowledged, it 'is not advanced enough to give an accurate picture of (its) potential contribution'. High energy losses in the 'electrolysis-to-wheel' supply chain are likely to keep this contribution relatively modest.

Batteries

By comparison with hydrogen fuel cells, battery operation in the road freight sector is a relatively mature technology. The batteries can be used in various capacities: in combination with an internal combustion engine (ICE) in a hybrid vehicle; in a plug-in hybrid vehicle (PHEV) where the battery can be externally charged; and in a battery-electric vehicle (BEV) powered solely by the battery. All three battery applications have been applied to freight vehicles, though almost entirely to small vans performing local delivery work. In the course of making numerous deliveries and collections these vehicles frequently brake and accelerate, a duty cycle that is well suited to electric operation, particularly as energy can be captured from the braking process and used to recharge the battery (so-called 'regenerative braking').

Sales of battery electric vans in the EU have been growing at around 15 per cent per annum, though in 2015 still accounted for only 0.5 per cent of new van sales, the vast majority of which were made in France (European Environment Agency, 2016a). La Poste, France's main postal service company, operates the largest electric vehicle fleet in Europe, comprising around 35,000 vans and scooters in 2017. This low market penetration at a European level can be attributed partly to the novelty of the technology, though as Leonardi, Cullinane and Edwards (2015: 282) note: 'Electric vehicles have been in use for deliveries for decades, with the British milk float an early and enduring example.' Other more serious issues have been inhibiting electric van sales:

- *Higher capital cost than petrol or diesel vans.* The higher purchase price must be set against much lower operating and maintenance costs and in some countries, like the UK, is substantially discounted by government grants. The average cost of lithium ion batteries fell by 73 per cent between 2010 and 2016 (Curry, 2017). As the capital cost differential has narrowed, electric vans can now be cheaper than their fossil-fuel equivalents on a full-life cost basis.
- *Limited delivery range between charges.* This so-called 'range anxiety' is easing as delivery distances between charges are lengthening and now exceed the requirements for many urban distribution operations. According to the Freight Transport Association, only a third of UK vans

travel more than 80 miles in a day, comfortably within the 100-mile range of all the electric vehicles on the market.

- *Weight of the battery*. The batteries in BEVs are quite heavy, limiting the payload that can be carried. An electric 3.5-tonne van would need to have its maximum gross weight raised to 4–4.5 tonnes to maintain the same payload as a diesel-powered equivalent (Browne, Rizet and Allen, 2014). Increasing the gross weight beyond 3.5 tonnes would then require the driver to have a goods vehicle licence. In the UK, the government is proposing to relax the weight and driving licence restrictions for electric and gas-powered vans.
- *Problems with recharging*. In most countries the density of recharging points is still limited, though steadily increasing. In interviews with operators of electric vans in France and the UK, Morganti and Browne (2018) also found 'queue anxiety' (having to queue for a recharging point) and 'grid unreliability' to be concerns, though both can be addressed by an increase in infrastructural capacity.

Despite these reservations, the electrification of local delivery operations is proceeding at a healthy rate as battery performance improves, the price differential with diesel and hybrid vehicles narrows and increasing numbers of cities establish low- and ultra-low-emission zones and offer privileges to operators of zero-emission vehicles. The main environmental advantage of BEVs after all is that they have no tailpipe emissions. The impact of their use on GHG emissions depends on the carbon intensity of the grid electricity they use which, as discussed earlier, varies widely by country. The preponderance of electric van sales in France is clearly associated with the country's high dependence on nuclear power (accounting for 77 per cent of French electricity generation in 2017) and correspondingly low level of CO_2 emissions per MWh. As the average carbon content of electricity drops in other countries, the decarbonization case for freight vehicle electrification will gradually strengthen. Heid et al (2017: 4) have constructed two scenarios for this electrification process. In their 'early adopter' scenario, galvanized by 'more aggressive adoption of low-emission zones by major cities', electric light- and medium-duty trucks will have a 15–34 per cent sales penetration by 2030, while in the 'late adopter' scenario it will lie somewhere between 8 and 27 per cent.

As discussed earlier, batteries are currently only able to bring low-carbon electricity to freight vehicles at the lower end of the size and weight spectrum with a local collection and delivery 'duty cycle'. Heavy trucks moving large quantities of freight over long distances would need extremely heavy and expensive batteries. Sripad and Viswanathan (2017) have calculated that a US Class 8 truck carrying an 11-tonne payload (three-quarters of the current average) over a distance of 600 miles (960 km) would need a lithium ion battery pack weighing 16 tonnes, implying that 'a greater fraction of the energy consumed to move the vehicle is spent on moving the battery pack rather than payload' (p.1671). They estimate the cost of this battery to be between $250,000 and $400,000, more than twice the current cost of a diesel-powered Class 8 truck ($120,000). Even a battery 'beyond current Li-ion systems' for a range of 300 miles could cost $100,000. The heavy battery weight presents a particular problem for US trucking because of the relatively low legal weight limit on heavy-duty vehicles. At 36 tonnes (80,000 lb) it is four tonnes below the limit for EU international haulage, 8 tonnes below the UK limit and less than half that of Finland. While recognizing that 'battery-induced payload loss' will remain a constraint and make electric long-haul trucks 'unfavorable for weight maximizers', Heid et al (2017) nevertheless see the total cost of operating these vehicles for some carriers reaching parity with diesel trucks by 2023. These might be niche applications, however, for companies moving very light loads or undertaking short shuttle movements.

Following the launch of two prototype battery-operated long-haul trucks in the autumn of 2017 it may be necessary to reassess the prospects for this carbon-reducing technology. In October 2017, Mitsubishi FUSO, a division of Daimler, displayed a heavy-duty electric truck which it claims can run for 220 miles on a single charge. Its gross weight of 23.2 tonnes would allow it to carry a payload of 13 tonnes within current US weight limits, only marginally below the current average load. The Tesla truck launched in November and portrayed as a 'gamechanger' for the trucking industry will, it is claimed, have a range of 500 miles on a single charge. At the time of writing, no indication has been given of the battery, vehicle and maximum payload weight or the likely cost of this new vehicle. Some commentators are anticipating a battery weight of 4–6 tonnes, roughly half that estimated by Sripad and Viswanathan (2017), which, when combined with the removal of the fuel tank and radical lightweighting, could preserve much of the existing carrying capacity of a Class 8 truck. It is possible therefore that there has been a breakthrough in battery technology which, once fully commercialized, could make it much easier than was previously thought to decarbonize long-haul trucking with low-carbon electricity.

Even without such a breakthrough, a viable alternative option now exists for electrifying heavy-duty trucks as discussed in the next section.

Electric road systems (ERS)

The electrification of highways has moved from being regarded as a far-fetched idea to a workable system and, in some countries, a policy priority in a remarkably short time. This is partly because the other means of switching long-haul road freight to low-carbon energy were looking increasingly problematic, but also because pilot projects are demonstrating that the technology is robust and likely to prove cost-effective. ERS now commands strong political support in Sweden and Germany; indeed, the two countries have signed a joint undertaking to work together in the development of this technology. A Swedish government study identified highway electrification as the most promising means of decarbonizing its long-haul trucking, something that will be required to achieve the country's goal of becoming 'fossil-fuel-free' by 2040. A German study by the Oeko Institute found that the development of an ERS would be the most cost-effective way of making long-haul trucking operations in Germany carbon neutral by 2050 (Table 6.2) (Mottschall, 2016). A UK study by Nicolaides, Cebon and Miles (2017) sees the electrification of long-haul road freight as an integral part of a broader strategy to electrify four categories of logistics operation (long-haul, urban delivery to shops, home delivery and auxiliary services), which it considers to be 'technically and financially feasible' (p.11). An earlier study by Highways England (2015) examined the feasibility of electrifying the trunk road network for all categories of traffic, although it recognized, with the support of the road freight sector, that commercial vehicles would be the first to benefit.

In 2016, the testing of ERS has moved from the test track[7] to stretches of public highway, the first of which was in Gavleborg, north of Stockholm. Other road trials are planned in Sweden, Germany and California. Three ERS systems are under development, two of which are conductive (ie requiring physical contact between the vehicle and the power source) and one inductive (ie where there is wireless transmission of the energy using a magnetic field):

1 *Overhead catenary system.* A pantograph mounted on the vehicle cab draws power from an overhead cable using a similar system to that of railway, tram and trolley-bus networks around the world. The trolley trucks used in the current trials have both electric and diesel or gas engines, allowing them to switch between energy sources when they change lanes or move between electrified and non-electrified roads. Installing these overhead cables does not require modifications to the road surface and so causes little disturbance to traffic flow during the construction phase.

Potential downsides of the catenary system are its higher visual impact, greater vulnerability to windy weather and vehicle collisions and its inability to power smaller, light-duty vehicles.

2 *Road-based conductive system.* A retractable metal arm under the vehicle connects with an electric rail embedded in the road surface. The rail is partitioned and a section only becomes electrically live as the vehicle passes. The installation of this system is much more intrusive and disruptive and is affected by routine maintenance of the road pavement. Opportunities for extending it to other categories of traffic are limited, but visually it is much less obtrusive than the catenary system. It is also less vulnerable to extreme weather and accidents.

3 *Road-based inductive system.* Transponders embedded in the road surface are activated as the truck passes to transmit the electrical energy to a receptor on the underside of the vehicle. This is the so-called Dynamic Wireless Power Transfer (DWPT) system favoured by Highways England (2015). It also requires infrastructural alterations to the road surface, but like system 2, is less vulnerable to physical impacts than the catenary system. Its main advantage is that it can be adapted to other categories of road traffic, potentially spreading the capital costs, and CO_2 savings, across many more road users.

The capital cost of road electrification is clearly an issue. One European study suggests that installing an overhead catenary line costs around €1.1–€2.5 million per km (Hey, quoted in Transport and Mobility Leuven, 2017). Using figures of a similar magnitude, a study by Fraunhofer IBP and IFEU (2015) calculated that it would cost between €30 and €65 billion to electrify 25,000 km of EU motorway using the catenary system, with a further €5 billion required annually to operate and maintain it. Assuming that EU electricity would be largely decarbonized by 2050, this continental network of electric highways would by then, according to IFEU's modelling, reduce CO_2 emissions from the long-haul movement of European road freight by 37 per cent.[8] The cost-effectiveness of ERS is enhanced by the concentration of large volumes of truck traffic on particular corridors. For example, heavy trucks in Germany emit 60 per cent of their CO_2 emissions when travelling along only 2 per cent of the road network and roughly a third of this intensively used network handles 60 per cent of all road tonne-kms (Akerman, 2016). Concentrating electrification on these heavily trafficked routes will maximize both the financial and carbon returns from the infrastructural investment.

Highway electrification can be combined with batteries to extend the carbon savings to other non-electrified parts of the road network. As the batteries will only be used for short feeder movements, they can be much smaller and lighter than those that would be required as the main or sole power source. They would be recharged while the truck is in motion travelling along the e-highway. Low-carbon biofuels or hydrogen fuel cells could also be used as a secondary power source for vehicles on non-electrified sections of the road network. If, however, we are on the eve of a step change in battery technology, as suggested by the recent launch of electric heavy-duty trucks, the case for highway electrification may be weakened and politicians may be reluctant to commit the large amounts of long-term funding it will require.

Summary

In climate change studies and plans, long-haul road freight is often portrayed as a 'hard to mitigate' sector of the economy because of its rapid growth rate, heavy dependence on fossil fuels and lack of viable decarbonization pathways. The three previous chapters explained how the growth rate can be slowed and possibly reversed; this one has shown how the potential exists to cut both energy use per truck-km and the carbon content of that energy by substantial amounts, in most cases using existing technologies and adopting operational practices that are tried and tested. Vehicle designers, manufacturers and operators have at their disposal a broad range of technologies and practices they can deploy, most of whose CO_2-reducing impacts are mutually reinforcing. While the pathways to more energy-efficient road freight movement are now quite clearly defined, the switch from fossil to low-carbon fuel is less certain and more controversial, with biofuels, hydrogen, batteries and highway electrification vying for political and industrial support.

Notes

1 https://ec.europa.eu/transport/modes/road/weights-and-dimensions_en.
2 In this paper the term 'mass energy performance' is used but this corresponds to 'energy intensity' as defined at the start of this chapter.
3 In this discussion the term 'low-carbon electricity' will include 'zero-carbon electricity'.

4 Hydrogen produced by either of these methods would yield little or no GHG savings in the absence of carbon capture of storage. As CCS is a very long-term prospect, hydrogen obtained from either of these sources will not be discussed here as a decarbonization option.
5 According to Bossel, when the hydrogen is liquefied the energy loss is even greater at 81 per cent.
6 Siemens operates a 2 km electrified test track on a former airfield near Berlin.
7 This estimate also assumes that by 2050, 43 per cent of long-haul vehicles would be electrically powered by overhead cable.

Transforming energy use in the maritime, air cargo and rail freight sectors 07

Energy consumed by ships

Moving goods by sea consumes only around a tenth as much energy per tonne-km as moving them by truck (International Energy Agency, 2016), but worldwide shipping accounts for around four times more tonne-kms than trucking (International Transport Forum, 2017b). As Wang and Lutsey (2013: 1) explain, 'Maritime shipping is highly fuel efficient, but its sheer volume and rapid growth make it a major consumer of energy and source of carbon emissions.'

The total amount of fuel consumed by ships is estimated in two ways. Using a top-down approach based on sales of bunker fuel, the International Energy Agency (2017b) calculates that it fluctuated around 260 million tonnes between 2007 and 2015. Smith et al (2014) and Olmer et al (2017), using a bottom-up approach based on the tracking of individual vessels, have produced significantly higher estimates and suggested that total fuel consumption declined from 352 million tonnes in 2007 to 298 million tonnes in 2015.[1] There was a marked convergence of the top-down and bottom-up estimates between 2007 and 2015, increasing confidence in the statistical analysis of fuel efficiency trends in the maritime sector.[2] In the remainder of this section, we will use the 'bottom-up' estimates.

According to Olmer et al (2017), shipping accounted for 2.6 per cent of global CO_2 emissions in 2015, down from 3.5 per cent in 2008. This decline in the maritime share of total global emissions was due to a divergence of trends: between 2007 and 2015, global emissions rose by 13 per cent while those from shipping declined by 15 per cent. This contraction of shipping's global carbon footprint was due mainly to a decline in its carbon intensity. For example, between 2007 and 2012 the average carbon intensity of container ships, bulk carriers and oil tankers, three categories of vessel that together account for around 86 per cent of all shipping capacity,[3] fell by, respectively, 18, 22 and 25 per cent (Smith et al, 2014). In each case, the growth of capacity (measured in deadweight tonnage (dwt)) exceeded the growth in demand (expressed in tonne-kms), suggesting that the reduction in carbon intensity was due more to improved energy efficiency than to higher load factors.[4]

In this chapter we shall focus on emissions from shipping operations rather than the wider maritime supply chain. The case for adopting a wider supply chain perspective is discussed in Woolford and McKinnon (2011). This recognizes the contribution that shippers, ports, freight forwarders and hinterland transport companies can make to the decarbonization of the maritime sector (McKinnon, 2014c; Gibbs et al, 2014). By carbon footprinting a global supply chain one gets an indication of the relative emissions from the ship, the ports and the hinterland transport. The example in Table 7.1

Table 7.1 CO_2 emissions from container movement from China (Wuhan) to UK (Glasgow) with road and intermodal hinterland options in the UK

Road feeder in UK	% of CO_2	Intermodal feeder in UK	% of CO_2
Road to Shanghai port	28.2%	Road to Shanghai port	31.2%
Port-Shanghai	0.2%	Port-Shanghai	0.3%
Deep-sea leg	58.5%	Deep-sea leg	64.7%
Port-Felixstowe	0.3%	Port-Felixstowe	0.3%
Road to Glasgow	12.8%	Rail to Coatbridge	3.1%
		Road to Glasgow	0.4%
Carbon intensity values:		road in China	120g / tonne-km
		road in UK	75g / tonne-km
		deep-sea container vessel	12g / tonne-km
		rail in UK	31g / tonne-km
		ports	16-18kg / container

relates to a 40-ft container movement from Wuhan in China to Glasgow in the UK and distinguishes road and intermodal hinterland movements within the UK. It shows that in both cases, the deep-sea shipping operation represents around 60 per cent of total emissions. Opportunities for decarbonizing the road and rail legs of the journey are discussed in, respectively, Chapter 6 and later in this chapter.

Port-related emissions typically represent a very small percentage of total supply chain emissions, as in the Wuhan-Glasgow example. Merk (2014) estimates that ports account for only 2 per cent of total shipping CO_2 emissions, though port emissions are projected to rise four-fold by 2050 on a business-as-usual basis. Several international initiatives are underway, most notably the World Ports Climate Initiative (organized by the International Association of Ports and Harbours (IAHP)), to help ports reduce their carbon footprints, mainly by a combination of energy efficiency measures and a switch to renewable energy. The majority of port-related emissions of CO_2, 58 per cent of them in the case of European and Asian ports, come from ships rather than landward activities (Merk, 2014). Decarbonization of the vessels will therefore make a major contribution to port decarbonization. This is not to belittle the efforts that port and terminal operators are making to cut emissions from those activities that they directly control. Carbon mitigation in the port sector is discussed at length in European Sea Ports Organization (2012), Winnes, Styhre and Fridell (2015), and Wilmsmeier and Spengler (2016).

Decarbonizing port operations will be relatively easy by comparison with the gargantuan task of bringing shipping emissions down to an acceptable level over the next few decades. A study for the European Parliament suggested that if decarbonization efforts are not intensified, maritime transport could represent 17 per cent of global GHG emissions by 2050 (Cames et al, 2015: 9). Smith et al (2014: 166) constructed four business-as-usual scenarios for the IMO in which total CO_2 emissions from shipping grow between 50 and 250 per cent over the period 2012–2050, this wide range reflecting differing assumptions about 'future economic and energy developments'. Even the lowest of these growth projections will require a rapid acceleration of the recent rate of decarbonization and transformation of energy use in this sector. This will involve both increasing energy efficiency and a major switch to lower-carbon energy.

Improving the energy efficiency of shipping

As with other freight modes, the energy efficiency of shipping can be increased by a combination of technological and operational improvements. The adoption rate for new technology in this sector is constrained by the relatively long lifespan of a ship and a replacement cycle extending over decades. At the start of 2017, the average ship was 20.6 years old and 53 per cent of vessels were in the 20+ age bracket (UNCTAD, 2017). The IMO emissions model assumes an average ship life of 25 years (Smith et al, 2014). The retrofitting of existing fleets permits shorter-term upgrades but major technical advances usually have to be incorporated into new vessels. Operational improvements can be implemented more rapidly, though sometimes require technical modifications to the vessel that extend the timescale and inflate the cost.

Numerous studies have examined the technical and operational options for decarbonizing shipping. The second IMO greenhouse gas report (Buhaug et al, 2009) provided one of the first detailed assessments of these options, estimating that changes to the hull and superstructure could cut CO_2/tonne-km by up to 20 per cent, to the power and propulsion systems by up to 15 per cent and to the 'concept, speed and capability' of new ships by as much as 50 per cent (p.73). Since then a substantial literature has accumulated on the decarbonization of shipping, much of it focused on energy efficiency. This has been comprehensively reviewed by Bouman et al (2017), who have compiled summary tables showing estimates of the carbon savings from 16 energy efficiency measures. Two measures in particular are reckoned to have been responsible for much of the reduction in carbon intensity observed between 2007 and 2015: increased vessel size and a reduction in vessel speed (ie slow steaming).

Increasing vessel size

As vessels get bigger their energy consumption increases at a slower rate than their carrying capacity. Bouman et al (2017: 410) observe that 'when cargo capacity is doubled, the required power and fuel consumption increases by about two-thirds, thus reducing fuel consumption per freight unit.' So long as the additional capacity is well utilized, energy use per tonne-km or per TEU-km can sharply decline. Wang and Lutsey (2013) found that CO_2 emissions per TEU-km were 31 per cent lower for 8,000+-TEU container ships than for those with a capacity of 2,000 TEU or less. They also noted that, in terms of carbon intensity, there were 'diminishing returns' as vessels increased in size. This was confirmed by Acciaro and McKinnon (2015) in a regression analysis of the relationship between fuel consumption and nominal vessel capacity for 2,030 container ship voyages. They found that 'average fuel consumption per TEU-km appears to stabilize above 8,000 TEU' (p.83). Bold claims have, nevertheless, been made for the carbon benefits of increasing the maximum size of container ships. Maersk (2014), for example, reported that its Triple E ships, introduced in 2013 with a carrying capacity in excess of 18,000 TEU, could cut carbon emissions per container by 50 per cent relative to the 'industry average on the Asia–Europe trade lane' because of a 16 per cent increase in capacity, 19 per cent reduction in engine power and a design speed two knots lower (Jiven (2011), quoted in McGill, Remley and Winther, 2013). The IMO's third GHG study expects the capacity of container ships to continue to expand over the next few decades, with the proportion of 14,500+-TEU vessels rising from 0.2 per cent in 2012 to 5 per cent in 2050 and the proportion of ships larger than 8,000 TEUs growing from 9.2 to 24 per cent. Other things being equal, concentrating capacity in larger vessels will significantly reduce their average carbon intensity. The size profiles of tankers and bulk carriers are expected to remain more stable.

Slow steaming

Reducing the speed of a vessel cuts fuel consumption by a disproportionately large amount, reflecting the fact that the 'fuel use and emissions per ton-mile are roughly proportional to the square of the speed' (Smith et al, 2014: 166). Slowing down a container ship by 10 and 20 per cent cuts fuel use and CO_2 emissions by, respectively, 15–19 per cent and 36–39 per cent (ICCT, 2011). In the container sector, slow steaming generally entails a 12.5 per cent reduction from a design speed of 24 knots to 21 knots, though some shipping lines have lowered the speed to 18 knots ('extra-slow steaming') or even 15 knots ('super-slow steaming') (Maloni, Paul and Gligor, 2013). This practice was initially implemented by container shipping lines in 2007 to economize on fuel during a period of steeply rising fuel prices and deepening global recession. Although primarily a response to economic pressures, slow steaming has proved to be an effective carbon-reduction measure. According to Cariou (2011) it resulted in CO_2 emissions from deep-sea container shipping being 11 per cent lower in 2010 than would otherwise have been the case. Having started with vessels that have the highest design speeds, ie container ships, slow steaming has now become fairly pervasive across the shipping industry. According to Smith et al (2014: 17), over the period 2007–2012 'the average reduction in at-sea speed relative to design speed was 12 per cent and the average reduction in daily fuel consumption was 27 per cent' (p.17). They point out, however, that 'a reduction in speed and the associated reduction in fuel consumption do not relate to an equivalent per cent increase in efficiency, because a greater number of ships (or more days at sea) are required to do the same amount of transport work' (p.17). While the net reduction in fuel use would have been less than 27 per cent, it would still have been substantial.

This raises the important issue of capacity requirements on shipping services subject to deceleration. In the early years of slow steaming it absorbed 'up to 40 per cent of potentially laid-up capacity' (Johnson, 2010, quoted by Mander, 2017: 231). Since then, as Mander explains, increases in vessel size have made it possible to move more freight on slower voyages. This expansion of ships has been motivated by a desire to exploit economies of scale rather than to compensate for slow steaming. In most sectors of the shipping industry over the past decade there has been sufficient excess capacity to accommodate slow steaming. If, however, as DNV–GL (2017) and others acknowledge, future growth in trade volumes exhausts this excess capacity, shipping lines will be confronted with a choice of reversing the slow steaming trend or acquiring enough additional capacity to maintain speeds at their current level. Opinions are divided on this issue.

Some believe that slow steaming has become a permanent feature of the shipping business. The fact that new container ships have been built with design speeds significantly slower than those of their predecessors suggests that the practice has become technologically embedded. Regulations on the energy efficiency of new ships, discussed below, are also helping to enforce a long-term commitment to slower speeds (Mander, 2017). Meanwhile, shippers have adjusted the scheduling of activities across their global supply chains to accommodate longer transit times. From interviews with senior shipping and export managers of 15 large shippers, I found that this adjustment had been easier than one might have expected, for several reasons (McKinnon, 2016a: 7–8):

- much of the freight carried by deep-sea container is not very time-sensitive;
- relative to order lead times of several months, additional deep-sea transit times of 3–4 days are relatively small;
- longer average transit times have been partly offset by improved service reliability;

- use of IT to improve the visibility of inbound container flows and the modification of internal processes has eased the impact of slow steaming;
- reduction of some port turnaround times has offset longer sea transit times.

In-transit inventory has unquestionably increased, though in most cases only marginally, and in a period of low interest rates the impact on inventory financing costs has been very small. According to Maloni, Paul and Gligor (2013), after allowance is made for these additional costs, slow steaming still yields large economic benefits. Their simulation modelling of container trade on the Asia–North America route showed that 'extra-slow steaming' cut total costs for carriers and shippers by 20 per cent, while reducing CO_2 emissions by 43 per cent. Both carrier and shipper can claim decarbonization benefits from slow steaming; the challenge has been to find an equitable way of splitting the net economic benefit between them. This could have been done by carriers reducing bunker fuel surcharges to compensate shippers for slow steaming, though Notteboom and Cariou (2013) detected little evidence of this happening. While the costs and benefits of its introduction could have been more fairly distributed, after a decade, slow steaming is now accepted as a commercial fact of life in the global shipping market. If, then, it is taken as a given in the future planning of ship capacity in periods of buoyant demand, more ships and containers will be required than if sailing speeds were to return to their pre-2007 levels. Building these extra ships and containers will release GHGs, but the 'embodied' emissions in this additional equipment will be more than offset by the CO_2 savings that will accrue from moving them more slowly during their working lives.

In an alternative future scenario, when demand is high and capacity in short supply, carriers will accelerate their services, using vessels more intensively at the expense of higher fuel consumption and CO_2 emissions. A sharp drop in bunker fuel prices might also induce a similar response (Cariou, 2011). Some shipping lines might break ranks and try to gain a competitive advantage by offering shorter transit times. Olmer et al (2017) found evidence of larger container ships (over 14,500 TEUs) and oil tankers increasing their speed by, respectively, 11 per cent and 4 per cent between 2013 and 2015, which could be the start of a relaxation of slow steaming. The latest IMO GHG report also talks of 'latent emission increases (suppressed by slow steaming and historically low activity and productivity)' that could materialize if the 'market dynamics… revert to their previous levels' (Smith et al, 2014: 17). Faber et al (2012: 9) also argued that 'carriers will likely return in large part to pre-2007 speeds when market conditions change and more capacity is required.' They go on to advocate the imposition of speed limits on ships to prevent this from happening. Cariou and Cheaitou (2012), on the other hand, show that, in the case of container shipping in Europe, this would have a high carbon mitigation cost and probably be counterproductive. Should commercial pressures to relax slow steaming strengthen, other forms of regulation and the use of market-based instruments could be more effective means of resisting them.

Three other sets of energy efficiency measures are expected to make a significant contribution to the decarbonization of shipping. These relate to the vessel's hull, its engine and propulsion system, and the way it is navigated.

Hull. Streamlining the hull reduces hydrodynamic resistance and the amount of energy needed to overcome it. As the amount of drag is a function of speed, slow steaming reduces the potential energy saving from hull redesign, though it can still be significant. Lindstad, Sandaas and Steen (2014) emphasize the carbon benefits of making hulls of bulk vessels more slender. Hulls can also be lightweighted and redesigned to reduce the amount of ballast water required. Special coatings can give them a smoother surface and reduce the build-up of slime, seaweed, shells and barnacles. It is estimated that 'in the absence of hull fouling control systems, within six months of active service a vessel could have up to 150 kg of marine life per square metre', significantly impairing fuel efficiency (Fathom Focus, 2013: 7). A more recent technical innovation, which has already been applied to several large passenger cruise liners, is the release of micro-bubbles to 'air lubricate' flat-bottomed hulls. It is claimed that this technology, 'with the right ship hull design', can cut CO_2 emissions by as much as 10–15 per cent (Wang and Lutsey, 2013; Raunek, 2017), though DNV-GL (2017) reckon that fuel savings will only be in the 3–5 per cent range.

Engine and propulsion system. This covers the design and control of ship engines, waste heat recovery and the nature of the propeller. The aim is to reduce the huge energy loss that occurs on a ship between the combustion of the fuel in the engine and the rotation of the propeller. In an average cargo ship operating under typical conditions only 27–28 per cent of the original energy is finally used to propel the vessel (Royal Academy of Engineering, 2013; Lindstad et al, 2015). Around 30 per cent is dissipated as waste heat, making it important to recover as much of this heat as possible. Waste heat recovery can cut fuel consumption by

6–8 per cent (Wang and Lutsey, 2013). Minimizing mechanical and frictional energy losses in the shaft and propeller can yield fuel savings of a similar magnitude.

Route optimization. By adopting a practice known as 'weather routeing' it is possible to adjust the route and speed of the vessel in relation to weather, waves and currents. After all, strong winds and rough seas can substantially increase hydrodynamic and aerodynamic resistance and hence fuel consumption. In stormy weather, routeing vessels to minimize this resistance can cut emissions and costs by between 11 and 19 per cent (Lindstad, Asbjørnslett and Jullumstrø, 2013). Wang and Lutsey (2013) estimate that, overall, weather routeing can reduce CO_2 emissions by 1–4 per cent.

It can be seen that there are many different ways of improving the energy efficiency of shipping, most of them mutually reinforcing, though many of them yet to be applied at scale. In an effort to accelerate their uptake the IMO has launched two schemes, one mandatory and relating to the energy efficiency of new vessels, the other voluntary and designed to encourage more energy-efficient management of existing fleets. Since 2013, all new vessels entering service have been assigned an Energy Efficiency Design Index (EEDI) based on their energy consumption and CO_2 emissions per capacity-km. This is essentially a fuel economy standard for the shipping industry which will steadily rise through time, putting pressure on the designers, builders and buyers of new vessels to make them more energy efficient. As a performance-based standard, EEDI does not prescribe technologies or design features, but instead aims to 'stimulate innovation and technical development of all elements influencing the energy efficiency of a ship' (IMO, 2011: 35). On the basis of the EEDI calculation, new vessels built after 2013 have had to be at least 10 per cent more fuel efficient than the average ship built between 1999 and 2009. This threshold will rise to 20 per cent for post-2020 vessels and 30 per cent for those deployed after 2025.

Although this tightening energy efficiency requirement may seem quite a bold move on the part of the IMO, techno-economic modelling by Smith et al (2016: 40) suggests that 'EEDI may have only a small role in shipping's decarbonization'. This is partly because it applies only to new ships and, as explained earlier, vessel replacement rates are very slow in the maritime sector. It also partly reflects the way in which EEDI is defined. Smith et al (2016) estimate that by 2050, shipping emissions may be only 3 per cent lower than they would have been without EEDI. They concede, however, that this regulation may stimulate a wider behavioural change in the industry which could amplify the direct improvements in energy efficiency.

IMO is also encouraging shipping lines to implement Ship Energy Efficiency Management Plans (SEEMP) on a voluntary basis. This requires them to calculate an 'energy efficiency operational indicator' (EEOI) for each vessel and commit to raising the EEOI scores by applying a set of 'best-practice measures in fuel-efficient operation' (IMO, 2011: 36). These measures include engine tuning and monitoring, improving trim, using advanced hull coatings and speed reduction. Most of the CO_2 savings to date have come from the last of these measures but not because of SEEMP; as discussed earlier, shipping lines have had a strong commercial motive to 'slow steam' their vessels over the past decade. SEEMP is the only IMO initiative to cut CO_2 emissions from shipping in the short term, and as such it has been criticized because there are 'large gaps' between what it 'actually requires of shipping companies and what has been deemed best practice in other industries' (Johnson et al, 2013: 188). It does not, for example, give any guidance on the level to which the energy efficiency of particular types, sizes and classes of vessel should be raised.

Repowering ships with low-carbon energy

Around three-quarters of the fuel consumed by shipping worldwide is 'heavy fuel oil' (HFO) (McGill, Remley and Winther, 2013). This is a residual fraction in the oil-refining process that is left when lighter, cleaner distillate fuels, such as kerosene, petrol and diesel, have been extracted. It is a thick, dirty, sulphur-rich fuel that needs to be pre-heated for combustion in a ship's engine. Its main advantages are that it is relatively cheap and widely available in large quantities through a bunker fuel supply network. It is, however, by far the most polluting of all transport fuels, causing shipping to emit 17 per cent of the world's anthropogenic NOx emissions, 10 per cent of its SOx emissions and enough particulate matter to hasten the deaths of tens of thousands of people each year (Holmes, Prather and Vinken, 2014; Corbett et al, 2007; Neslan, 2016). Over the past decade, efforts have been made to reduce the high levels of pollution from ships, particularly of SOx, in Emission Control Areas (ECAs) such as the Baltic and North Seas and the east and west coastlines of North America. At a global level the IMO is reducing the maximum permitted sulphur content in marine fuel from 3.5 per cent to 0.5 per cent in 2020. These initiatives have generated much discussion and research on the use of alternative fuels in the maritime sector to cut emissions of noxious gases. Their impact on GHG emissions from shipping has so far been a secondary concern, though the switch to lower-carbon energy sources features prominently in future roadmaps for the decarbonization of shipping (eg Smith et al, 2016; DNV-GL, 2017). It is also worth noting that there is a conflict between efforts to minimize noxious emissions from shipping for health reasons and to decarbonize it. As Eide et al (2013: 4185) explain:

> The agreed regulations on shipping emissions of SOx and NOx will, as an unintended side effect, affect climate in the following ways: firstly, because the indirect cooling effects of NOx and SOx will diminish, and secondly, because some of the measures needed to comply with the new rules may increase fuel consumption and emissions of CO_2.

To meet the tightening restrictions on sulphur emissions, the shipping industry has been shifting to lower-sulphur versions of HFO and to marine diesel oil (MDO), the latter accounting for around a quarter of the fuel consumed by ships (McGill, Remley and Winther, 2013). Both of these fuels have much lower sulphur content, but very similar carbon intensity to normal HFO at around 3.1–3.2 tonnes of CO_2 per tonne of fuel (RightShip, 2013). On a 'well-to-propeller' basis,[5] however, low-sulphur HFO has a carbon intensity around 2 per cent higher than normal HFO (Chryssakis, 2014). Efforts to 'desulphurize' marine fuel are also in conflict with the decarbonization of shipping in two other respects, both relating to the use of devices called 'scrubbers' on vessels to remove sulphur from the exhaust gases. According to DNV-GL (2017), the extra energy used to power these scrubbers increases GHG emissions by 5 per cent. It also argues that the decision to invest in scrubbers may lock shipping lines into longer-term use of HFO and discourage them from switching to lower-carbon fuels. In what they describe as an 'emissions paradox... vessels with scrubbers are... very unlikely to adopt a different fuel type in the future, therefore slowing the transition to low-carbon fuels' (DNV-GL, 2017: 6). This is only one of several factors inhibiting the switch to low-carbon energy in the maritime sector, as we shall see below.

The low-carbon energy options can be divided into six categories: natural gas, biofuel, wind and solar, electricity, hydrogen and nuclear.

Natural gas

Both LNG and LPG can be used to power ships. LNG is generally seen as the more appropriate gaseous fuel for shipping, as LPG tends to be more expensive and in higher demand in other sectors of the transport market. Nevertheless, LPG's potential as a ship fuel may have to be reassessed in the light of recent research (Brinks and Chryssakis, 2017). Interest in natural gas has been driven mainly by the tightening of sulphur regulations, as both LNG and LPG emit virtually no sulphur and only a fraction of the NOx and PMs released by burning HFO. Unlike the use of scrubbers to meet these regulations, switching to natural gas also helps to cut CO_2 emissions relative to HFO, by 20 per cent for LNG and 17 per cent for LPG (DNV-GL, 2017). As with the application of natural gas in the trucking sector, however, some of the CO_2 benefits are eroded by a methane slip problem. Methane leaks from both the fuelling system and incomplete combustion in the engine. More research is needed to assess the extent to which this problem weakens the case for a switch to LNG/LPG for decarbonization purposes and how it can be alleviated in the maritime sector.

LNG has been used to power ships for many years, particularly in vessels carrying LNG as a cargo. The ship classification societies have rules in place for the use of LNG in new ships and existing vessels converted to burn this fuel. It has been concluded, therefore, that 'there is little of a technical nature that would prevent adoption of LNG as a marine fuel' (Royal Academy of Engineering, 2013: 29).

Apart from the methane slip problem, the main barrier to the use of LNG is the lack of bunkering infrastructure for this fuel. There has been significant investment in LNG fuelling points in ECAs such as the Baltic, but many are designed for smaller vehicles providing short-sea services. Through its Connecting Europe Facility, the EU is providing financial support for the development of LNG terminals. It will, however, take substantial investment and a long time to build up a network of LNG bunkering terminals both for short-sea and deep-sea vessels. Lloyd's Register (2012) considered the use of LNG to be a 'viable option' in deep-sea ships and forecast that it would account for between 3.2 and 8.4 per cent of global HFO demand by 2025. This was contingent on extensive development of an LNG bunkering capability at a strategic network of ports. Chryssakis et al (2014: 11) predicted that 'over the next four decades… LNG has the potential to become the fuel of choice for all shipping segments, provided the infrastructure is in place.' It envisages between 40 and 70 per cent of ships using LNG or LPG by 2050 (DNV-GL, 2017).

Biofuel

Various biofuels, such as biodiesel, biomethanol and biogas, have been successfully trialled on ships either as the sole fuel or blended with other fuels (eg biodiesel with MDO and biogas with LNG). Their use in the maritime sector, however, is at a very early stage, although if it were scaled up there could be substantial carbon benefits. Biodiesel and biomethanol can both cut GHG emissions by 50 per cent, and biogas by 90 per cent relative to HFO (Chryssakis et al, 2014). Their use in the maritime sector is, however, subject to similar constraints and concerns as in the road freight sector, as discussed in Chapter 6. Given the huge energy demands of the shipping industry, a substantial switch from HFO to biofuels would require vast amounts of agricultural land. MacKay, for example, estimated that powering all the world's shipping with biofuel derived from vegetation would require a land area equivalent to twice that of the UK (quoted by the Royal Academy of Engineering, 2013: 27).

Wind and solar power

Ships can capture wind power in several ways: soft sails, fixed sails, cylindrical rotors (known as Flettner rotors), kites and wind turbines (Mofor, Nuttall and Newell, 2015). These 'wind assistance' technologies have been successfully trialled and can generate significant amounts of energy, though the supply is weather-sensitive and, according to a marginal abatement cost analysis by DNV (2010), has a relatively high carbon mitigation cost. Studies reviewed by Bouman et al (2017), nevertheless show much higher CO_2 reduction potential for wind than for solar power. The amount of solar power that can be generated on a ship is tightly limited by the amount of deck area available for photovoltaic panels and dependent on the amount of sunlight on the routes the vessel typically sails. Overall, these forms of renewable energy can help to reduce fossil fuel use on particular types of vessel but are likely to make only a minor contribution to the decarbonization of shipping.

Electricity

The second GHG report from the IMO (Buhaug et al, 2009) barely mentions electrification as a means of decarbonizing shipping. Five years later, Chryssakis et al (2014) reported that 'recent developments in ship electrification hold significant promise for more efficient use of energy.' In 2015 a battery-powered 120-car ferry went into service on the west coast of Norway (Mofor, Nuttall and Newell, 2015) and in late 2017, the world's first battery-powered cargo ship was launched in China, with a 2,000-metric-ton carrying capacity and 80 km range (Quanlin, 2017). Until recently batteries were seen as having a supplementary role on ships within hybrid propulsion systems, in powering ancillary equipment for heating, air conditioning, lighting etc, and storing renewable energy from solar panels and rotors onboard the vessel. For the foreseeable future this may be their main contribution to decarbonization and it may be relatively small. However, the ability to power cargo ships with batteries has now been demonstrated and greatly increases the potential for using low-carbon electricity from power grids in shipping operations. A long-established practice called 'cold ironing' has allowed ships to connect to a shore-side electrical supply while in port, enabling them to switch off their main and auxiliary engines. In the few ports with a cold-ironing capability, this has been done mainly to cut emissions of noxious pollutants. Depending on the carbon intensity of the electricity, it can also cut CO_2 emission from vessels while in port. The use of cold ironing to recharge ship batteries for the propulsion of small vessels on short-range freight services opens up new opportunities for decarbonization. The weight problem which discourages the use of batteries in heavy trucks is less of a problem in ships, which have a much greater gross weight and are not subject to legal weight limits.

Hydrogen

The problems of using hydrogen as an energy carrier for low-carbon electricity were discussed in Chapter 6 in the context of road freight decarbonization. They are just as acute and constraining in the maritime sector. Energy losses and costs in the production, storage and delivery of hydrogen are high. A new hydrogen 'bunkering' infrastructure would have to be created. Compressed hydrogen requires six to seven times more storage space than HFO (Chryssakis et al, 2014), making it very difficult to retrofit an existing vessel to run on the gas and imposing a significant payload penalty. Marine applications of the fuel cells that would be needed to convert hydrogen into electrical power are still at a very early stage in their development (DNV-GL, 2017). Overall, therefore, the outlook for hydrogen as an alternative, low-carbon energy source for shipping does not look promising.

Nuclear

Nuclear power is already used in naval vessels and could conceivably be introduced on commercial vessels, where it would permit near-zero-carbon operation. An exhaustive review of its possible application in the maritime sector by the Royal Academy of Engineering (2013: 40) confirmed its technical feasibility but also highlighted the numerous barriers, including 'international regulation, public perception, initial capital cost and financing, training and retention of crews, refuelling and safe storage of spent fuel and setting up and maintenance of an infrastructure.' Few would therefore dissent from the view of Buhaug et al (2009: 136) that 'installing nuclear reactors onboard is not foreseen to be an interesting option for international shipping for environmental, political, security and commercial reasons.'

Summary

The transformation of energy use in the maritime sector will involve a complex mix of interventions. Energy consumption per tonne-km can be substantially reduced by a synergistic combination of technical and operational measures. Energy efficiency improvements on their own, however, will not be enough to achieve the deep reductions in GHG intensity that will be required, given the huge predicted growth in maritime cargo volumes (Smith et al, 2016). They will need to be accompanied by a major shift to lower-carbon energy sources. LNG currently seems to be the strongest candidate to replace HFO and MDO because it is plentiful, relatively cheap and can be used on existing vessels after appropriate retrofitting. However, it will take time to establish an adequate bunkering network worldwide, the net CO_2 saving at 20 per cent is fairly modest, and on a well-to-propeller basis the net GHG savings will be much smaller because of methane leakage problems. It will need, therefore, to be supplemented by other low-carbon energy sources in combinations that will vary with the size and type of vessel and the nature of the shipping operation.

Bouman et al (2017) assess the current state of knowledge on both energy efficiency improvement and the switch to low-carbon energy in the maritime sector. They review 150 studies, 60 of which contain quantitative data, and arrive at a few general estimates of the potential for decarbonization in this sector. This exercise was complicated by the studies using differing categorizations of various measures and, in some cases, producing widely varying estimates of likely CO_2 savings. Using median values for the combined impact of all the energy efficiency and alternative energy options, Bouman et al calculate a CO_2 emission reduction potential of 39 per cent by 2030 and 73 per cent by 2050 relative to what the studies regarded as the business-as-usual emission levels in those years. Again using median values, this large body of research evidence suggests 'an emission reduction of between 50 and 60 per cent per freight unit transported up to 2050' (p.417).

Energy use in the air cargo sector

Air freight is by far the most carbon-intensive freight transport mode, with an average CO_2/tonne-km figure eight to ten times that of trucking and 80–100 times that of shipping (Sims et al, 2014). On the other hand, only a very small fraction of global freight traffic moves by air. To put it into perspective, in 2015 the movement of air cargo worldwide (expressed in tonne-kms) was roughly equivalent to the movement of freight by all modes in the UK in that year (Airbus, 2016; Department for Transport, 2017d).

CO_2 estimates for air cargo are based on top-down calculations. It would be preferable if, as in the case of shipping, one could conduct a bottom-up analysis based on the activities and fuel consumption of individual aircraft carrying freight, but this is currently not possible. The carbon footprinting of global air freight is complicated by the fact that just over half of it flies in the bellyholds of passenger aircraft (Airbus, 2016). Jardine (2009) estimated that freight typically accounted for 15–30 per cent of emissions in wide-body passenger jets and 0–10 per cent in narrow-bodied planes. This split depends, however, on the methodology used to allocate CO_2 emissions between the passengers and freight carried on the same plane (eg IATA, 2014b). Currently, different organizations apply different allocation rules, though efforts are being made to harmonize them (GLEC, 2016).

On the basis of available data it is not possible to monitor changes in the carbon intensity of air cargo through time. There is also very little published research specifically addressing the energy efficiency of air freight operations and opportunities for switching air cargo services to lower-carbon fuels. Literature on the decarbonization of aviation tends not to distinguish between passenger and freight services and where it does it is mainly concerned with the passenger side of the business.

ICAO, the UN organization responsible for international aviation, and IATA, the trade body representing 280 airlines worldwide, have both committed to stabilizing CO_2 emissions from international aviation (both passenger and freight) by 2020 and making its subsequent growth carbon neutral. By 2050, IATA aims to have reduced total CO_2 emissions from aviation by 50 per cent against a 2005 baseline, despite the huge predicted growth in air transport over the next few decades. IATA (2017a) is forecasting a 3.6 per cent compound annual growth rate in air passenger traffic in the next 20 years, causing air passenger volumes to double by the mid-2030s. Air freight tonne-kms are forecast to increase at a slightly faster rate of 3.8–4.0 per cent per annum over this period (Boeing, 2016; Airbus, 2017). An increasing proportion of this additional air freight will be carried in the bellyholds of passenger aircraft; Airbus (2017) expects the bellyhold share of tonne-kms to rise from 52 to 61 per cent. To some extent the growth of bellyhold freight capacity can be seen as a 'by-product' of the expansion of air passenger capacity, particularly on long-haul flights. The relationship between the upward trends in air travel and air cargo is more synergistic than this, however, because as freight represents 9 per cent of the revenue of combined passenger/freight airlines (IATA, 2015), the investment case for new aircraft is often critically dependent on the freight revenue.

Data provided by ICAO (2016) shows just how difficult it is going to be to meet the aviation industry's carbon commitments while accommodating rising passenger and freight demand. It predicts that, on a business-as-usual basis, the total amount of fuel consumed by international aviation will rise from around 200 million tonnes in 2016 to between 450 and 1,090 million tonnes in 2050, with 850 million tonnes as the mid-range projection.[6] ICAO has an 'aspiration' to see the energy efficiency of aviation increase by 2 per cent per annum. If this were achieved and combined with improved air traffic management (ATM), fuel consumption in 2050 would be around 440 million tonnes, still more than double the 2016 figure. By 2050 a huge 'emissions gap' would have opened up between the flatline carbon neutrality baseline beyond 2020 and the emissions from burning 440 million tonnes of kerosene. ICAO (2016: 20) suggests that this emissions gap could be closed by 2050 but this would require 'nearly complete replacement of petroleum-based fuels with sustainable alternative jet fuel', in this case biofuel, derived from plants such as Jatropha or organic waste. This in turn 'would require approximately 170 new large biorefineries to be built every year from 2020 to 2050, at an approximate capital cost of US $15 billion to $60 billion per year if growth occurred linearly' (p.20).

It is recognized by ICAO, IATA and the major airlines that this is not a realistic scenario and, therefore, that carbon neutrality can only be achieved by buying carbon credits and relying on others to cut emissions on behalf of the airline industry. ICAO has devised an elaborate 'carbon offset and reduction scheme for international aviation' (CORSIA), which will take effect in 2020 and involve national governments. CORSIA has nevertheless been criticized for lacking 'environmental integrity', being voluntary until 2027 and generally being very unlikely to close the emissions gap (Carbon Market Watch, 2016). Wider concerns about the use of carbon offsetting were discussed in Chapter 2 of this book.

To what extent might it be possible to reduce this dependence on offsetting by improving the energy efficiency of air freight operations and switching to alternative fuels?

Making air cargo operations more fuel efficient

Energy costs represent a high proportion of an airline's costs. When the oil price is around US $50 per barrel, fuel accounts for 17 per cent of the average airline's operating costs, rising to over 30 per cent at $100 per barrel (Airbus, 2016). For purely financial reasons, therefore, aircraft designers, manufacturers and operators have a strong incentive to maximize energy efficiency and minimize the industry's exposure to fluctuating oil prices. ICAO (2013) estimated that since the 1960s, jet aircraft had become 80 per cent more energy efficient and projected a slightly slower rate of future efficiency improvement, of 0.57–1.50 per cent per annum, forward to 2050.

Most of the past gains in energy efficiency have come from improved engine technology and streamlining of the aircraft fuselage. To achieve a step change in fuel efficiency, a 'fundamental shift in design is required', such as a move to open-rotor engines, propfans and blended wing airframes (Bows-Larkin, 2015: 691). This echoes the view (ACARE, 2008: 61) almost a decade ago that 'although there is scope for further improvement by evolving existing technologies, further substantial improvement will require the introduction of breakthrough technologies and concepts into everyday service.' In a subsequent report they argued that further energy efficiency improvements of 15–20 per cent from engines and 20–25 per cent from airframes would be required, effectively 'doubling the historic rate of improvement' (ACARE, 2011: 67).

Such radical changes will not only be very costly, they will take a long time to achieve. Air transport, like shipping, is essentially 'a long-life-cycle industry' (ICAO, 2007). It can take 10 years to design a new aircraft, which will then be manufactured for around 20–30 years, with each aircraft having a typical lifespan of 25–40 years. ICAO will be introducing a CO_2 and particulate matter standard for new planes built after 2020. Unlike fuel economy standards for trucks and EEDI in the shipping industry, this CO_2 standard will relate to the use of alternative fuels as well as energy efficiency. Like EEDI, however, the net effect of this new standard on total aviation CO_2 emissions over the next few decades may be very limited given the long aircraft development and use cycles.

The diffusion of more energy- and carbon-efficient technologies into the air cargo sector proceeds at different rates. Innovations normally impact first on passenger jets and therefore benefit bellyhold freight. The Boeing 787, for instance, is 20–25 per cent more fuel efficient and has 47 per cent more revenue-earning cargo space than previous aircraft of its type. At a later stage, technical advances are incorporated into purpose-built air freighters, many of which are purchased by integrated express carriers. Much general air cargo flies in former passenger aircraft converted into freighters at a later stage in their life. In 2014 the average age of an all-cargo plane in Europe was 19.6 years, much higher than the average of 10.3 years for the entire European fleet (European Environment Agency, 2016b). Overall, therefore, it takes longer for air freight operations to benefit from fuel- and CO_2-reducing advances in aircraft technology.

The weight and streamlining of the airframe have a major influence on energy efficiency. Lightweighting has been achieved by the use of lighter alloys and composites. The Airbus A380, for example, has twice the proportion of composites in its airframe as the A320, introduced 20 years previously. The switch to 'fly by wire', involving the replacement of hydraulic controls with wiring, has also reduced aircraft weight. The use of 3D printing allows aircraft manufacturers to reduce the weight of many onboard components (Orcutt, 2016). Retrofitting 'winglets' to the ends of aircraft wings improves aerodynamics, raising fuel efficiency by an average of 4–6 per cent (Flight International, 2008).

Improvements in energy efficiency are not confined to the aircraft. Improved air traffic management (ATM) can make a significant contribution to energy and CO_2 reduction. ATM includes the airborne routeing of the aircraft as well as its taxiing on the ground. Flying across Europe's congested airspace carries a significant environmental and economic penalty. ACARE (2008: 64) estimated that within Europe, 'between 13 per cent and 15 per cent of fuel is consumed through excessive holding either on-ground or in-flight and through indirect routeing and non-optimal flight profiles.' By far the most important ATM initiative in Europe aims to move away from nationally based air traffic control to a 'Single European Sky' by 2020 (IATA, 2013). This EU initiative should cut CO_2 emissions per flight by 10 per cent and enable European ATM to handle a three-fold growth in air traffic (ICAO, 2016). There is a danger, however, that 'more efficient use of airspace and increases in airport capacity', while reducing fuel use per passenger and tonne-km, will 'serve to maintain or raise growth rates, increasing absolute energy consumption' (Bows-Larkin, 2015: 692). In projections of the future carbon intensity of aviation, therefore, allowance must be made for such a rebound effect.

Flying freight with low-carbon energy

The aviation industry's decarbonization plans place heavy emphasis on the use of alternative jet fuels (AJF). To date, several airlines, such as KLM, have successfully demonstrated that commercial aircraft can be operated safely and efficiently with a blend of biofuel and kerosene. The main barrier to wider uptake of biofuels in this sector is what ICAO (2016: 153) calls 'the tremendous price gap between conventional fuels and biofuels for aviation', which suppresses demand and 'in turn limits the investment in biorefineries that is needed in order to scale up production'. Hind (2014) suggests that in Europe a tonne of biofuel would cost around three times that of kerosene. High cost is not the only constraint, however. Gegg, Budd and Ison (2014) also see biofuel use in aviation being inhibited by 'limited availability of suitable feedstocks, uncertainty surrounding the definition of the sustainability criteria, and a perceived lack of both national and international political and policy support for aviation biofuel.' Other concerns about life-cycle GHG emissions, land use implications and food market distortions, discussed in Chapter 6, also apply to aviation-related biofuels.

The decarbonization of freight transport with low-carbon electricity, which offers large potential for land-based freight modes, is unlikely to be a feasible option for air freight in the foreseeable future. Aviation requires power sources with a high energy density and on this criterion, batteries fall well short of kerosene. However, greater use is being made of batteries to power ancillary equipment. For example, they give the Boeing 787 five times as much onboard electrical power as other comparable aircraft. Recent advances in battery technology have raised the possibility of large hybrid jets being developed which combine the use of batteries with liquid fuel in the propulsion system, particularly during take-off (Calder, 2017). It will be a long time, however, before electrification has a significant impact on the carbon intensity of air cargo operations.

Summary

Much of the discussion in this section has related to aviation in general rather than just air freight. This is partly because there is a dearth of air freight-specific environmental data, but it can also be justified on the grounds that over half the freight moved by air shares planes with passengers. Complaints that the aviation sector is not doing enough to curb its carbon emissions therefore apply as much to freight as to passenger services. In a report for the European Parliament, Cames et al (2015: 40) predict that if it continues to fall behind other sectors in its efforts to decarbonize, aviation could account for 22 per cent of total global CO_2 emissions by 2050. Some commentators now take the view that, given the limited options for improving the energy efficiency of aircraft and switching them to lower-carbon fuel within the required time frame, the only remaining option is to restrain the growth in demand (Peeters, 2017; Bows-Larkin et al, 2016). While this may be too pessimistic a view, it suggests that major users of air cargo services may have to give more serious thought to the opportunities outlined in Chapters 3 and 4 to reduce freight demand and shift it to alternative lower-carbon modes.

Energy use in the rail freight sector

Globally, rail freight operations have a low average carbon intensity (14g CO_2/tonne-km), account for a small proportion of total freight movement (6.9 per cent of tonne-kms in 2015) and emit a relatively small share of total energy-related CO_2 emissions (0.7 per cent in 2015) (IEA/UIC, 2017). The average energy and carbon intensities of rail freight have also declined by roughly 40 per cent over the past 40 years. One might therefore consider rail freight operators to be under less pressure than their counterparts in the road, maritime and aviation sectors to achieve further improvements in energy and carbon efficiency. They cannot be complacent, however, about their environmental advantage over their main modal competitor – trucking. In some countries, road transport has recently been improving energy efficiency and environmental performance at a faster rate, eroding rail's carbon advantage. The trucking industry has the advantage of replacing its vehicles more frequently and thus more quickly adopting new energy-saving technology. Locomotives, like ships and aircraft, typically have a working life extending over several decades. In 2012 the average age of the world's 47,000 electric locomotives was 27 years (SCI Verkehr, 2012), while in 2014, (SCI Verkehr, 2012) 36 per cent of the diesel locos in the United States were more than 25 years old (Humphrey, 2015). Older rolling stock is, however, amenable to retrofitting with energy-saving devices in the short to medium term.

It is difficult to generalize about energy use in rail freight operations globally because of wide international variations in the major determinants of energy efficiency, many of them related to infrastructure. Infrastructure influences the length, weight and speed of freight trains. In European countries trains typically have a maximum length of 750–850 metres, whereas in the United States they can exceed 3,000 metres. Maximum axle weights and loading gauge[7] restrictions also vary widely. These capacity restrictions are particularly important because, as Gucwa and Schäfer (2013) have shown, the energy intensity of rail is much more sensitive to loading than it is for other freight modes. Their analysis found that 'an increase in cargo capacity on a transportation system by 10 per cent leads to roughly 8.3 per cent reduction in rail energy intensity and to a 5.5 per cent reduction in truck and shipping intensity' (p.47). They also acknowledge the importance of average speed and note that Indian freight trains run 25 per cent slower than those in the United States, while Canadian ones run 20 per cent faster than US ones.

The management of rail infrastructure, and in particular the relative priority given to freight and passenger services, is also very relevant. In rail networks dominated by passenger services, freight trains are regularly diverted into sidings to allow the faster passenger trains to pass. Large amounts of energy are wasted slowing down, stopping and then accelerating a train, often weighing in excess of 1,000 tonnes, back to its normal running speed. In countries whose railways are used primarily for moving freight and where passenger services are relatively sparse, such energy losses are much lower.

Another infrastructural variable which differentiates national rail networks is the proportion of electrified track. Only a tiny fraction of the US rail network is electrified, in sharp contrast to the EU, where 52 per cent is electrified, and Switzerland, where almost all lines are electrified. Research by Bombardier (quoted in UIC, 2016) estimated that energy costs were approximately 60 per cent lower for electrically powered trains than for their diesel equivalents, though this comparison is clearly sensitive to the prevailing diesel fuel and electricity prices in particular countries.

The nature of a country's rail infrastructure clearly imposes restrictions on the extent to which energy and carbon efficiency can be improved. There are, however, many carbon-reducing measures that can be applied virtually anywhere.

Improving the energy efficiency of freight trains

Potential energy efficiency improvements accrue from:

- *Technological enhancements to locomotive engines*: for example, use of 'permanent magnet motors' can improve the energy efficiency of electric locos by 3 per cent, while the companies Siemens and Alstom claim that hybridization of diesel locos can cut fuel consumption by as much as 40 per cent (UIC, 2016; OBB et al, 2014).

- *Anti-idling devices*: locomotives spend a significant amount of their time stationary with the engine idling, partly to power ancillary equipment. Stodolsky (2002) estimated that an idling engine consumed 3.5–5 gallons (16–23 litres) of fuel per hour. Through the use of automatic 'stop-start' systems and the installation of auxiliary power units, he estimated that US railroads could achieve fuel savings of up to 10 per cent.
- *Fuel-efficient driving*: driver training programmes and the use of electronic Driver Advisory Systems (DAS) have been shown to yield significant fuel savings. Eco-driving is not therefore confined to the trucking industry. In one trial of 1,620-tonne freight trains travelling between Munich and Hamburg, DAS cut fuel consumption by 10 per cent (OBB et al, 2014).
- *Aerodynamics*: aerodynamic profiling has tended to be less important in the rail freight sector than to trucking, partly because freight trains typically travel more slowly than long-haul trucks but also because the close coupling of rail wagons minimizes drag. For one type of train, however, aerodynamics does have a big impact on energy efficiency – the intermodal train comprising containers, swap-bodies or trailers. The gaps created by vacant slots on intermodal trains can cause significant turbulence and energy loss. Lai, Barkan and Önal (2008) estimate that optimizing the positioning of containers on trains using one of America's busiest intermodal routes between Chicago and Los Angeles would save 15 million gallons (68 million litres) of fuel a year, worth $28 million and emitting around 170,000 tonnes of CO_2.
- *Lightweighting*: railway rolling stock has traditionally been heavy, using materials that give it the required structural strength. New composite materials can provide the necessary strength and are much lighter. It has been estimated that composites could reduce the tare weight of rolling stock by 30–40 per cent and fuel consumption by 4 per cent, though in Europe, regulations currently constrain their use in the rail sector (PhysOrg, 2016). An EU-funded project called Refresco[8] has examined how the use of composites in the rail sector can be extended from components to the bodies of rail vehicles.

Speed reduction has become an important source of fuel savings in the shipping and trucking sectors and could also be applied to rail freight. This, however, would risk making rail less competitive with road and conflict with modal shift objectives which can potentially yield larger carbon savings (McKinnon, 2016a).

Switching rail freight operations to lower-carbon energy

A large and increasing proportion of rail freight services around the world are in the enviable position of having a direct connection to grid electricity and are therefore benefiting from the steady decarbonization of electricity generation. Electricity increased its share of the global rail energy mix from 38 to 44 per cent between 2005 and 2015, while the carbon content of this electricity declined (IEA/UIC, 2017). The worldwide share of railway electricity generated by renewable energy rose by 65 per cent over the same period. These statistics apply to all rail services; it is not known what proportion of rail freight operations are currently electrified, though it is likely to be over one-third. This proportion will gradually increase as the degree of network electrification continues. The use of a new generation of hybrid locomotives also makes it easier for freight trains to combine electric and diesel propulsion on a single haul (OBB et al, 2014). Some rail freight operators, such as DB Schenker, are already buying zero-carbon electricity from the grid to be able to offer their customers a carbon-neutral rail freight service.

The carbon intensity of diesel-powered rail freight services has been subject to conflicting trends. The blending of biodiesel with conventional diesel has been cutting emissions, at least on a tank-to-wheel (TTW) basis. Meanwhile, the need to meet tightening regulations on noxious emissions, particularly SOx, has forced rail operators to switch to lower-sulphur fuels at the expense of higher CO_2 emissions (McKinnon, Allen and Woodburn, 2015). This is a similar environmental trade-off to that being experienced in the trucking and shipping industries, where some of the efforts to improve air quality are unfortunately inhibiting decarbonization.

Summary

This section on rail freight is much shorter than those for trucking, shipping and air cargo. This is partly because it represents a small share of total freight-related CO_2 emissions (4–5 per cent) and presents fewer decarbonization challenges than these other modes. It also reflects the fact that much less research has been done and literature published on energy use by the rail freight sector. One big environmental advantage that rail commands over the other freight modes is its direct connection to grid electricity, allowing rail companies to 'outsource' much of the responsibility for decarbonization to electricity suppliers. This is not an option available to US railroads, although per tonne-km they are already highly energy efficient and, for several decades, have been increasing their carbon efficiency at an impressive rate (Association of American Railroads, 2017).

If the modal shift aspirations outlined in the previous chapter are realized, the amount of freight carried by rail is likely to rise steeply over the next few decades. Rail energy use will not increase in proportion, partly because energy efficiency will continue to improve, but also because changes in the rail freight commodity mix, particularly the substitution of lighter manufactured products for heavier coal and oil, will lighten the loads. However, rail must not rest on its 'CO_2 laurels' (Reidy, 2017), as the gap in carbon intensity between rail and road is likely to narrow over a period when this gradually becomes a more important modal selection criterion for shippers.

Notes

1. Also using a bottom-up approach, DNV-GL (2017) estimated that shipping consumed 233m tonnes of fuel in 2016, though this figure may not be directly comparable with those of Smith et al (2014) and Olmer et al (2017).
2. Cargo vessels consume over 90 per cent of the fuel used in the maritime sector (McGill, Remley and Winther, 2013). This section is only concerned with ships carrying freight.
3. Expressed as dead-weight tonnage of capacity multiplied by nautical miles sailed (Olmer et al, 2017).
4. Smith et al (2014: 165) acknowledge that there is no data on cargo load factors. It is not possible therefore to independently monitor the trend in average vessel utilization on a tonne-km basis.
5. Well-to-propeller is the maritime equivalent of well-to-wheel in the road freight sector
6. ICAO data only relates to international air transport and excludes air travel within countries. Peeters (2017) estimates that domestic air transport accounts for over 30 per cent of total CO_2 emissions from air transport. Total fuel consumption and CO_2 emissions from aviation are therefore greater than these figures suggest.

7 Loading gauge defines the maximum width and height of railway rolling stock that can be accommodated within the infrastructure at tunnels, bridges and stations.

8 Refresco Project (2018) [online] http://www.refresco-project.eu/ [accessed 6 February 2018].

Decarbonizing logistics at the national level

08

The case of the United Kingdom

Literature on the decarbonization of logistics tends to be polarized between company-focused studies and global analyses of what needs to be done to meet climate change targets. Relatively little work has been done at a national level, despite the fact that it is at this level that most of the public policy initiatives on climate change are devised and implemented. Many of the studies that have adopted a national perspective are concerned with particular transport modes, technologies or activities. Few have attempted to give a broader overview of the key issues and interrelationships (eg Guérin, Mas and Waisman, 2014). In this concluding chapter, I illustrate how it is possible to build up this bigger picture using the TIMBER framework discussed in Chapter 1, applied to the United Kingdom. This is not simply because it is my home country and the one in which much of my research has been based. My choice has also been influenced by the availability of relevant statistics, which few countries can match, and the UK's long history of government- and industry-sponsored efforts to improve the environmental performance of logistics. As we will see, however, even with these favourable circumstances it is still difficult to assess the likelihood of a country's logistics sector meeting long-term climate change targets.

A brief history of sustainable logistics initiatives in the UK

The UK pioneered much of the early work on sustainable logistics. The British Government conducted a major inquiry into the effects of trucks on the environment in the late 1970s and shortly afterwards set up an independent 'Lorries and the Environment Committee' to advise it on environmental improvement measures. Between the early 1990s and 2010, the government ran Energy Efficiency Best Practice and Freight Best Practice Programmes providing companies with free advice on how they could improve the energy efficiency of their freight transport operations. In 1999, it published its first 'Sustainable Distribution' strategy document (Department of the Environment, Transport and the Regions, 1999), outlining a national policy to reduce the environmental impact of logistics. This was updated nine years later (Department for Transport, 2008). Also in that year an independent Committee on Climate Change was established to provide impartial advice on the decarbonization of economic activity, including freight transport.

These government initiatives have been supplemented by industry-led schemes to improve the environmental sustainability of logistics, the most relevant of which to the decarbonization of logistics has been the Logistics Carbon Reduction Scheme (LCRS),[1] launched by the UK Freight Transport Association in 2010 to secure industry commitment to cutting carbon emissions. In 2011, the LCRS declared a target of reducing the carbon intensity of freight transport operations by 8 per cent between 2010 and 2015 (FTA, 2011). The 41 companies belonging to the scheme over this five-year period managed to cut their CO_{2e} emissions per vehicle-km by 7 per cent, coming close to meeting this target (FTA, 2016). In the meantime, the number of companies belonging to the scheme has swollen to 130, operating a total of 88,181 trucks and vans. The UK government has welcomed these industry efforts, though has questioned their generality. In its first 'Freight Carbon Review' it stated that 'available evidence shows that some parts of the freight industry are making substantial efforts, through a wide range of measures, to reduce their carbon emissions. It is less clear what is happening across the wider industry' (Department for Transport, 2013: 7).

Freight-specific carbon reduction initiatives fall within the broader remit of the UK government's climate change policy. This was enshrined in the 2008 Climate Change Act, which committed the government at the time and future governments to cutting CO_2 emissions by 80 per cent by 2050 against a 1990 baseline. The inclusion of international transport in this target makes it one of the most ambitious yet declared by a national government. The UK government has set a series of five-yearly 'carbon budgets' up to 2032 to ensure that the country gets onto an emission-reduction trajectory that will meet the 80 per cent target by 2050. This will require a 57 per cent reduction between 1990 and 2032 (Committee on Climate Change, 2017).

Figure 8.1 Scenarios for the decarbonization of UK road freight by 2050

Tonne-km	High Mobility 2030 projection 165%								Medium Past 10 year trend projected 127%								Low Stable at 2007 level							
Mode shift	Road share of tkm 64% → 50%								Road share of tkm 64% → 50%								Road share of tkm 64% → 50%							
Empty running	High 22% of HGV kms				Low 17% of HGV kms				High 22% of HGV kms				Low 17% of HGV kms				High 22% of HGV kms				Low 17% of HGV kms			
Payload weight	Low 70% load same max wt		High 70% load wt up to 60t		Low 70% load same max wt		High 70% load wt up to 60t		Low 70% load same max wt		High 70% load wt up to 60t		Low 70% load same max wt		High 70% load wt up to 60t		Low 70% load same max wt		High 70% load wt up to 60t		Low 70% load same max wt		High 70% load wt up to 60t	
Fuel efficiency	+20	+40	+20	+40	+20	+40	+20	+40	+20	+40	+20	+40	+20	+40	+20	+40	+20	+40	+20	+40	+20	+40	+20	+40
Carbon intensity	–30	–30	–30	–30	–30	–30	–30	–30	–30	–30	–30	–30	–30	–30	–30	–30	–30	–30	–30	–30	–30	–30	–30	–30
CO_2 in 2050 (% change)	–44	–58	–58	–69	–48	–61	–61	–70	–57	–68	–68	–76	–60	–70	–70	–77	–66	–75	–75	–81	–68	–76	–76	–82
SCENARIO	**S1**	**S2**	**S3**	**S4**	**S5**	**S6**	**S7**	**S8**	**S9**	**S10**	**S11**	**S12**	**S13**	**S14**	**S15**	**S16**	**S17**	**S18**	**S19**	**S20**	**S21**	**S22**	**S23**	**S24**

SOURCE McKinnon and Piecyk (2009)

In the absence of separate sectoral carbon reduction targets, it is assumed that the freight transport industry will be expected to meet this target. McKinnon and Piecyk (2009) constructed a series of 24 logistical scenarios for the UK in 2050, eight of which envisaged the road freight sector achieving, or coming close to, the target through the 'combination of a series of radical, but probably feasible' shifts in key parameters (Figure 8.1).

The Committee on Climate Change (CCC), whose role is to provide the government with independent monitoring and advice on Britain's transition to becoming a 'low-carbon economy', has examined opportunities for freight transport decarbonization in each of its five-year carbon budget assessments. It has focused more on supply-side opportunities, emphasizing the contribution that new technologies can make, although in 2015 it commissioned a study of demand-side trends likely to cut carbon emissions from the road freight sector up to 2032 (Greening et al, 2015). Reference will be made to the CCC's work in the course of this chapter.

The remainder of the chapter will review the potential for decarbonizing logistics in the UK within the TIMBER framework. In June 2016, the UK electorate voted in a referendum to leave the European Union. At the time of writing, the nature of Britain's withdrawal and its likely impact on economic performance and environmental policy are very uncertain. It is therefore too early to say how supply chains and their carbon intensity will be affected. No reference will therefore be made to the possible effects of Brexit on the decarbonization of UK logistics.

CO_2 emissions from logistics in the UK: 2006 and 2016

Few countries collect as much freight data as the UK or have as rigorous a system of freight data collection. Despite this, it is still very difficult to accurately carbon footprint the country's freight transport system using official data. This was first done on a comprehensive basis as part of a major study, 'Transport and Climate Change' undertaken for the Commission for Integrated Transport (2007). The calculation, which I did for the Commission, covered only domestic freight movements and excluded freight journeys with an origin, destination or both outside the country. It estimated that freight transport emitted just over 34 million tonnes of CO_2 in 2006 (McKinnon, 2007a).

The 2006 figure included an estimate of emissions from vans[2] transporting freight, using data collected in surveys of company-owned vans in 2003 and

2004. The 2004 survey revealed that only a third of the distance travelled by company-owned vans was 'in connection with the collection and delivery of goods' (Department for Transport, 2004). Vans were used as much for commuting as for moving freight. Despite the statistical evidence that vans are heavily used as a mode of personal travel it is common for them simply to be treated as a freight mode (eg Pye et al, 2015). The UK van surveys were discontinued in 2008 and other UK studies of van traffic conducted since then (eg AEA Technology, 2010; Clarke et al, 2014) have not provided the data required to update the CO_2 estimate on a consistent basis. This is unfortunate because over the past decade, the steep rise of online retailing has propelled the growth in van traffic (Braithwaite, 2017). Back in 2004, vans carried only 7 per cent of all road tonne-kms (Department for Transport, 2004). It is likely that over the past decade this percentage will have significantly increased and with it delivery vans' share of freight-related emissions, which was 12 per cent in 2006. Neither increase can currently be quantified using the available data.

Calculating total carbon emissions from UK freight transport is also complicated by a major discrepancy in official government estimates of the total annual distance travelled by trucks. This distance is measured in two ways: by a survey of truck operators (called the 'Continuing Survey of Road Goods Transport' (CSRGT)) and by the monitoring of traffic flows on the road network (called the 'Road Traffic Survey' (RTS)). In 2006, traffic counts yielded a truck-km estimate 25 per cent higher than that derived from the operator survey; by 2016 the gap had widened to 40 per cent. This discrepancy is highly significant to the carbon footprinting exercise because road haulage is by far the dominant mode of freight movement and annual truck-kms are a key parameter in the carbon calculation. A government study attributed this statistical gap to several factors, mainly the exclusion of foreign-registered vehicles from the CSRGT, under-reporting of haulage activity in this survey, inaccurate vehicle classification in the RTS, and inaccuracies in the grossing up of CSRGT mileage data to annual estimates (Department for Transport, 2010). It is likely that the traffic-based estimate of vehicle-kms is the more accurate and so it is the one used here. The estimates of fuel efficiency, however, come from the CSRGT, as do the tonne-km estimates required for carbon intensity calculations. It is assumed that the average fuel efficiency and carbon intensity of road haulage operations derived from the CSRGT also apply to the much higher truck mileage recorded by the RTS.

It should also be noted that different methods are used to calculate emissions from different transport modes. The so-called energy-based approach, outlined in Chapter 2, can be used for trucking operations because the annual CSRGT collects fuel consumption data from a large sample of trucks. It is presumably also the method used by the railway companies submitting fuel and CO_2 data to the Office of Rail and Road (2017). CO_2 estimates for trucking are obtained by multiplying the fuel efficiency of different vehicle classes by the aggregated distances they travel. No comparable energy data is available for waterborne freight movements or pipeline flows. The measurement of their carbon emissions therefore employs the activity-based method, which involves multiplying their total tonne-kms by their average modal carbon intensity, expressed as gCO_2 per tonne-km. While the tonne-km figures are considered to be reasonably accurate, it is difficult to derive an average carbon intensity value that is representative of the mix of freight operations undertaken by a particular mode. This task is made even more difficult where the official modal emission factor covers both passenger and freight movements and where the freight services are particularly heterogeneous, as is the case with domestic shipping.

Despite all these caveats, it is possible to compare CO_2 emissions from the UK freight transport system in 2006 and 2016 on a reasonably consistent basis. This system excludes vans and comprises lorries with a gross weight in excess of 3.5 tonnes (ie heavy goods vehicles (HGVs)), freight trains, waterborne services and pipelines.[3] For waterborne freight services and pipelines, the same carbon intensity values were used for both years. This means that any changes in the average carbon intensity of these modes over the past decade are not reflected in the comparison. As pipelines represent only 5 per cent of tonne-kms and their carbon intensity tends to be fairly stable, this is a minor issue. Failing to allow for changes in the carbon intensity of domestic shipping is a more serious shortcoming but unavoidable given the data currently available. Only changes in the carbon intensity of road haulage and rail freight will affect the comparison, but this is considered acceptable as these modes accounted for four-fifths of all tonne-kms in 2016 (excluding vans).

CO_2 emissions from the UK freight transport system were approximately 18 per cent lower in 2016 than in 2006, declining from 30.1 to 24.7 million tonnes[4] (Figure 8.2). Much of this reduction was due to an 8 per cent decline in the total amount of freight moved. Four per cent more tonne-kms were moved by road in 2016, but this was more than offset by the 42 per cent and 22 per cent reductions experienced by, respectively, waterborne and rail freight services (Figure 8.3). Much of the decline in waterborne traffic was attributable to the contraction of Britain's North Sea oil business, while

Figure 8.2 CO_2 emissions from freight transport in the UK in 2006 and 2016: tonnes

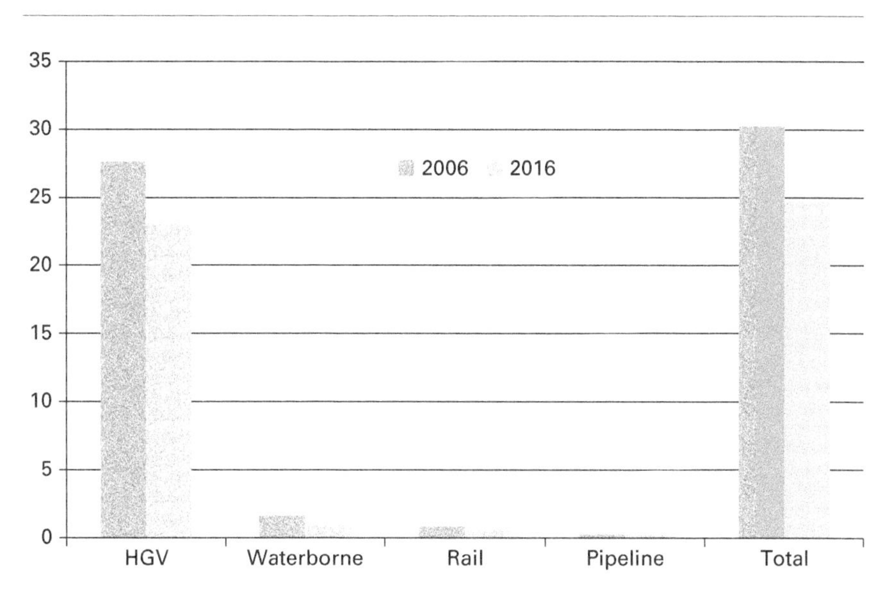

Figure 8.3 UK freight modal split 1993–2016: per cent of tonne-kms

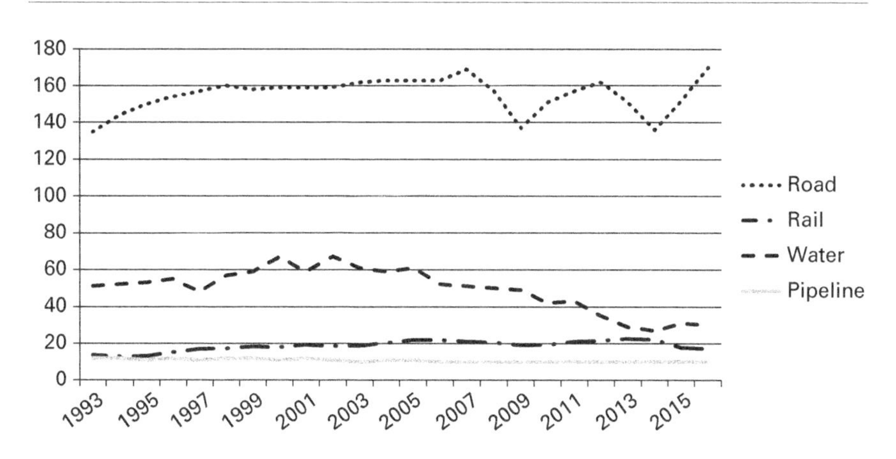

SOURCE Department for Transport (2017a)

the drop in rail tonnage was due mainly to the loss of coal traffic – both trends beneficial from a wider climate change standpoint. Nevertheless, they helped to increase the road share of the freight market from 66 per cent to 75 per cent, running counter to the government's long-term policy of shifting freight from road to rail and water.

The reduction in total CO_2 emissions from UK freight movement was clearly not the result of a modal shift to lower-carbon modes. One can also reject several other possible causes. According to CSRGT, it was not due to an improvement in the average fuel efficiency of UK-registered lorries (Department for Transport, 2017c). For both rigid and articulated lorries, the 2006 and 2016 averages were almost identical at, respectively, 3.3–3.4 km and 2.8–2.9 km per litre. Nor was there a reduction in the empty running of lorries; on the contrary, the proportion of HGV-kms run empty increased from 27 per cent to 30 per cent between 2006 and 2016, reversing an earlier downward trend and contradicting the collective views of an expert Delphi panel, which in 2008 predicted, on average, that empty running would drop to 21.9 per cent by 2020 (Piecyk and McKinnon, 2010, 2013).

Trends in the rail freight sector can also be eliminated as possible causes of the overall decline in freight-related CO_2 emissions. The carbon intensity of UK rail freight operations actually went up over the 10-year period, from 27.8 gCO_2/tonne-km to 33.5 gCO_2/tonne-km (Office of Rail and Road, 2017). This increase appears to have been the result of the changing mix of commodities transported by rail. The steep decline in coal traffic and growth of lighter manufactured goods carried on intermodal services reduced the average density of UK rail freight. Since the carbon intensity metric is weight-based, reductions in average product density are likely to inflate it, other things being equal. This highlights an anomaly in the way that the carbon performance of different transport modes is assessed, as discussed in Chapter 4, and does not necessarily mean that the carbon efficiency of rail freight services has deteriorated. The higher carbon intensity value for rail in 2016 had little impact on the overall freight CO_2 trend because rail's share of tonne-kms was small (7.5 per cent in 2016) and diminishing (down from 8.9 per cent since 2006).

The one key parameter which, in carbon terms, showed a marked improvement over the decade was the average lading factor on HGVs. The proportion of available carrying capacity on laden trips (maximum payload weight multiplied by distance travelled) that was actually used rose from 57 per cent in 2006 to 68 per cent in 2016. This one-fifth improvement in the weight-based load factor may also have been supplemented by greater utilization of vehicle cube. Between 2000 and 2010 the proportion of lorry loads subject to a volume constraint increased from 38 per cent to 68 per cent (Department for Transport, 2011) and this upward trend may have continued, though no published data is available to assess this. Nor has any research yet been done to explain why the average loading of lorries has increased so much in such a short time. It may have been the result of a combination of some or all of the measures outlined in Chapter 5.

While freight transport's carbon footprint has been shrinking, emissions from warehousing have been rising. Government data for the industry category 'warehousing and transport support' (SIC 52) indicates that these emissions rose by 30 per cent between 2006 and 2015, from 2.0 to 2.6 million tonnes (Department for Transport, 2017d). These figures can be combined with freight transport data for 2006 and 2015 to obtain CO_2 estimates for UK logistics as a whole (excluding van traffic). This indicates that total logistics emissions fell from 32.1 million tonnes in 2006 to 27.4 million tonnes in 2015, a drop of 17 per cent. The increase in warehouse-related emissions was much more than offset by the decline in transport emissions, though warehousing's share of the total rose from 6.2 per cent to 9.5 per cent.

The main messages from this statistical comparison of 2006 and 2015/2016 data are that:

- Total CO_2 emissions from freight transport operations in the UK have significantly diminished, mainly because of a reduction in total tonne-kms handled by rail and waterborne freight services. This is consistent with the long-term decline in the freight transport intensity of the UK economy observed by McKinnon (2007a). Between 2006 and 2016 the amount of freight movement generated by each £1 billion of GDP fell by roughly 18 per cent.
- Carbon emissions have fallen despite significant increases in road's share of the freight market, the proportion of truck-kms run empty and the rising average carbon intensity of rail freight.
- The average fuel efficiency of trucking operations remained fairly stable.
- The marked reduction in the carbon intensity of UK road haulage, and hence the freight sector as a whole, has been due mainly to improved loading of the nation's truck fleet (on laden trips).
- HGVs increased their share of total freight-related CO_2 emissions from 91.4 per cent to 93.2 per cent (Figure 8.2).
- Total CO_2 emissions from UK logistics declined by 17 per cent despite a 30 per cent increase in the amount of warehousing-related emissions.
- This analysis provides only a partial view because it excludes the movement of freight in vans and takes no account of changes in the carbon efficiency of waterborne freight services over the past decade.

Looking forward, the UK government has projected a business-as-usual trend for CO_2 emissions for HGVs over the period 2015–2035. No similar projections have been made for other modes. It is predicted that CO_2 emissions from HGVs will 'fall gradually out to 2025, reflecting fuel efficiency improvements across the HGV fleet driven by government-backed, industry-led action as well as incremental improvements in new HGV efficiency year on year' (Department for Transport, 2017a: 18). Beyond 2025, the growth in truck traffic will offset the carbon efficiency gains, causing a slight rise in total emissions. The government acknowledges that this will make it 'increasingly challenging to meet our climate change targets within the road freight sector' (p18).

It is against this background that we can now examine the potential for decarbonizing logistics in the UK and achieving a downward divergence from the business-as-usual trend. The assessment will be made using the TIMBER framework.

Impact of TIMBER factors on the decarbonization of UK logistics

Technology

Road freight

The first major review of CO_2-reducing technologies for HGVs in the UK was undertaken by Ricardo almost a decade ago (Atkins, 2010). It estimated their potential impact on CO_2 emissions and indicated when they would be available for adoption. In all but two cases (engine heat recovery and the use of fly wheels), these technologies, listed in Table 8.1, were considered at the time to be ready for commercialization. It was anticipated that a large proportion of technology-related carbon savings will accrue from wider uptake of these currently available systems, devices and designs. A subsequent study identified six low-carbon technologies as being the most promising, in each case estimating the likely CO_{2e} savings and financial payback period (Ricardo-AEA, 2012). Dual-fuel, ie combining gas and liquid fuel, and fully gas-powered engines were considered to offer the greatest carbon abatement potential with payback periods of one to four years. This was before the scale of the 'methane slip' problem, discussed in Chapter 6, was fully recognized. More recent research (Stettler et al, 2016)

Table 8.1 Carbon-reducing HGV technologies recommended for the UK

Engine heat recovery	Low rolling resistance tyres
Combustion efficiency improvements	Predictive cruise control
Mechanical turbo-compounding	Single wide tyres
Pneumatic booster	Flywheel hybrid
Aerodynamic fairings	Aerodynamic trailers/bodies
Automatic manual transmission	Alternative fuelled vehicles
Electrical turbo-compounding	CNG/biomethane vehicles
Full hybrid	

SOURCE Atkins (2010)

has indicated that incomplete combustion of methane can cause total GHG emissions from dual-fuel trucks to be 50–127 per cent higher than those of an equivalent diesel vehicle. Another technological priority on the list, aerodynamic profiling, remains a very cost-effective means of decarbonization. According to Ricardo-AEA (2012), it can offer 6–9 per cent emission savings with a one- to three-year payback. The other three technologies highlighted in this study, predictive cruise control, lightweighting and anti-idling devices, could each yield emission reductions of 1–2 per cent.

A government survey of 700 operators of HGVs in 2015 suggested that the adoption rate for some of the major carbon-reducing technologies, such cab roof air deflectors, automatic manual transmission and vehicle telematics, was relatively high (Department for Transport, 2017a) (Figure 8.4). For many others, however, it was a third or less, indicating much untapped potential to use these CO_2-reducing technologies. The figures may also underestimate that potential because operators participating in the survey do not necessarily apply the technologies across their entire fleets.

Britain is the home of the so-called teardrop trailer, which slopes at both the front and back to minimize the drag caused by air turbulence at both ends of the vehicle. They are now quite a common sight on UK roads and have been shown to offer large improvements in fuel efficiency. On its website, the main manufacturer of these trailers, Don Bur (2017), cites the results of 33 company case studies indicating average fuel savings of just over 11 per cent.

Another distinctive truck design feature in the UK is double-decking. This is a logistics phenomenon largely confined to the UK and made possible by the relatively high clearances over most of the trunk road network. This

Figure 8.4 Percentage uptake of carbon-reducing truck technologies in the UK: survey of 700 operators

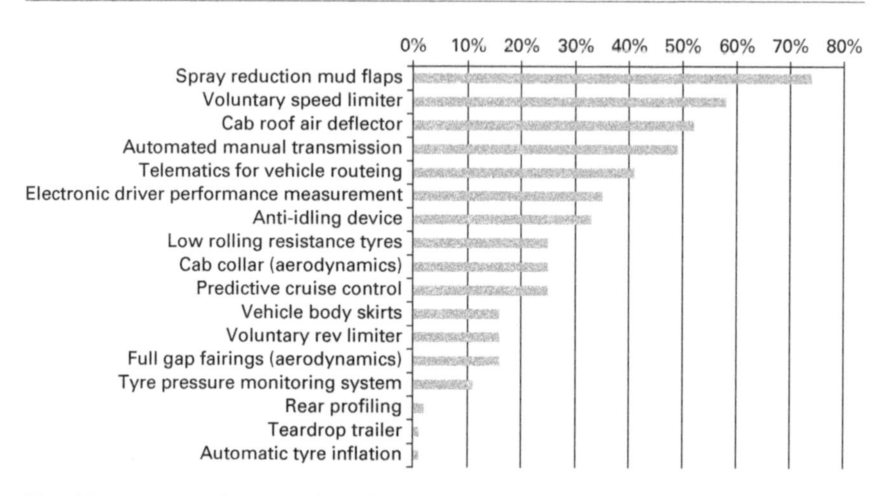

SOURCE Department of Transport (2017a)

allows trucks to be up to five metres tall, in contrast to their counterparts on the European mainland which are limited to 4.0–4.2 metres. Indeed, there is no legal height limit on trucks operating in the UK, though by 'custom and practice', five metres is the accepted maximum. Extra vertical clearance can be gained by running trailers on smaller wheels. By inserting a second deck it is possible to carry two rows of pallets or roll cages,[5] increasing the carrying capacity by 80–90 per cent, depending on the trailer configuration and density of the commodity (Freight Best Practice Programme, 2007). As double-deck vehicles must operate within the general truck weight limit (of 44 tonnes), they cater mainly for lower-density loads. Double-decking has proved a very effective means of decarbonizing road freight operations in the UK. Tesco, Britain's largest supermarket chain, halved carbon emissions per case delivered to its stores in the UK between 2007 and 2012 and attributed most of this reduction to the increased use of double-deck trailers. As trailers do not have to be registered in the UK, there are no official statistics on the numbers of double-deck trailers. I estimated that there were approximately 6,500–7,500 double-deck trailers operating in the UK in 2010 (McKinnon, 2010c) and this figure is likely to have greatly increased since then, partly reflecting a doubling of their share of tonne-kms carried by articulated lorries from 3.9 per cent in 2010 to 7.9 per cent in 2013 (Department for Transport, 2015e). I also calculated that in the absence of double-decking, total CO_2 emissions from UK-registered articulated lorries would have been 4.3 per cent higher in 2008, equivalent in CO_2 terms to adding an extra 151,000 new cars to Britain's roads in that year (McKinnon, 2010c).

The carbon benefit derived from the double-decking of road trailers can be increased in several ways:

1. *Replacement of powered decks by fixed decks.* Most of the original double-deck trailers had hydraulically powered decks. These typically weigh around four tonnes, imposing a significant fuel penalty and reducing vehicle carrying capacity. Many companies, most notably Tesco, have been replacing them with much lighter fixed decks, made possible by the installation of lifting equipment at factories, warehouses and shops (TransDEK, 2015).
2. *Aerodynamic profiling.* Because of their greater height, double-deck trailers create more wind resistance. The latest generation of double-decks has sloping fronts to improve streamlining, though at the expense of some internal load space.
3. *Greater mixing of light and heavy loads,* often facilitated by the horizontal collaboration discussed in Chapter 5, can allow double-decks to be used for a broader range of commodities with more variable density.

Because UK road infrastructure can accommodate tall trailers, truck operators have gained additional volumetric carrying capacity vertically whereas their counterparts in countries permitting longer vehicles, such as Sweden and Finland, have gained it horizontally. The British government is, however, currently running a 10-year trial of 'longer semi-trailers' (LSTs) which are either one or two metres longer than the standard 13.6-metre trailer. The results of this trial are discussed under the Regulation heading.

Rail freight

In 2016, 93 per cent of UK rail freight was hauled by diesel locomotives and only 7 per cent on electrified services. The 'workhorse' of diesel-operated services is the Class 66 locomotive. Over the past 20 years it has gradually replaced the previous generation of locomotives, such as the Class 86, introduced back in the 1960s. As the lifespan of a locomotive is typically in excess of 30 years, the Class 66 fleet is likely to dominate the UK rail freight fleet for the foreseeable future, despite having what AECOM/ARUP (2016: 69) considers 'relatively poor fuel consumption'. A new Class 70 locomotive (called 'Powerhaul') was introduced by Freightliner, one of the UK rail freight companies, in 2009, and is estimated to be around 18 per cent more fuel efficient than the Class 66 (Railway Technology, 2013). It achieves this greater fuel efficiency partly by regenerative braking and the use of auxiliary power units. These more fuel-efficient, lower-CO_2 locos account for only around 7–8 per cent of the UK rail freight fleet and any future switch from Class 66's is likely to be slow.

The 7 per cent of rail freight that is electrically hauled relies on a type of locomotive introduced back in the 1980s (Class 90). This percentage of electrically powered freight services is very low considering that 37 per cent of the UK rail network is electrified. It is constrained by the fact that very few industrial premises or terminals have an electrified rail connection, necessitating a switch to diesel haulage at the start and/or end of the journey. Depending on the journey length, it is often much quicker and cheaper to run a diesel loco on the end-to-end journey. One UK rail freight operator, DRS, has recently acquired a small fleet of hybrid locomotives (Class 88), combining electric and diesel–electric haulage. Wider uptake of this technology will help rail freight companies improve energy efficiency and give them more access to low-carbon electricity. The decarbonization of UK electricity is discussed later in this chapter.

In all countries, the updating of locomotive and wagon fleets is a slow process. In the UK, however, there is an additional constraint. Britain's relatively small railway loading gauge means that locos and wagons must be tailored to this infrastructural constraint. Foreign manufacturers, however, have limited incentive to produce locomotives to UK specifications as the UK market is relatively small, certainly by comparison with North America and the rest of the EU. It is likely that over the next 10–20 years carbon efficiency gains from rail technology will come mainly from retrofitting energy-saving devices onto the existing loco fleet, for example:

- use of auxiliary power units to remove the need for the main engine to idle when stationary just to support heating, communication systems etc;
- driver advisory systems (DAS) to give drivers in-cab advice on the adjustment of speed and braking to maximize fuel efficiency;
- devices to support preventative maintenance, which can cut fuel losses.

More radical innovations are on the horizon. A new type of short, high-speed freight train, capable of carrying up to three ISO containers, has been proposed. According to its promoters, 'TruckTrain is designed to extend the commercial reach of rail beyond the limits set by orthodox train technology and to be fully cost competitive with road freight over short-, medium- and long-distance sectors' (www.trucktrain.co.uk). It is claimed that by promoting modal shift to rail and achieving higher levels of fuel efficiency, TruckTrain will yield significant carbon reductions. It is not known if and when this radical rail freight concept will be commercialized. It is worth noting, too, that in its assessment of the potential for freight modal shift in the UK and related CO_2 savings, AECOM/ARUP (2016) sees a relatively small role for 'alternative locomotive technology'.

Infrastructure

Infrastructural improvement can assist efforts to cut carbon emissions, primarily in two ways: by easing traffic congestion and by promoting a shift to lower-carbon modes. The first of these ways relates mainly to the road network, partly because it carries three-quarters of freight tonne-kms in the UK, but also because it is much more prone to congestion. The adverse effects of traffic congestion on truck fuel efficiency and CO_2 emissions were discussed in Chapter 6. Infrastructure investment has also proved quite an effective means of promoting the use of rail for freight movement in the UK.

Road infrastructure

A comparison of 20 European countries in 2012 revealed that the UK incurred the highest road congestion costs (National Infrastructure Commission, 2017: 145). The UK government concedes that there has been long-term under-investment in road infrastructure, with traffic growth far exceeding the increase in available road space. This has resulted in traffic densities on UK motorways being approximately 2.5–3 times greater than in France and Germany.

The impact of traffic congestion on road freight movements in the UK was monitored by a series of seven surveys over the period 2002–2009, comprising a total of 55,820 journey legs (McKinnon et al, 2009). Twenty-six per cent of these legs were subject to a delay and 35 per cent of these delays (ie 9 per cent of the total) were attributed mainly to traffic congestion. The incidence of congestion-related delays varied widely among the sectors. The length of the delay caused by traffic congestion averaged 24 minutes, by comparison with an unweighted average delay time for all causes of 41 minutes. When the frequency and duration of delays are combined and the results weighted by the number of journeys surveyed, traffic congestion was found to be responsible for around 23 per cent of total delay time in the UK road freight system. No attempt was made in these studies to measure the carbon impacts of these congestion-related delays. Research by Palmer and Piecyk (2010), however, shed some light on this issue. They analysed the effects on CO_2 emissions of varying the start times of trucks on 56 routes across the UK trunk road network, taking account of traffic levels on the roads along which they travelled. This found an average difference of 5 per cent in CO_2 emissions between best (off-peak) and worst (peak) start times. If road traffic continues to grow and road capacity does not expand to accommodate it, congestion and congestion-related emissions will increase, widening this gap in CO_2 emissions between peak and off-peak operation congestion. The UK logistics sector has, to some extent, reduced its exposure to traffic congestion by rescheduling deliveries into the evening and night, taking advantage of the increase in commercial and industrial premises operating on a 24-hour cycle. Between 1985 and 2005, the proportion of HGV-kms run between 8 pm and 6 am in the UK rose from 8.5 per cent to 19.8 per cent, though over the past decade this has increased only marginally (to 21 per cent in 2016) (Black, 1995; Department for Transport, 2006b; 2017d). As discussed in Chapter 6, there are numerous logistical constraints on the rescheduling of freight deliveries.

2017 marks the half-way point in the implementation of what the UK government claims is the 'biggest-ever upgrade to (UK) motorways and key A roads', an investment of £15.1 billion in 'strategic roads' and adding a further 221 'lane miles of extra capacity to (the) busiest motorways' (Department for Transport, 2014b). This road construction programme is being supplemented by an extension of the UK's 'managed motorways' scheme which integrates several technological developments such as MIDAS (motorway incident detection system), variable messaging and road-based telematics to improve traffic flow, thereby increasing the effective capacity of the network and minimizing delays caused by congestion (Highways Agency, 2013). At the local level, the government has been working with municipal authorities to ease congestion at around 70 'local pinch points' in the road network. A separate 'local pinch point' fund has been established to fund these upgrades. Government documents provide no indication of the likely effects of these infrastructural upgrades on the average carbon intensity of road freight movements in the UK. As they are likely to be easing congestion, at least in particular areas and corridors, the carbon impact should be positive relative to a do-nothing scenario.

The longer-term outlook is uncertain. The government is forecasting that road traffic will grow between 19 and 55 per cent between 2010 and 2040

(Department for Transport, 2015d). The wide gap between these estimates reflects uncertainty about the future course in a range of variables including GDP, population, employment, car ownership, oil prices and fuel efficiency. On the basis of these forecasts and planned road improvements, it is anticipated that by 2040, between 8 and 17 per cent of vehicle miles will be run on congested roads. The government's main congestion metric, seconds lost per vehicle-km, could vary within an even wider range from 18.3 to 34, representing at one extreme no change between 2010 and 2040, and an 86 per cent increase at the other. Uncertainty in the government's modelling of future road congestion trends is compounded by uncertainty about the future roll-out and traffic impact of vehicle automation, platooning and infrastructural connectivity, none of which are formally integrated into the traffic forecasting model. These developments should permit much more intensive use of available road space, minimize the operational impact of congestion and 'reduce the need for expensive physical enhancements to road capacity' (National Infrastructure Commission, 2016: 147). Carbon emissions per truck-km across this 'smart infrastructure' are likely to be significantly lower than today.

Much road construction in the UK is opposed by environmental groups who argue that adding extra capacity generates additional traffic and complain that official appraisals of road schemes fail to take adequate account of the environmental impact of the 'induced' traffic. A study by the government's Standing Advisory Committee on Trunk Road Assessment (1994: 205) concluded that 'induced traffic can and does occur, probably quite extensively, though its size and significance is likely to vary widely in different circumstances'. A recent review of 54 road schemes completed between 2002 and 2010 has estimated that the additional traffic using this extra capacity has to date emitted around 8 million tonnes of CO_2 (from all categories of traffic) (Sloman, Hopkinson and Taylor, 2017). The sponsoring organization recommends that 'increasing road capacity needs to become the option of last resort rather than the default, as is currently the case. Otherwise we face a dead-end of increasing congestion, needless environmental damage and sprawling development' (CPRE, 2017: 3). It goes on to argue that the UK's current road-building plans are in conflict with its long-term climate change commitments. This is a controversial issue that, as far as logistics decarbonization is concerned, cannot be resolved on the basis of currently available evidence. Regrettably, it is not an issue that has been highlighted by the National Infrastructure Commission (2017) in its consultation on '*congestion, capacity and carbon*'.

Rail infrastructure

Upgrading rail infrastructure can contribute to the decarbonization of UK freight transport in two ways. It can improve the competitiveness of rail freight services and thereby lure traffic from higher-carbon road haulage operations. It can also reduce the carbon intensity of moving freight by rail. A much-quoted statistic asserts that, in the UK, rail emits around 75 per cent less CO_2 per tonne of cargo moved than trucks (eg Rail Delivery Group, 2014; Department for Transport, 2016a). There is no question that rail commands a large carbon advantage over road, though as discussed earlier, its average carbon intensity is increasing while that of road is declining. As companies' decision making around modal choice is likely to become increasingly sensitive to modal differences in carbon emissions, rail cannot be complacent and 'rest on its laurels as a leader in sustainable transportation' (Rail Technology Magazine, 2008).

We will focus on six interrelated ways in which UK rail infrastructure can be improved to attract more traffic, all of which can also reduce the carbon intensity of rail freight services.

1 Increase capacity/relieve bottlenecks

It was recently acknowledged that 'over the past 20 years, passenger and freight usage on Britain's railway network has increased significantly (by 116 per cent and 34 per cent respectively), exceeding government and industry projections and expectations' (Institute of Mechanical Engineers, 2017: 5). This has created 'an increasingly congested and capacity-constrained network' (p.5), on which the number of train paths available for freight services is tightly limited on particular corridors at particular times. Capacity for freight on the main spinal route for freight traffic, the West Coast Main Line (WCML), which carries 40 per cent of rail tonne-kms, is currently very limited. It has been identified as a bottleneck inhibiting the development of a green rail freight corridor between Glasgow and Duisburg and constraining the related CO_2 savings (Aditjandra et al, 2012). The available freight capacity will greatly expand when the new High-Speed 2 (HS2) line diverts passenger trains, though this will not happen until at least 2026 between London and Birmingham and 2033 on sections to Manchester and Leeds. In the meantime, Network Rail (2017), the UK rail infrastructure provider, has a programme of capacity upgrades on key freight corridors.

2 Accelerating freight services

This can be done by upgrading the track to allow freight trains to travel faster (as locomotive performance improves) and reducing the time that freight trains spend in passing loops. As passenger trains get priority on the UK rail network and freight trains are slower, the latter are frequently diverted into loops to allow passenger trains to pass. As discussed in Chapter 7, this not only lengthens transit times, it also consumes more energy and emits more CO_2 as the freight train has to stop, restart and regain its earlier momentum.

3 Enhancing the loading gauge

The loading gauge is the amount of clearance around the track at stations, bridges and tunnels. There are different loading gauges across the UK rail network, but most of them are relatively low by international standards, restricting the dimensions of rolling stock. The low loading gauge has become a more serious constraint as intermodal units, in particular ISO shipping containers, have expanded. Much of the freight-specific investment in UK rail infrastructure in recent years has gone into gauge enhancement on routes handling deep-sea container traffic to and from the ports of Southampton and Felixstowe. The changing mix of commodities moved by rail makes expansion of the loading gauge critical to the future development of Britain's freight railway. Intermodal traffic is predicted to grow at 5 per cent per annum between 2013 and 2043 (Network Rail, 2017). The lower density flows handled by intermodal services require more cubic capacity and hence higher loading gauges.

4 Electrifying the track

Network Rail (2009: 5) estimated that 'electric vehicles, on average, emit 20 to 30 per cent less CO_2 emissions than their diesel counterparts, depending upon on the energy mix used for generation.' Between 2012 and 2017, the average carbon intensity of UK mains electricity dropped by 49 per cent (Committee on Climate Change, 2017; MyGrid, 2017), further strengthening the carbon case for rail electrification. It can also increase the average speed of freight trains, increase their maximum payload weight, raise track utilization and improve overall productivity. Network Rail (2017: 40) is committed to 'achieving a critical mass of electrified network, to make utilization of electric locos viable, including diversionary route capability to ensure that electric freight services can run during times of disruption to primary routes.' The current electrification of the Great Western line from London to Cardiff helps to fulfil this ambition. However, in 2017, the government cancelled electrification schemes in Northern England, the Midlands and Wales, partly because of their escalating cost, casting doubt on the future of rail electrification in the UK. It is important to distinguish the proportion of the network electrified, currently 34 per cent in the UK (Office of Rail and Road, 2017), from the proportion of freight traffic electrically hauled, only 7 per cent (AECOM/ARUP, 2016). More could be done, for example with the use of hybrid locomotives, to increase this proportion across the existing electrified network.

5 Development of more intermodal freight terminals

Very few factories and warehouses in the UK are rail-connected, making it necessary for companies using intermodal services to access the rail network. In 2017 there were only 36 intermodal terminals in the UK (excluding the ports), some of which serve wide areas. In a small country like the UK, where the average length of haul for road freight is only 92 km (Department for Transport, 2017c), intermodal services are only competitive where road feeder journeys are short and quick. The density of intermodal terminals will therefore need to increase to sustain the recent growth of intermodal traffic.

6 Accommodating longer and/or heavier trains

Increasing the maximum length and axle weight limit on freight trains permits greater load consolidation, cutting train-kms, fuel consumption, emissions and operating costs. Tonnes lifted per train-km in the UK did increase by 80 per cent between 2003/4 and 2013/14 (Network Rail, 2017), though the potential exists for greater load consolidation. For the lower-density, intermodal traffic that is rapidly increasing its share of UK rail freight, it is beneficial to extend train length to gain additional cube. Network Rail is aiming to make 775 metres the 'baseline' length of freight trains across the network, significantly longer than most current UK trains. This will require a lengthening of passing loops and so-called 'chords' in rail intersections. For dense product flows, such as steel, construction materials and aggregates, allowing maximum axle weight to increase beyond the current 25-tonne limit would reduce their carbon intensity but require significant infrastructural investment.

As with the road network, physical upgrades to rail infrastructure are being supplemented by advances in IT, creating what Network Rail (2017: 38) calls the 'Digital Railway' in which advances in signalling, communication and train management not only improve efficiency and service quality, but can also help to decarbonize rail freight operations.

Infrastructure features prominently in the government's 'Rail Freight Strategy' (Department for Transport, 2016a), particularly as it is seen as a way of unlocking the large potential CO_2 savings that AECOM/ARUP (2016: 66) estimate will accrue by 2030 from 'capacity and gauge enhancements' (662,000 tonnes), development of the 'strategic freight network' (484,000 tonnes) and the construction of new intermodal terminals (217,000 tonnes).

Market trends

The provision of logistics services in the UK is very competitive and efficient by international standards. The UK was one of the first countries in the world to deregulate its road haulage industry (in 1970), creating a free market within which a number of logistical innovations such as contract distribution and pallet-load networks were pioneered. It was also the first European country to privatize its rail freight sector (between 1995 and 1997). The country, therefore, has a liberal and mature market for logistics services. To what extent do these favourable market conditions translate into carbon efficiency?

In an attempt to answer this question, we will examine six logistical market trends relevant to the decarbonization of logistics:

1. freight transport intensity of the economy;
2. freight modal split;
3. international transit traffic and cabotage;
4. collaborative initiatives;
5. online trading of logistics capacity;
6. growth of online retailing.

1 *Freight transport intensity of the economy*

In logistical terms, a distinguishing feature of the UK economy has been its declining freight transport intensity. Until the mid-1990s there was a close correlation between economic growth and the growth of freight movement, measured in tonne-kms. Between 1950 and the mid-1990s, every £1 million of GDP (at constant prices) generated roughly 200,000 tonne-kms of freight traffic. Over the past 20 years, this number of tonne-kms has approximately halved. I explored possible reasons for this decoupling of GDP and tonne-km growth trends at a time when this trend was becoming pronounced (McKinnon, 2005). I concluded that it was largely due to the service sector increasing its share of GDP, the offshoring of manufacturing industry to low-labour-cost countries and the stabilization of long-term trends that had been driving freight traffic growth, particularly the centralization of inventory and wider sourcing of supplies. Evidence of this stabilization can be found in the trend in one critical freight transport parameter: the average length of haul. This is the average distance moved by each unit of freight. It almost doubled in the UK over a 40-year period, rising from 63 km in 1964 to 123 km in 2000. Since then it has trended downwards and is currently around 110 km (Department for Transport, 2017c). This suggests that the spatial structure of UK supply chains is now relatively stable after decades when freight hauls were lengthening, partly in response to the development of the motorway network (McKinnon and Woodburn, 1996).

Reducing freight transport intensity was presented in Chapter 3 as a means of decarbonizing logistics. This decline in the freight intensity of the UK economy might therefore be considered to be beneficial in carbon terms; however, this is not necessarily so. The two main trends depressing the ratio of tonne-kms to GDP – the growth of the service sector and offshoring of production activities – are likely to be increasing carbon emissions elsewhere. The growth of services, whose share of UK GDP increased from 67 per cent to 79 per cent between 1997 and 2017 (Office of National Statistics, 2017b), has been closely associated with the near doubling of van traffic between 1990 and 2016 (Department for Transport, 2017c; Braithwaite, 2017). Vans have been by far the fastest-growing category of road traffic in the UK. As discussed earlier, van traffic was excluded from the tonne-km calculation because of a lack of data and so is not included in the analysis of freight transport intensity. It has, nevertheless, been a major source of additional CO_2 emissions. CO_2 emissions from all vans, not only those carrying freight, rose from 13.2 million tonnes in 1993 to 17.3 million tonnes in 2014 (Department for Transport, 2005; 2017a), partly offsetting the reduction in freight-related emissions discussed earlier in this chapter.

The offshoring of manufacturing, mainly to low-labour-cost countries, has not only transferred production-related emissions to these countries; it has also displaced emissions from associated logistical activity. In the process, total logistics-related emissions will have increased, partly as a consequence of domestic freight transport, warehousing and handling operations being less energy efficient in many of the countries to which production was offshored, and partly because of the lengthening of international supply links to and from these countries. So the carbon footprint of UK logistics operations has to some extent been reduced at the expense of higher emissions in other parts of the world, where their global warming impact per tonne of CO_2 is just as great.

2 Freight modal split

Figure 8.5 uses index values to show how the UK freight modal split, measured in tonne-kms,[6] changed over the period 1993–2015. This reveals that:

- Rail tonne-kms increased by 60 per cent between 1993 and 2013, boosted by the privatization of the rail freight sector in the mid-1990s and reconfiguration of the coal supply chain, which replaced short hauls of domestically produced coal from English mines to power stations with much longer hauls of imported coal from ports in Scotland and the Northeast of England. Since 2013, rail freight tonne-kms have declined by 23 per cent, mainly because of an 84 per cent drop in coal traffic. Domestic intermodal, including container movements to the ports, is now the biggest category of traffic (39 per cent), followed by construction materials (25 per cent).
- Waterborne tonne-kms have declined since the early 2000s. Most of this decline has been in so-called 'one-port traffic' to and from offshore locations like oil rigs, and largely reflects the decline in Britain's North Sea oil industry as the oil and gas reserves are depleted.

Figure 8.5 Tonne-km trends in UK freight transport modes: index values 1993 = 100

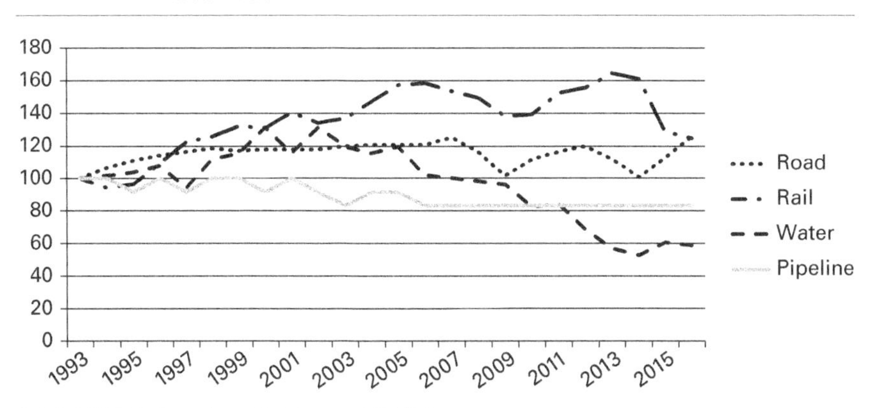

DATA SOURCE Department for Transport (2017d)

- Over the past decade road tonne-kms have fluctuated around 10 per cent above the 1993 baseline. The decline in rail and waterborne traffic mainly for reasons unrelated to the relative competitiveness of road have significantly increased road's share of the freight market, to 75 per cent of total tonne-kms in 2016.

A superficial review of these tonne-km trends would suggest that, in carbon terms, freight modal split in the UK is heading in the wrong direction. In reality, the allocation of freight between modes is being distorted by major structural changes in the freight market, which ironically are intimately linked to wider decarbonization of the UK economy. Most of the decline in the rail and waterborne share of total tonne-kms is the result of declining coal and North Sea oil traffic. Rail's non-coal traffic has been rising and the government considers the long-term outlook for the rail freight business to be promising (Department for Transport, 2016a). Rail tonne-kms are projected to grow at 2.9 per cent per annum between 2011 and 2043 (Network Rail, 2013), with the intermodal, construction, biomass and automotive sectors generating most of the future growth. As intermodal traffic has a much lower density than the coal traffic it is replacing, it will be difficult, however, to restore rail's earlier modal share in terms of tonne-kms. This illustrates the limitations of using a weight-based modal split metric (Woodburn, 2007). The most important metric in the present context is CO_{2e} savings from a switch from road to rail, and that is forecast to be around 2.3 million tonnes in 2030 (AECOM/ARUP, 2016).

The increasing contribution of intermodal services to logistics decarbonization in the UK is illustrated by two encouraging trends. The first is the growing involvement of rail in retail logistics. For example, supermarket chains Tesco and ASDA have switched significant volumes of longer-haul traffic from road to rail on Anglo-Scottish routes (Monios, 2015). Tesco's use of rail freight services removes approximately 26 million truck-kms from the UK road network annually and reduces CO_2 emissions by 80 per cent on the routes affected (Direct Rail Services, 2017). Second, as Woodburn (2017) observes, rail increased its share of containerized port hinterland movements from 14.7 to 16.6 per cent between 2007 and 2015 while raising train load factors by a quarter. He estimates that this cut CO_{2e} emissions by 34,000 tonnes over an eight-year period.

3 *Transit traffic and cabotage*

Because of its peripheral location, very small quantities of international freight traffic traverse the UK by comparison with more centrally located

European countries like Germany and Austria. Transit traffic is confined to surface freight flows between the Irish Republic and mainland Europe. Cabotage[7] penetration of the UK freight market is correspondingly low; in 2014 it accounted for only 1.5 per cent of road tonne-kms, about a quarter of the equivalent German figure (Eurostat, 2017b). It was estimated in 2010 that only 3.5 per cent of HGV-kms in the UK were run by foreign-registered vehicles (Department for Transport, 2014a). Total carbon emissions from freight transport in the UK are therefore little affected by international transit flows or the activities of foreign carriers.

4 Collaborative initiatives

There has been considerable interest in horizontal collaboration in the UK, mainly in the fast-moving consumer goods (FMCG) sector. Several successful bilateral collaborations, such as those involving Kimberly-Clark and Kelloggs (*Logistics Manager*, 2008) and Nestlé and United Biscuits (White et al, 2017), which cut transport CO_2 emissions by, respectively, 270 and 250 tonnes annually, have been widely publicized and presented as examples for other companies to follow (ECR UK, 2008). As discussed in Chapter 5, so-called 'Starfish' studies have assessed the potential benefits of multilateral collaboration within the FMCG sector. The first study used transport data collected from 27 large FMCG companies to examine seven collaborative scenarios and found that, in the case of less-than-truckload (LTL) consignments, they could offer CO_2 savings of between 4 and 20 per cent (Palmer and McKinnon, 2010). Several of the companies participating in this study subsequently developed collaborative arrangements, though the resulting benefits were not tracked or publicized. In a more recent survey of 50 FMCG businesses, IGD/ECR UK (2015) found that 32 per cent of suppliers and 42 per cent of retailers considered 'sharing transport' to be the greatest benefit of collaboration. A government-commissioned study of 'freight industry collaboration' (White et al, 2017) conducted a 'stakeholder survey' of 47 organizations to determine the potential for further collaboration in the UK and what can be done by the government and others to realize it. The research also modelled the impact on CO_2 emissions if collaboration could cut the empty running of trucks by 1–4 per cent. The authors concluded that 'many of the easy wins were already taken' and that there was now a need for 'more fundamental changes to encourage operators to work together' (p.4). For 'horizontal logistics collaboration' to be successful and yield significant environmental benefits, a number of prerequisites need to be met, as investigated by Sanchez-Rodrigues, Harris and Mason (2015).

5 Online trading of freight capacity

Extensive use is made of online freight exchanges in the UK, with much of this business channelled through web-based trading platforms such as Timocom, Trade Extensions and Returnloads.net. The market for logistics e-marketplaces in the UK is considered mature and offers both carriers and shippers a variety of online services to improve the utilization of their fleets (Wang and Pettit, 2016). Returnloads.net has estimated that 'in 2016 loads totalling over 16.5 million miles were covered on the platform, resulting in a potential saving of 25,514 tonnes of CO_2' (Department for Transport, 2017c). However, no data is available in the UK on the current level of uptake, the rate at which this market is growing or the extent to which these online platforms are helping to reduce the carbon intensity of road freight operations at a national level.

6 Growth of online retailing

The UK has among the highest levels of online retail penetration in the world (Braithwaite, 2017; Saleh, 2017). In November 2017 the online market accounted for 17 per cent of total retail sales and was growing at 16 per cent per annum (Office of National Statistics, 2017a). This transformation of Britain's system of retail distribution clearly has implications for logistics-related CO_2 emissions. Quantifying its effect on these emissions is fraught with difficulty, however, as discussed by Edwards, McKinnon and Cullinane (2011). Early research in the UK suggested that purchasing a small non-food item, such as a book, online and having it delivered to the home instead of making a car or bus trip to the shop to buy it could substantially reduce 'last mile' CO_2 emissions per item (Edwards, McKinnon and Cullinane, 2010). Assuming that only one item was purchased and using average UK values for the fuel efficiency, loading and distance travelled by the various modes, shopping by car would have emitted 4,274 gCO_2 per item, shopping by bus 1,265 g and home delivery only 181 g. These CO_2 differentials narrowed as the number of items purchased per trip/order increased and allowance was made for the probability of the home delivery failing and items being returned. Subsequent research, comparing online and conventional retailing on a 'life cycle' basis, has shown that reductions in transport emissions on the last link in the supply chain to the home can be partly offset by higher emissions elsewhere in the retail system (van Loon et al, 2014; van Loon et al, 2015). To be meaningful the carbon calculation must extend beyond logistics to assess the effects of online retailing on, among other things, personal travel behaviour, patterns of consumption and packaging. It appears that in the UK the growth of internet retailing is reducing shopping-related travel. Between 2002 and 2016 the number of shopping trips per person fell by 18 per cent and average distance travelled on these trips by 19 per cent (Department for Transport, 2017d), admittedly against a general decline in the level of personal travel for all purposes.

It is hard to predict the likelihood of online shoppers modifying their behaviour to reduce the carbon intensity of e-fulfilment. A multi-year survey of public attitudes to online retailing in the UK comprising around 1,300 respondents in 2015 and 2016 found that three-quarters 'would consider favouring an environmentally sustainable delivery channel', though only 30 per cent would be prepared to pay more to use it, and even then only very small amounts of money (less than £10 a year in most cases) (IMRG/Blackaby, 2016: 37). There has been a proliferation of fulfilment channels and options in recent years allowing the consumer to 'click and collect' their orders from many different types of pick-up points using a variety of transport modes. Each of these options has a different carbon rating, making it very difficult to estimate the net effect on CO_2 emissions of the overall switch to online retailing. The growth of online sales volumes should in theory be improving van load factors and hence lowering emissions per order, but this trend is being counteracted by two other factors:

- The compression of order lead times as online retailers increasingly use speed of delivery as a competitive differentiator. This is likely to be depressing vehicle fill rates.
- The fragmentation of deliveries across a larger spread of possible delivery locations (homes plus collection points) as retailers develop their so-called 'omni-channel' capabilities (KPMG, 2016).

In this country, at the vanguard of the online retail revolution, policy makers need to gain a better understanding of its net impact on carbon emissions, much of which is logistics-related.

Behaviour

The UK has had a long tradition of training truck drivers to drive fuel-efficiently. This is partly in response to the relatively high price of diesel fuel in the UK. For almost 25 years the UK has had amongst the highest diesel prices in the EU, mainly because of the high level of tax imposed on this fuel (McKinnon, 2007c). In November 2017, the diesel fuel price in the

UK was 11 per cent above the EU average (European Commission, 2017b). The 'stick' of high fuel taxes has been accompanied by several 'carrots', the most significant of which was government financial support for a Safe and Fuel-Efficient Driving (SAFED) programme between 2003 and 2005, which subsidized training in eco-driving skills for 6,375 truck drivers. This training programme was subsequently commercialized and extended to around 82,000 HGV drivers by 2009. By that year a total of 154,000 drivers, almost half of all UK truck drivers, had received some form of training in fuel-efficient driving (RoSPA, 2010). Those receiving SAFED training were estimated to reduce their fuel consumption by between 2 and 12 per cent (Department for Transport, 2009), averaging around 7 per cent (Greening et al, 2015). The longevity of the training benefits was shown to depend critically on subsequent driver monitoring, feedback, coaching and motivation (AECOM, 2016). It has since become increasingly common in the UK for companies, particularly larger operators, to supplement driver training with the installation of onboard monitoring equipment to check that fuel-efficient driving practices are 'embedded'. Of a sample of HGV operators surveyed by the UK government in 2015, 35 per cent had installed 'electronic driver performance monitoring' (Department for Transport, 2017c).

Fuel efficiency is a key element in the Fleet Operator Recognition Scheme (FORS) in London, a voluntary accreditation scheme for truck, van, bus and coach operators to which over 4,500 businesses have subscribed (Transport for London, 2014). This provides driver training on fuel-efficient driving, expert advice on fuel-saving technologies, and various toolkits on, for example, anti-idling and avoiding congestion.

Despite these positive developments, having reviewed the available empirical data collected by AECOM (2016) and others, the Department for Transport (2017a: 22) judged 'evidence on current efficient driver training uptake rates' to be 'inconclusive'. It concedes that there is quite a high level of eco-driver training in the UK, but fears that the results may be biased as it tends to be the more progressive operators who respond to surveys. The case for promoting universal eco-driver training and monitoring in the freight sector is certainly strong, as it has been shown to have the 'lowest implementation cost' of a broad range of truck decarbonization measures (Greening et al, 2015).

Energy system

The average carbon intensity of energy used in the UK logistics sector is being reduced in several ways:

1 decarbonization of the UK electricity supply;
2 extension of electrification within the logistics sector;
3 switch from diesel to lower-carbon fuels.

1 Decarbonization of the UK electricity supply

The average carbon intensity of UK electricity has dropped sharply from 540 gCO_2/kwh in 2012 to 275 gCO_2/kwh at the end of 2017 (Committee on Climate Change 2017; MyGrid, 2017). This was achieved by radically altering the energy mix in a remarkably short time. Between 2012 and 2016, coal's share of electricity generation shrank from 39 per cent to 9 per cent and will be phased out completely by 2025 (Department of Energy and Climate Change, 2013; Department for Business, Energy and Industrial Strategy, 2017). The low-carbon power sources, ie renewables and nuclear, increased their combined share from 30 per cent to 46 per cent, while gas, a significantly less carbon-intensive fossil fuel than coal, also displaced much of the coal-fired capacity, raising its share from 28 per cent to 42 per cent. The aim is to have three-quarters of Britain's electricity generated by renewables and nuclear and the remainder gas powered by 2030, pushing the carbon intensity down to under 100 gCO_2/kwh by then (Committee on Climate Change, 2017).

2 Extending electrification within the logistics sector

If the decarbonization of the UK power sector proceeds as planned, those logistical activities that are currently electrified will indirectly have their carbon intensity reduced. This includes most warehousing and handling operations, electrified rail freight services and the tiny proportion of local deliveries currently undertaken by electric vans. In addition to purchasing lower-carbon electricity from the grid, some warehouse operators are 'micro-generating' zero-carbon electricity using site-based solar panels and wind turbines, though this is comparatively rare. As discussed earlier, only 7 per cent of rail freight services are electrically powered. This percentage will have to rise dramatically if rail is to take full advantage of the decarbonization of grid electricity.

Only around 800 of the 371,830 new vans registered in the UK in 2015 were electric (0.2 per cent) (European Environment Agency, 2016a; SMMT, 2017). Demand is currently very low for the variety of reasons discussed in Chapter 6, including the limited distance range on a single charge (causing so-called 'range anxiety'), the weight of the battery, the price premium against diesel vans, the lack of recharging points, and recharging times (Morganti and Browne, 2018). These barriers are being gradually overcome. According to the Freight Transport Association, a third of UK van deliveries are over a distance of 80 miles or less, significantly below the typical 100-mile range of an electric van, and the delivery range of new vans is being steadily extended. The weight of the battery typically reduces the payload of a small van by 5–15 per cent and a large van by 25 per cent (LowCvP, 2016). In recognition of the payload penalty imposed by a heavy battery, the government is increasing the maximum legal weight of electrically powered vans, thereby allowing drivers with ordinary licences to drive them. The government also makes a 20 per cent financial contribution towards the cost of a new electric van, up to a maximum of £8,000, to help narrow the capital cost differential between electric and diesel vehicles. The Committee on Climate Change (2015: 219) expects electric vans and small HGVs to 'become cost-effective by the mid-2020s'. The number of charging locations is also predicted to grow rapidly over the next few years from roughly 4,900 at the end of 2017 to 7,900 by 2020, with average recharging times reducing as battery performance improves.

The electrification of long-haul trucking in the UK is considered a distant prospect. Highways England (2015), which is responsible for the English trunk road network, has studied the feasibility of using an inductive system of Dynamic Wireless Power Transfer (DWPT) embedded in the road surface to power electric vehicles while on the move. This technology is favoured in the UK because it can power all classes of vehicle, 'can be installed under the road without any additional visible infrastructure... (does) not introduce additional safety risks (collision or electrical safety) and potentially minimize(s) the need for maintenance' (p.22). The government, nevertheless, seems open-minded about whether DWPT or Overhead Wire Power Transfer (OWPT) would be the most appropriate technology for the UK. It is observing international trials on both (Department for Transport, 2017a) but has no plans for e-highway trials comparable to those currently underway in Germany and Sweden.

The possible use of hydrogen as a low-carbon energy carrier for HGVs has been investigated by E4Tech et al (2017). In their 'critical path' scenario, they see the use of hydrogen, produced mainly by electrolysis, 'becoming increasingly prevalent' among HGVs between 2030 and 2040, with '90 per cent of HGVs hydrogen-fuelled' by 2050 (p.2). This is a view shared by Pye et al (2015: 28), whose pathway to the 'deep decarbonization' of HGVs in the UK 'is based on a shift to hydrogen-fuelled vehicles in the long term, with compressed natural gas (CNG) playing an important transitioning role.' Element Energy (2015) also foresees a widespread uptake of hydrogen by trucks between 2030 and 2050 and explores the infrastructure required to provide it. Others, such as Cebon (quoted in Auto Industry Newsletter, 2016), dismiss hydrogen as a cost-effective means of getting low-carbon electricity into the UK lorry fleet, mainly on the grounds of high energy losses and the expense of creating a new hydrogen supply network.

3 Switching from diesel to lower-carbon fuels

As the low-carbon electrification of freight transport is a longer-term option, the switch to lower-carbon fuels is being prioritized in the short to medium term. To date, however, there has been very limited repowering with these fuels across the UK logistics sector. They are mainly liquid biofuel (biodiesel in the case of road freight vehicles), liquid petroleum gas (LPG) for vans and liquid and compressed natural gas (LNG and CNG) and biomethane, principally for trucks.

The blending of biofuel with petrol and diesel has, since 2008, been subject to a Renewable Transport Fuel Obligation (RTFO), described as 'one of the government's main policies for reducing greenhouse gases from road transport in the UK' (Department for Transport, 2016b). This requires fuel suppliers to blend a specified proportion of biofuel into the fuel that they sell; in 2015–16 it was 4.75 per cent. A tradeable permit system is used to ensure that the biofuel entering the system meets sustainability criteria, as defined in 2011, and to support the development of the biofuel sector. In 2016, biodiesel, used in trucks, vans and also cars, accounted for only 2.4 per cent of all road transport fuel consumed in the UK (Department for Business, Energy and Industrial Strategy, 2017), most of it converted from recycled vegetable oil. As Bailey (2013) explains, there are a host of economic and environmental problems inhibiting increased use of biofuels in the UK transport sector.

Relatively small numbers of vans are powered by LPG in the UK. Relative to petrol vans, they offer 14 per cent savings in tailpipe CO_2 emissions and 20 per cent on a full life-cycle basis, but have broadly comparable CO_2 emissions to diesel-powered vans, which are by far the most commonly used for delivery purposes. The use of LNG and CNG has been increasing (FTA, 2016), though from a very low base. Much of the CNG used is in dual-fuel trucks, which combine natural gas with diesel to give them freedom of movement, while the density of gas refuelling points remains relatively low. In February 2016, there were only 43 gas refuelling stations across the UK (24 LNG, 16 CNG and three joint), split evenly between public and private installations (Element Energy, 2014). The FTA has been pressing for the establishment of a 'national gas refuelling infrastructure' (FTA, 2012). As the density of refuelling points increases, the proportion of trucks running solely on LNG or CNG will increase and reliance on dual-fuelling will diminish. This may be the longer-term solution to the 'methane slip' problem, discussed in Chapter 6, which afflicts dual-fuel lorries and, as the government accepts (Department for Transport, 2017a: 11), 'can offset any CO_2 savings derived from natural gas'.

The development of a gas refuelling network will also support the distribution of biogas. On a well-to-wheel (WTW) basis, biomethane produced by the anaerobic digestion of organic waste currently offers the greatest decarbonization potential, though its use in the logistics sector is minimal. This is because production capacity is limited and most of the available gas is used for other activities, particularly electricity generation and heating, in contrast to other European countries, such as France and Sweden, where it has been developed more as a transport fuel (FTA, 2012).

In summary, of the trends reviewed under the Energy heading, the decarbonization of grid electricity is proceeding at the fastest rate. By comparison, the uptake of alternative, lower-carbon fuels in the UK freight transport sector is proving relatively slow and problematic. This suggests that greater priority should be given to the electrification of freight transport in the short to medium term to help accelerate its decarbonization.

Regulation

As mentioned in the introduction, the regulatory regime for road haulage in the UK has been very liberal by international standards since the country deregulated this industry back in 1970. There is no quantitative licensing or tariff regulation, and own-account operators are free to carry goods for others. Unlike in much of Europe, there is no restriction on the movement of trucks at the weekend, though in some urban areas, most notably in London, there are restrictions on their movement during the night. Road haulage operations are governed by EU regulations on cabotage, drivers' hours and truck speed limits, though 'construction and use' regulations for trucks operating within the UK are subject to the EU 'subsidiarity' principle and specified by the British government. This has allowed the government to deviate from the C&U regulations that apply in many other EU countries. For example, in 2001 it raised the maximum weight of trucks running on six or more axles to 44 tonnes. I estimated that, within three years, this regulatory change had enabled operators of these heavier vehicles to cut CO_2 emissions by roughly 136,000 tonnes per annum (McKinnon, 2005). The six-axle, 44-tonne truck is now the standard 'workhorse' of the UK haulage industry, in contrast to the five-axle, 40-tonne truck which dominates road freight operations across the European mainland. The UK logistics sector has now fully adapted to this regulatory change. As discussed earlier, the UK also differs from most other EU countries in having no legal limit on trailer height. This, combined with greater height clearances under tunnels and bridges, has permitted the extensive use of high cube/double-deck trailers in the UK. As discussed earlier, this has significantly reduced the average carbon intensity of road haulage in the UK (McKinnon, 2010b).

Another much more recent regulatory innovation in the UK, which again does not conform to the EU modular system of truck dimensions, is a longer semi-trailer (LST). The government launched a 10-year trial of LSTs in 2012, issuing permits for a quota of 1,800 trailers, 20 per cent of which are one metre longer (14.6 m) and 80 per cent two metres longer (15.65 m), the latter increasing cubic capacity by 15 per cent (four more pallet slots). The maximum truck weight remains fixed at 44 tonnes. This means that once allowance is made for the additional tonne of extra tare weight in the longer trailer, the maximum payload weight is reduced by one tonne. Up to the end of 2015, the resulting load consolidation reduced truck-kms by roughly 5 per cent and this is considered to 'provide a rough proxy for all emission savings' (Risk Solutions, 2017: 10). The government has recently extended the trial by five years and increased the number of permits by 1,000, but, despite the reported economic and environmental benefits, has not committed to making this regulatory change permanent. As less than 3 per cent of the UK articulated trucks fleet will be allowed to operate with these longer trailers during the trial phase, the overall carbon impact of these regulatory changes will be minimal at least until the early 2020s.

The LST is the only recent concession that the UK government has been prepared to make to the wider international trend towards 'high-capacity transport'. Following a major study of several configurations of longer truck (Knight et al, 2008), the government at the time rejected the case for a length increase to 25 or 34 metres, primarily because of the threat that this might pose to the viability of some rail freight services, but also because of concerns about infrastructure, safety and public disapproval. Modelling by Greening et al (2015), however, suggests that 'larger and heavier' lorries will have to be used after 2025 to meet the emission reduction target for road freight transport. They project that their use will account for around a quarter of the CO_2 reductions from 'logistics measures' by 2035 (p.28).

At the other end of the size range, the government is considering an increase in the maximum weight of vans using alternative, low-emission forms of energy. This change would be 'technologically neutral' and apply to electric as well as gas-powered vehicles to relieve the payload penalty they suffer because of the additional weight of the battery or gas tank (Department for Transport, 2017a). In addition to more than offsetting this weight penalty, the proposed increase from 3.5 tonnes to 4.25 tonnes would also allow drivers with car licences to operate these heavier vans, removing the need to have an HGV licence to drive these cleaner, lower-carbon vehicles.

At the local level, public intervention in the freight sector is concerned much more with local air quality, road safety and congestion than with CO_2 emissions, although on balance, it is likely to have the effect of cutting these emissions per tonne-km. For example, the Low-Emission Zone for trucks in London, introduced in 2008 and soon to be replicated in other UK cities, relates solely to NOx and PM emissions but incentivizes operators to upgrade to newer and hence more fuel-efficient vehicles.

The regulatory heading can be extended to include various financial mechanisms that the UK government uses to help to decarbonize the freight transport sector. Of these, the most effective has unquestionably been the duty imposed on road fuel, which is very high by international standards, though has been declining in real terms in recent years as the rate has been frozen since 2011–12. The relatively high cost of road fuel should be promoting energy efficiency in the trucking sector and encouraging greater use of lower-carbon modes. The fuel used by diesel locomotives is taxed at only 19 per cent of the duty levied on truck fuel, giving the rail freight sector a significant fiscal advantage. This can be supplemented by Modal Shift Revenue Support (MSRS) grants where it is demonstrated that the environmental benefits of sending freight by rail rather than road are great enough (Department for Transport, 2015a). Separate MSRS schemes exist for intermodal rail freight services and rail bulk and inland waterway services. Another set of 'Waterborne Freight Grants' are available for the transfer of road freight from road to coastal and short-sea shipping, again where it can be demonstrated that the environmental benefits exceed a predefined threshold (Department for Transport, 2015b). In the valuation of these benefits, the government assigns a monetary value to the savings in CO_2 and other externalities per HGV-mile displaced from the road network. The intention is to use the grant to tip the financial balance in favour of rail or waterway where this is justified by the monetary value of the environmental benefits. The latest awards are expected to remove 114,000 lorry journeys from the road network between May 2017 and March 2019 and to yield an environmental benefit-to-cost ratio of 7.1–7.5 to 1 (Department for Transport, 2017b).

In summary, UK government regulatory and fiscal policies are exerting decarbonization pressures on the freight sector from several directions.

This pressure will have to intensify, however, for the country to meet the 80 per cent carbon reduction target in the 2008 Climate Change Act.

Conclusion

This review suggests that on balance the TIMBER factors, summarized in Table 8.2, are promoting the decarbonization of UK logistics. Several business-as-usual trends, such as the declining ratio of freight tonne-kms

Table 8.2 Summary of TIMBER factors promoting and inhibiting the decarbonization of logistics in the UK

External factor	Supportive	Inhibiting
Technology	*Road:* Uptake of many carbon-saving technologies still relatively low. Potential for wider adoption of support for platooning and vehicle automation. *Rail:* Greater use of hybrid locomotives, DAS and auxiliary power units.	Fragmentation and tight margins in road haulage industry limiting resources for vehicle upgrades and renewal. Uncertain residual values for greener trucks and vans.
Infrastructure	*Road:* Major upgrade of trunk road network. Congestion relief at 70 local 'pinch points'. Smart motorway management techniques. Network of gas refuelling points for trucks. *Rail:* Increasing capacity on some routes. Loading gauge enhancement on key routes. Development of more intermodal terminals.	Worsening road traffic congestion on key routes and in urban areas. Cancellation of rail electrification schemes. Growth of passenger services limiting freight capacity on key rail links and nodes. Lack of progress on trialling electrification of highways.

(continued)

Table 8.2 *(Continued)*

External factor	Supportive	Inhibiting
Market	Decoupling of freight and GDP growth trends. Decline of rail coal traffic – growth of intermodal. Diffusion of horizontal collaboration. Growing use of online freight exchanges. Online retailing reducing car shopping CO_2.	Large retailers tightening order lead times. Growth of van traffic for online delivery – shortening delivery times reducing load factors.
Behaviour	Expansion of eco-driver training programmes for trucks and trains. Wider telematic monitoring of driver behaviour.	Truck driver shortage – diverting attention from eco-driving schemes.
Energy	Rapid decarbonization of electricity generation. Increasing use of biodiesel and biomethane. Extension of gas-refuelling and battery-recharging networks.	Methane slip problem. High price of electric vehicles. Inadequate supply of biofuels.
Regulation	Extension of longer semi-trailer (LST) trial. Relaxation of weight and driving licence restrictions on electric vans. Continuation of modal shift support schemes.	Opposition to liberalization of high-capacity transport.

to GDP and the upgrading of vehicle technology, are lowering both the average carbon intensity of logistics and its overall carbon footprint. These trends are being reinforced by government policies, such as revenue support for modal shift and industry-led initiatives such as the Logistics Emission Reduction Scheme. Some developments, on the other hand, are having an inhibiting effect. For example, worsening traffic congestion, the curtailing of the rail electrification programme and political resistance to allowing longer and heavier lorries on UK roads will all constrain the rate at which logistics operations in the UK can decarbonize.

This chapter has illustrated how the TIMBER framework can be used to broaden the perspective on logistics decarbonization at the national level and show public policy makers and managers how it is influenced by macro-level trends beyond the control of individual companies. Much of the current analysis in government, industry and academic circles tends to concentrate on a small subset of these trends, mainly those relating to low-carbon fuels and vehicle technology. This is exemplified by the study of Pye et al (2015) cited earlier which envisages the long term decarbonization of the UK road freight sector being achieved primarily by a shift to hydrogen-fuelled vehicles with little consideration of alternative and supplementary options. As this chapter, and indeed the whole book, has tried to demonstrate, this 'fuel and vehicle' school of logistics decarbonization is taking much too narrow a view of the subject and is closing off many options which collectively can make a large contribution to the process.

In the UK, as in other countries, a diverse mix of technological and operational means exist to decarbonize logistics, some under corporate control, others requiring government intervention. As shown in Figure 8.6, they vary enormously in both their carbon abatement potential and the ease with which they can be implemented. The relative positioning of these developments is based partly on previous versions of this graph by Cebon (Auto Industry Newsletter, 2016), International Energy Agency (2017b) and the Smart Freight Centre (2017) but also on a fresh assessment of the evidence amassed for this book. New interventions have been added and several recalibrated specifically for the UK logistics industry and public policy arena. The final product offers both positive and negative messages. The positive news is that most interventions are clustered on the right and therefore judged to be implementable. On the other hand, some of the measures offering the greatest carbon abatement potential also have the highest technical, regulatory and operational barriers to overcome. For those measures in the bottom-left-hand quadrant, the potential contribution to logistics decarbonization seems limited, at least in the short to medium term. In the longer term most, if not all of these options will probably need to be deployed for the logistics sector to achieve the 80 per cent reduction in CO_2 emissions by 2050 (against a 1990 baseline) decreed by the UK Climate Change Act for the economy as a whole. Preparing the sector for this very low-carbon world will certainly not be easy but, as the well-being of future generations is at stake, those planning and managing logistics must rise to the challenge.

Figure 8.6 Comparison of decarbonization options in terms of carbon abatement potential and ease of implementation

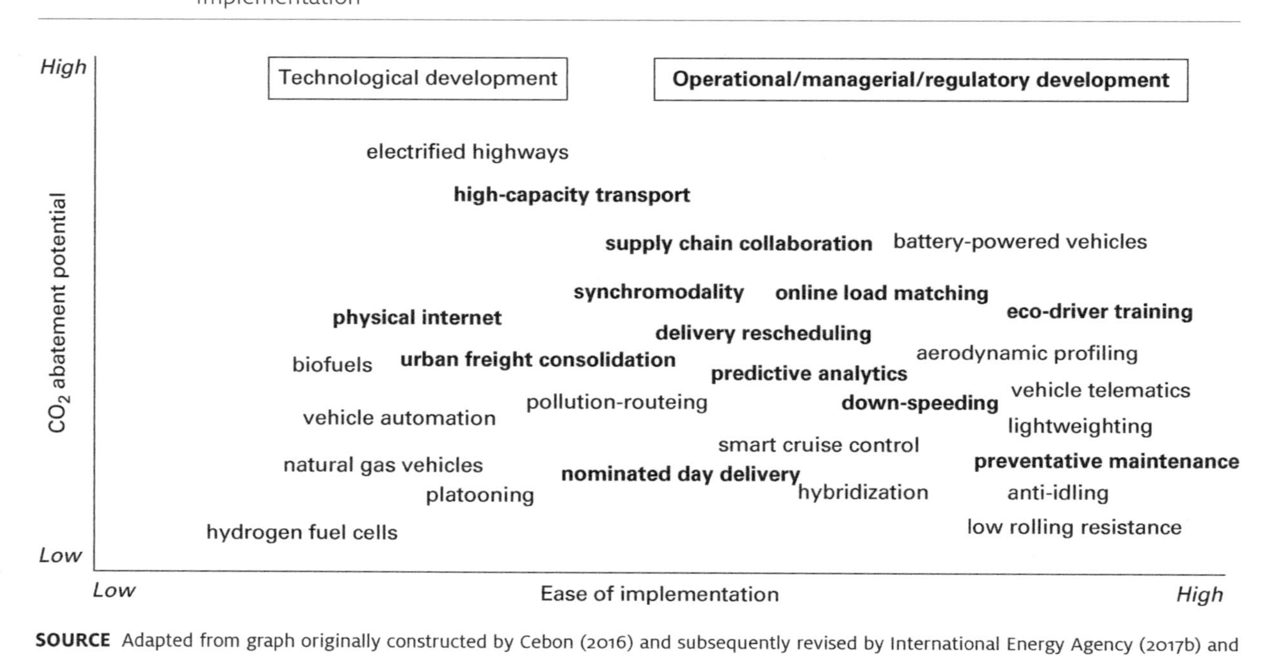

SOURCE Adapted from graph originally constructed by Cebon (2016) and subsequently revised by International Energy Agency (2017b) and Smart Freight Centre (2017).

Notes

1 In 2017 it was renamed the Logistics Emission Reduction Scheme.

2 Vans are defined as having a gross weight of less than 3.5 tonnes.

3 Domestic airfreight is excluded as it represents a tiny proportion of tonne-kms. No tonne-km data has been collected for pipelines since 2013, though they carried 10 bn tonne-kms every year between 2006 and 2013, and it can be assumed that this figure has remained reasonably stable up to 2016.

4 List of data sources for the calculation.

5 For low-density products, such as foam for furniture or car seats, which can be stacked to heights of four metres or more, it is not necessary to insert a second deck to fill these high-cube trailers.

6 Road relates to lorries with a gross weight of 3.5 tonnes or higher. Waterborne transport combines coastal shipping and inland waterway movements. As the UK has very little canal and riverborne traffic, this trend relates primarily to coastal shipping operations.

7 Cabotage is domestic haulage work undertaken by foreign registered carriers.

Bibliography

A T Kearney (1997) *The Efficient Unit Loads Report*, Brussels: ECR Europe

A T Kearney (2008) *5th Quinquennial Survey of European Logistics Costs*, Brussels: European Logistics Association.

A T Kearney (2015) *US Reshoring: Over before it began?* AT Kearney Consultants [online] https://www.atkearney.com/documents/10192/7070019/US+Reshoring.pdf

A T Kearney (2017) *The Widening Impact of Automation* (Research Report) A T Kearney Consultants [online] https://www.atkearney.com/digital-transformation/article?/a/the-widening-impact-of-automation-article

ACARE (2008) *Strategic Research Agenda 1: Volume 2 – environment*, Brussels: Advisory Council for Aeronautics Research in Europe

ACARE (2011) *Aeronautics and Air Transport: Beyond vision 2020 (Towards 2050)* Brussels: Advisory Council for Aeronautics Research in Europe

Acciaro, M and McKinnon, A C (2015) Carbon emissions from container shipping: an analysis of new empirical evidence, *International Journal of Transport Economics*, **42** (2), pp 75–92

Aditjandra, P T et al (2012) Investigating freight corridors towards low carbon economy: evidence from the UK *Procedia – Social and Behavioral Sciences*, **48**, pp 1865–76, https://doi.org/10.1016/j.sbspro.2012.06.1161

AEA Technology (2010) *Light Goods Vehicle – CO_2 Emissions Study: Final report*, London: Department for Transport

AEA Technology (2011) *Reduction and Testing of Greenhouse Gas (GHG) Emissions from Heavy Duty Vehicles – Lot 1: Strategy*, Report to the European Commission – DG Climate Action, Brussels

AEA Technology (2012) *Employment-Based Energy Consumption in the UK*, Harwell: AEA Technology plc

AECOM (2016) *Eco-Driving for HGVs: Final report*, London: Department for Transport

AECOM/ARUP (2016) *Future Potential for Modal Shift in the UK Rail Freight Market*, London: Department for Transport

Afionis, S et al (2017) Consumption-based carbon accounting: does it have a future? *Wiley Interdisciplinary Reviews: Climate Change*, **8** (1), https://doi.org/10.1002/wcc.438

Airbus (2016) *Mapping Demand: Global market forecast*, Blagnac: Airbus SAS

Airbus (2017) *Growing Horizons: Global market forecast 2017–2036*, Blagnac: Airbus SAS

Akerman, P (2016) *eHighway: Electrified heavy-duty road transport*, presented at The Future Role of Trucks for Energy and Environment, organized by the Joint Research Centre of the European Commission and the International Energy Agency, Brussels [online] https://www.iea.org/media/workshops/2016/thefutureroleoftrucks/7_Akerman_PA_eHighway_IEA_JRC_workshop.pdf

Alderson, W (2006) Marketing efficiency and the principle of Postponement, in *A Twenty-First Century Guide to Aldersonian Marketing Thought*, ed Wooliscroft, Tamilia and Shapiro (pp 109–13), Boston, MA: Springer, https://doi.org/10.1007/0-387-28181-9_8

Alexander, P et al (2017) Losses, inefficiencies and waste in the global food system, *Agricultural Systems*, **153** (Supplement C), pp 190–200, https://doi.org/10.1016/j.agsy.2017.01.014

ALICE (2014) *ALICE Recommendation to Horizon 2020 Work Programmes,* Brussels: EU European Technology Platform for Logistics [online] http://www.etp-logistics.eu/wp-content/uploads/2015/07/ALICE-Recomendations-HORIZON2020-WP-2016-2017-v141218_DEF-2.pdf

Alises, A, Vassallo, J M and Guzmán, A F (2014) Road freight transport decoupling: a comparative analysis between the United Kingdom and Spain, *Transport Policy*, **32**, pp 186–93, https://doi.org/10.1016/j.tranpol.2014.01.013

Allen, J et al (2012) The role of urban consolidation centres in sustainable freight transport, *Transport Reviews*, **32** (4), pp 473–90, https://doi.org/10.1080/01441647.2012.688074

Allwood, J M et al (2011) Material efficiency: a white paper, *Resources, Conservation and Recycling*, **55** (3), pp 362–81, https://doi.org/10.1016/j.resconrec.2010.11.002

Alternative Fuels Data Center (2017) Hydrogen production and distribution, *US Department of Energy* [online] https://www.afdc.energy.gov/fuels/hydrogen_production.html

American Trucking Association (2015) *Automated Driving and Platooning Opportunities and Issues* (White Paper) American Trucking Association

Anderson, K (2015) Duality in climate science, *Nature Geoscience*, **8**, pp 898–900, https://doi.org/10.1038/ngeo2559

Arvidsson, N (2013) The milk run revisited: a load factor paradox with economic and environmental implications for urban freight transport, *Transportation Research Part A: Policy and Practice*, **51** (Supplement C), pp 56–62, https://doi.org/10.1016/j.tra.2013.04.001

Asian Development Bank (2013) *The Clean Development Mechanism: A field guide for transport projects*, Manila: Asian Development Bank

Association of American Railroads (2017) *Freight Railroads Help to Reduce Greenhouse Gas Emissions*, Washington DC: AAR

Atkins, P (2010) *Technology Roadmap for Low Carbon HGVs*, London: Ricardo

ATRI (2016) *An Analysis of the Operational Costs of Trucking: 2016 update*, Arlington, VA: American Transportation Research Institute

Auto Industry Newsletter (2016) Hydrogen fuel cells: yes or no? *Auto Industry Newsletter* (14 December) [online] http://autoindustrynewsletter.blogspot.co.uk/2016/12/hydrogen-fuel-cells-yes-or-no.html

Auvinen, H et al (2014) Calculating emissions along supply chains: towards the global methodological harmonisation, *Research in Transportation Business & Management*, **12**, pp 41–46, https://doi.org/10.1016/j.rtbm.2014.06.008

Bailey, R (2013) *The Trouble with Biofuels: Costs and consequences of expanding biofuel use in the United Kingdom*, London: Chatham House

Baker, P and Marchant, C (2015) Reducing the environmental impact of warehousing, in *Green Logistics: Improving the environmental management of logistics*, ed A McKinnon et al (3rd edn), London: Kogan Page Ltd

Balachandran, B (2015) Indian Tyre industry: trend analysis of sales, net profit, EPS and radialisation level in India, *International Journal of Research*, **2** (6), pp 755–65

Ballot, E, Russell, M and Montreuil, B (2014) *The Physical Internet: The network of logistics networks*, Paris: PREDIT

Barth, M, Wu, G and Boriboonsomsin, K (2015) *Intelligent Transportation Systems for Improving Traffic Energy Efficiency and Reducing GHG Emissions from Roadways*, UC Riverside: National Center for Sustainable Transportation

Barton Assocs (2013) *Trade and Competitiveness Assessment of Mandated Speed Limiters for Heavy Trucks Operating in Canada*, Ottawa: Transport Canada

BASt (2016) *German Field Trial with Longer Trucks*, Bergisch Gladbach: BASt / Federal Highway Research Institute

Bastianoni, S, Pulselli, F M and Tiezzi, E (2004) The problem of assigning responsibility for greenhouse gas emissions, *Ecological Economics*, **49** (3), pp 253–57, https://doi.org/10.1016/j.ecolecon.2004.01.018

Becker, A et al (2016) A method to estimate climate-critical construction materials applied to seaport protection, *Global Environmental Change*, **40**, pp 125–36, https://doi.org/10.1016/j.gloenvcha.2016.07.008

Bektaş, T and Laporte, G (2011) The pollution-routing problem, *Transportation Research Part B: Methodological*, **45** (8), pp 1232–50, https://doi.org/10.1016/j.trb.2011.02.004

Birtchnell, T et al (2013) *Freight Miles: The impact of 3D printing on transport and society*, report of a workshop held as part of ESRC Project ES/J007455/1, Lancaster University, University of Lancaster

Black, I (1995) *Modelling the Links between Economic Activity and Goods Vehicle Movements*, CCLT Working Paper, Cranfield University

Boeing (2016) *World Air Cargo Forecast 2016-17*, Seattle: Boeing

Böge, S (1995) The well-travelled yogurt pot: lessons for new freight transport policies and regional production, *World Transport Policy and Practice*, **1** (1), pp 7–11 https://doi.org/10.1108/EUM0000000004024

Bond, T C et al (2013) Bounding the role of black carbon in the climate system: a scientific assessment, *Journal of Geophysical Research: Atmospheres*, **118** (11), pp 5380–52, https://doi.org/10.1002/jgrd.50171

Boon, W and Wee, B van (2017) Influence of 3D printing on transport: a theory and experts judgment based conceptual model, *Transport Reviews*, pp 1–20, https://doi.org/10.1080/01441647.2017.1370036

Boriboonsomsin, K (2015) *Reducing the Carbon Footprint of Freight Movement Through Eco-Driving Programs for Heavy-Duty Trucks*, UC Riverside: National Center for Sustainable Transportation

Bossel, U (2004) The hydrogen 'illusion', *Cogeneration and On-Site Power Production*, (March–April), pp 55–59

Boston Consulting Group (2011) *Made in America Again: Why manufacturing will return to America again*, Boston

Boston Consulting Group (2015a) BCG launches platform that could save the container industry billions of dollars (press release 17 November 2015) [online] https://www.bcg.com/d/press/17november2015-bcg-launches-xchange-22133

Boston Consulting Group (2015b) Reshoring of manufacturing to the US gains momentum, *BCG Perspectives* [online] https://www.bcgperspectives.com/content/articles/lean-manufacturing-outsourcing-bpo-reshoring-manufacturing-us-gains-momentum/

Bouman, E A et al (2017) State-of-the-art technologies, measures, and potential for reducing GHG emissions from shipping: a review, *Transportation Research Part D: Transport and Environment*, **52** (Part A), pp 408–21, https://doi.org/10.1016/j.trd.2017.03.022

Bows-Larkin, A (2015) All adrift: aviation, shipping, and climate change policy, *Climate Policy*, **15** (6), pp 681–702, https://doi.org/10.1080/14693062.2014.965125

Bows-Larkin, A et al (2016) Aviation and climate change: the continuing challenge, in *Encyclopaedia of Aerospace Engineering*, New York: John Wiley and Sons

Braithwaite, A (2017) *Implications of Internet Shopping Growth on the Van Fleet and Traffic Activity*, London: RAC Foundation

Brinks, H and Chryssakis, C (2017) *LPG as a Marine Fuel*, Oslo: DNV-GL

British Standards Institution (2011) *Guide to PAS 2050: How to carbon footprint your products, identify hotspots and reduce emissions in your supply chain*, London: BSI

Browne, M, Rizet, C and Allen, J (2014) A comparative assessment of the light goods vehicle fleet and the scope to reduce its CO_2 emissions in the UK and France, *Procedia – Social and Behavioural Sciences*, **125**, pp 334–44, https://doi.org/10.1016/j.sbspro.2014.01.1478

BSL (2017) *2016 Report on Combined Transport in Europe*, Paris: BSL

Buhaug, O et al (2009) *Prevention of Air Pollution from Ships: Second IMO GHG Study 2009*, London: International Maritime Organisation

Calder, S (2017) Could you be flying on a battery-powered plane by 2027? *Independent*, 28 November

Cames, M et al (2015) *Emission Reduction Targets for International Aviation and Shipping* (Study for the ENVI Committee), Brussels: European Parliament

Campaign for Better Transport (2017) *Rising to the Challenge*, London: CBT

Campbell, T et al (2011) *Could 3D Printing Change the World?* (Strategic Foresight Initiative), Washington DC: Atlantic Council

Camuzeaux, J R et al (2015) Influence of methane emissions and vehicle efficiency on the climate implications of heavy-duty natural gas trucks, *Environmental Science & Technology*, **49** (11), pp 6402–10, https://doi.org/10.1021/acs.est.5b00412

Caplice, C and Sheffi, Y (1994) A review and evaluation of logistics metrics, *The International Journal of Logistics Management*, **5** (2), pp 11–28, https://doi.org/10.1108/09574099410805171

Carbon Market Watch (2016) The CORSIA: ICAO's market-based measure and implications for Europe [online] https://carbonmarketwatch.org/wp-content/uploads/2016/10/Post-Assembly-Policy-Brief-Web.pdf

Carbon Pricing Leadership Coalition (2017) *Report of the High-Level Commission on Carbon Prices*, Washington DC: International Bank for Reconstruction and Development/World Bank

Carbon Tax Center (2017) Where carbon is taxed [online] https://www.carbontax.org/where-carbon-is-taxed/

Carbon Trust (2006) *Carbon Footprints in the Supply Chain*, London: Carbon Trust

Cariou, P (2011) Is slow steaming a sustainable means of reducing CO_2 emissions from container shipping? *Transportation Research Part D: Transport and Environment*, **16** (3), pp 260–64, https://doi.org/10.1016/j.trd.2010.12.005

Cariou, P and Cheaitou, A (2012) The effectiveness of a European speed limit versus an international bunker-levy to reduce CO_2 emissions from container shipping, *Transportation Research Part D: Transport and Environment*, 17 (2), pp 116–23, https://doi.org/10.1016/j.trd.2011.10.003

CDP (2017) *Putting a Price on Carbon: Integrating climate risk into business planning*, Carbon Disclosure Project

CDP/WRI/WWF (2015) Sectoral Decarbonisation Approach (SDA): A method for setting corporate emission reduction targets in line with climate science, *Science-based Targets* [online] http://sciencebasedtargets.org/wp-content/uploads/2015/05/Sectoral-Decarbonization-Approach-Report.pdf

CEN (2012) *Methodology for Calculation and Declaration of Energy Consumption and GHG Emissions of Transport Services (Freight and Passenger)*, Brussels: European Committee of Standardization

Chan, F T S and Zhang, T (2011) The impact of Collaborative Transportation Management on supply chain performance: a simulation approach, *Expert Systems with Applications*, **38** (3), pp 2319–29, https://doi.org/10.1016/j.eswa.2010.08.020

Cheah, L W (2010) *Cars on a Diet: The material and energy impact of passenger vehicle weight reduction in the US* (PhD Thesis), Boston: Massachusetts Institute of Technology

Chryssakis, C (2014) *Alternative Fuels for Shipping: Pathways to 2050*, Oslo: DNV

Chryssakis, C et al (2014) *Alternative Fuels for Shipping* (Position Paper 1), Oslo: DNV-GL

Clark, P U et al (2016) Consequences of twenty-first-century policy for multi-millennial climate and sea-level change, *Nature Climate Change*, **6** (4), pp 360, https://doi.org/10.1038/nclimate2923

Clarke, G et al (2014) *Van Travel Trends in Great Britain*, London: RAC Foundation

Clean Air Asia (2013) *China Green Freight Policy and Institutional Analysis Report*, Manila: Clean Air Asia and World Bank

Clean Cargo Working Group (2014) Global marine trade lane emission factors, *BSR* [online] http://www.bsr.org/reports/BSR_CCWG_Trade_Lane_Emissions_Factors.pdf

Coady, D et al (2015) *How Large are Global Energy Subsidies?* IMF Working Paper WP/15/105, Washington DC: International Monetary Fund

Coley, D, Howard, M and Winter, M (2011) Food miles: time for a re-think? *British Food Journal*, **113** (7), pp 919–34, https://doi.org/10.1108/00070701111148432

Comer, B et al (2017) *Prevalence of Heavy Fuel Oil and Black Carbon in Arctic Shipping: 2015–2025*, Washington DC: International Council for Clean Transportation

Commission for Integrated Transport (2007) *Transport and Climate Change*, London: HMSO

Committee on Climate Change (2015) *Sectoral Scenarios for the Fifth Carbon Budget: Technical report*, London: Committee on Climate Change

Committee on Climate Change (2017) *Meeting Carbon Budgets: Closing the policy gap*, London: Climate Change Committee

Constantinescu, C, Mattoo, A and Ruta, M (2015) *The Global Trade Slowdown: Cyclical or structural* (Working Paper 15/6), Washington DC: International Monetary Fund

Copacino, W and Rosenfield, D B (1985) Analytic tools for strategic planning, *International Journal of Physical Distribution & Materials Management*, **15** (3), pp 47–61, https://doi.org/10.1108/eb014610

Corbett, J J et al (2007) Mortality from ship emissions: a global assessment, *Environmental Science & Technology*, **41** (24), pp 8512–18, https://doi.org/10.1021/es071686z

Cornelisson, J and Janssen, R (2017) *Truck Platooning Real-Life Cases NL*, Delft: TNO

Coyle, M (2007) *Effects of Payload on the Fuel Consumption of Trucks*, London: Department for Transport

CPRE (Council for the Protection of Rural England) (2017) *The End of the Road: Challenging the road building consensus*, London: CPRE

Craig, A J, Blanco, E E and Sheffi, Y (2013) Estimating the CO_2 intensity of intermodal freight transportation, *Transportation Research Part D: Transport and Environment*, **22** (Supplement C), pp 49–53, https://doi.org/10.1016/j.trd.2013.02.016

Crawford, R H (2009) Life cycle energy and greenhouse emissions analysis of wind turbines and the effect of size on energy yield, *Renewable and Sustainable Energy Reviews*, **13** (9), pp 2653–60, https://doi.org/10.1016/j.rser.2009.07.008

Cristea, A et al (2013) Trade and the greenhouse gas emissions from international freight transport, *Journal of Environmental Economics and Management*, **65** (1), pp 153–73, https://doi.org/10.1016/j.jeem.2012.06.002

CSIR (1997) *Damaging Effects of Overloaded Heavy Vehicles on the Roads*, Pretoria: South African Department for Transport

Curry, C (2017) *Lithium-ion Battery Costs and Market*, London: Bloomberg New Energy Finance

Curry, T et al (2012) *Reducing Aerodynamic Drag and Rolling Resistance for Heavy Duty Trucks: Summary of available technologies and applicability to Chinese trucks*, San Francisco: Report by Bradley Assocs for ICCT

Dantzig, G B and Ramser, J H (1959) The truck dispatching problem, *Management Science*, **6** (1), pp 80–91, https://doi.org/10.1287/mnsc.6.1.80

Davenport, T H (2016) Prescriptive analytics project delivering big benefits at UPS, *Data Informed* (19 April) [online] http://data-informed.com/prescriptive-analytics-project-delivering-big-dividends-at-ups/

Davies, A (2016) Uber's self-driving truck makes its first delivery: 50,000 Budweisers, *WIRED* [online] https://www.wired.com/2016/10/ubers-self-driving-truck-makes-first-delivery-50000-beers/

Davis, L W and Gertler, P J (2015) Contribution of air conditioning adoption to future energy use under global warming, *Proceedings of the National Academy of Sciences*, **112** (19), pp 5962–67, https://doi.org/10.1073/pnas.1423558112

DB Schenker (2015) *Get Your Business Rolling with Innovative Rail Logistics Solutions Between China and Europe*, Berlin

De Jong, G et al (2010) *Price Sensitivity of European Road Freight Transport – for a Better Understanding of Existing Results* (Report for Transport and Environment), Delft: CE Delft

DECC (2015) *Updated Short-Term Traded Carbon Values Used in UK Public Policy Appraisal*, London: Department for Energy and Climate Change

DEFRA (2017) *UK Government GHG Conversion Factors for Company Reporting*, London: Department of the Environment, Food and Rural Affairs [online] https://www.gov.uk/government/publications/greenhouse-gas-reporting-conversion-factors-2017

Dekker, R, Bloemhof, J and Mallidis, I (2012) Operations research for green logistics: an overview of aspects, issues, contributions and challenges, *European Journal of Operational Research*, **219** (3), pp 671–79, https://doi.org/10.1016/j.ejor.2011.11.010

Delgado, O, Rodriquez, F and Muncrief, R (2017) *Fuel Efficiency Technology in European Heavy Duty Vehicles: Base and potential for the 2020–2030 time frame*, Berlin: International Council for Clean Transportation

Department for Business, Energy and Industrial Strategy (2017) *Digest of United Kingdom Energy Statistics 2017*, London: DBEIS

Department for Transport (2004) *Survey of Van Activity*, London: Department for Transport

Department for Transport (2005) *Transport Statistics Great Britain*, London: Department for Transport

Department for Transport (2006a) *Fuel Management Guide*, London: Department for Transport

Department for Transport (2006b) *Road Traffic Statistics*, London: Department for Transport

Department for Transport (2008) *Delivering a Sustainable Transport System: The logistics perspective*, London: Department for Transport

Department for Transport (2009) *SAFED for HGVs: A guide to safe and fuel efficient driving for HGVs*, London: Department for Transport

Department for Transport (2010) *Guidance on Measuring and Reporting Greenhouse Gas (GHG) Emissions from Freight Transport Operations*, London: Department for Transport

Department for Transport (2011) *Road Traffic Statistics 2010*, London: Department for Transport

Department for Transport (2013) *Freight Carbon Review*, London: Department for Transport

Department for Transport (2014a) *Discrepancies between Road Freight and Road Traffic HGV Traffic Estimates: Methodological note*, London: Department for Transport

Department for Transport (2014b) *Road Investment Strategy: 2015–2020*, London: Department for Transport

Department for Transport (2015a) *Guide to Mode Shift Revenue Support Scheme*, London: Department for Transport

Department for Transport (2015b) *Guide to the Waterborne Freight Grant Scheme,* London: Department for Transport

Department for Transport (2015c) *Quiet Deliveries Good Practice Guidance: Key principles and processes for community and resident groups*, London: Department for Transport

Department for Transport (2015d) *UK Road Traffic Forecasts 2015*, London: Department for Transport

Department for Transport (2015e) *Unpublished Data from the Continuing Survey of Road Goods Transport*, London: Department for Transport

Department for Transport (2016a) *Rail Freight Strategy: Moving Britain ahead* London: Department for Transport

Department for Transport (2016b) *Renewable Transport Fuels Obligation 2015–26*, London: Department for Transport

Department for Transport (2017a) *Freight Carbon Review*, London: Department for Transport

Department for Transport (2017b) *Mode Shift Revenue Support and Waterborne Freight Grant Applications and Background Information*, London: Department for Transport

Department for Transport (2017c) *Road Freight Statistics 2016*, London: Department for Transport

Department for Transport (2017d) *Transport Statistics Great Britain 2017*, London: Department for Transport

Department of Energy and Climate Change (2013) *Digest of United Kingdom Energy Statistics 2013*, London: DECC

Department of the Environment, Transport and the Regions (1999) *Sustainable Distribution: A strategy*, London: HMSO

DG CLIMA (2017) *Reducing CO_2 Emissions from Heavy Duty Vehicles*, European Commission [online] https://ec.europa.eu/clima/policies/transport/vehicles/heavy_en

DG Move (2015) *An Overview of the EU Road Transport Market in 2015*, Brussels: European Commission

Dhooma, J and Baker, P (2012) An exploratory framework for energy conservation in existing warehouses, *International Journal of Logistics Research and Applications*, **15** (1), pp 37–51, https://doi.org/10.1080/13675567.2012.668877

Dicken, P (2015) *Global Shift: Mapping the changing contours of the global economy* (7th edn) London: Paul Chapman

Direct Rail Services (2017) Intermodal – Tesco: Less CO_2 for a better environment [online] https://www.directrailservices.com/intermodal-%E2%80%93-tesco.html

DNV (2010) *Pathways to Low Carbon Shipping: Abatement potential towards 2030*, Oslo: DNV

DNV-GL (2017) *Low Carbon Shipping Towards 2050*, Hovik, Norway: DNV-GL

Doherty, B (1999) Paving the way: the rise of direct action against road-building and the changing character of British environmentalism, *Political Studies*, **47** (2), pp 275–91, https://doi.org/10.1111/1467-9248.00200

Doll, C et al (2008) *Long-Term Climate Impacts of the Introduction of Mega-Trucks: Study for the community of European railways and infrastructure companies (CER)*, Karlsruhe: Fraunhofer ISI

Dominguez-Faus, R (2016) *The Carbon Intensity of NGV C8 Trucks (revised)*, Davis: Institute of Transportation Studies, University of California Davis

Dominguez-Faus, R (2017) *Climate, Energy Transition and the Use of Natural Gas in Freight Transportation: Pros and cons*, presented at the Webinar Series, Energy Sector Management, HEC Montreal, Montreal [online] https://www.youtube.com/watch?v=Ynd4hpuFdZg

Don Bur (2017) Teardrop case studies, *Don Bur Ltd* [online] http://www.donbur.co.uk/gb-en/info/teardrop-trailer-case-studies.php

Dong, C et al (2017) Investigating synchromodality from a supply chain perspective, *Transportation Research Part D: Transport and Environment*, https://doi.org/10.1016/j.trd.2017.05.011

E4Tech, UCL Energy Institute and Kiwa Gastec (2017) *Scenarios for Deployment of Hydrogen in Contributing to Meeting Carbon Budgets and the 2050 Target: Final report*, London

EASAC (2018) *Negative Emission Technologies: What role in meeting Paris Agreement targets?* Halle: European Academies Science Advisory Council

ECB (2016) *Understanding the Weakness in Global Trade: What is the New normal?* (Occasional Paper Series) Frankfurt: European Central Bank,

Economist (2017a) Greenhouse gases must be scrubbed from the air, *Economist* (16 November)

Economist (2017b) 3D printers start to build factories of the future, *Economist* (29 July)

Economist (2017c) New rail route between China and Europe will change trade patterns, *Economist* (16 September)

ECR UK (2008) Transport collaboration guide, *Institute of Grocery Distribution* [online] https://www.2degreesnetwork.com/groups/2degrees-community/resources/igd-guide-transport-collaboration/

Eddington, R (2006) *The Eddington Transport Study: Understanding the relationship – how transport can contribute to economic success* (Vol 1), London: Department for Transport

Edwards, J B, McKinnon, A C and Cullinane, S L (2010) Comparative analysis of the carbon footprints of conventional and online retailing: a 'last mile' perspective, *International Journal of Physical Distribution & Logistics Management*, **40** (1/2), pp 103–23 https://doi.org/10.1108/09600031011018055

Edwards, J, McKinnon, A and Cullinane, S (2011) Comparative carbon auditing of conventional and online retail supply chains: a review of methodological issues, *Supply Chain Management: An International Journal*, **16** (1), pp 57–63, https://doi.org/10.1108/13598541111103502

Edwards-Jones, G et al (2008) Testing the assertion that 'local food is best': the challenges of an evidence-based approach, *Trends in Food Science & Technology*, **19** (5), pp 265–74, https://doi.org/10.1016/j.tifs.2008.01.008

Egis (2010) *Greenhouse Gas Emissions Mitigation in Road Construction and Rehabilitation*, Washington DC: World Bank

Eglise, R and Black, I (2015) Optimising the routeing of vehicles, in *Green Logistics: Improving the environmental sustainability of logistics*, ed A McKinnon et al (3rd ed), London: Kogan Page Ltd (pp 229–42)

Eide, M S et al (2013) Reducing CO_2 from shipping – do non-CO_2 effects matter? *Atmos. Chem. Phys.*, **13** (8), pp 4183–4201, https://doi.org/10.5194/acp-13-4183-2013

Element Energy (2014) *A Fuel Roadmap for the UK*, London: Low Carbon Vehicle Partnership

Element Energy (2015) *Transport Energy Infrastructure Roadmap: Hydrogen road*, London: Low Carbon Vehicle Partnership

Elhedhli, S and Merrick, R (2012) Green supply chain network design to reduce carbon emissions, *Transportation Research Part D: Transport and Environment*, **17** (5), pp 370–79, https://doi.org/10.1016/j.trd.2012.02.002

Ellen MacArthur Foundation and McKinsey & Co (2014) *Towards a Circular Economy: Accelerating the take-up across global supply chains*, London

Ellen MacArthur Foundation and McKinsey Center for Business and Environment (2015) *Growth Within: A circular economy vision for a competitive Europe*, London.

Energy Transitions Commission (2017) *Better Energy Greater Prosperity: Achievable pathways to low-carbon energy systems*, ETC, London

EPA (2017) *Global Greenhouse Gas Emissions Data*, United States Environmental Protection Agency [online] https://www.epa.gov/ghgemissions/global-greenhouse-gas-emissions-data

Esper, T L and Williams, L R (2003) The value of collaborative transportation management (CTM): its relationship to CPFR and information technology, *Transportation Journal*, **42** (4), pp 55–65

Establish Inc (2010) *Logistics Cost and Service*, presented at the CSCMP Annual Global Conference

EU CO3 Project (2013) Innovation: horizontal collaboration [online] http://www.co3-project.eu/innovation/

European Commission (2004) Proposal for a European Parliament and Council directive on intermodal loading units COM(2004) 361, European Commission, Brussels, [online] http://eur-lex.europa.eu/legal-content/EN/TXT/?uri=LEGISSUM:l24271

European Commission (2011) *White Paper: Roadmap to a single European transport area – towards a competitive and resource-efficient transport system*, Brussels

European Commission (2013) *Marco Polo: The results* [online] https://ec.europa.eu/transport/marcopolo/files/infographics-marco-polo-results.pdf

European Commission (2014) *EU Energy, Transport and GHG Emissions: Trends to 2050 – Reference scenario 2013*, Brussels: European Commission

European Commission (2017a) *Sustainability Criteria* [online] https://ec.europa.eu/energy/en/topics/renewable-energy/biofuels/sustainability-criteria

European Commission (2017b) *EU Weekly Oil Bulletin* [online] https://ec.europa.eu/energy/en/data-analysis/weekly-oil-bulletin

European Court of Auditors (2016) *Rail Freight Transport in the EU: Still not on the right track*, Brussels: European Commission

European Environment Agency (2010) *Load Factors for Freight Transport (TERM 030)* [online] http://www.eea.europa.eu/data-and-maps/indicators/load-factors-for-freight-transport/load-factors-for-freight-transport-1

European Environment Agency (2016a) *Electric Vehicles in Europe*, Copenhagen: European Environment Agency

European Environment Agency (2016b) *European Aviation Environmental Report 2016*, Copenhagen: European Environment Agency

European Parliamentary Research Service (2014) *Reshoring of EU Manufacturing* (Briefing), Brussels

European Wind Energy Association (2008) *Wind Energy: The facts*, Brussels: EU Intelligent Energy Programme

European Sea Ports Organisation (2012) *Green Guide: Towards excellence in port environmental management and sustainability*, Brussels: ESPO

Eurostat (2014) *Volume of Freight Transport Relative to GDP* [online]http://ec.europa.eu/eurostat/tgm/table.do?tab=table&plugin=1&language=en&pcode=tsdtr230

Eurostat (2017a) *Road Freight Transport by Journey Characteristics* [online] http://ec.europa.eu/eurostat/statistics-explained/index.php/Road_freight_transport_by_journey_characteristics

Eurostat (2017b) *Road Freight Transport Statistics – Cabotage* [online] http://ec.europa.eu/eurostat/statistics-explained/index.php/Road_freight_transport_statistics_-_cabotage

Eurostat (2017c) *Freight Transport Statistics: Modal split* [online] http://ec.europa.eu/eurostat/statistics-explained/index.php/Freight_transport_statistics_-_modal_split

Eurostat (2017d) *Road Freight Transport Methodology: Reference manual for the Implementation of Council Regulation No. 1172/98 on statistics on the carriage of goods by road*, Luxembourg: Eurostat

Evans, J A et al (2014) Assessment of methods to reduce the energy consumption of food cold stores, *Applied Thermal Engineering*, **62** (2), pp 697–705, https://doi.org/10.1016/j.applthermaleng.2013.10.023

Exergia, E3M and COWI (2015) *Study of Actual GHG Data for Diesel, Petrol, Kerosene and Natural Gas*, Report for DG Ener, Brussels: European Commission

Fabbes-Coste, N and Colin, J (2007) Formulating a logistics strategy, *Global Logistics*, ed D Waters (5th edn), London: Kogan Page Ltd

Faber, J et al (2012) *Regulated Slow Steaming for Maritime Transport: An assessment of the options, costs and benefits*, Delft: CE Delft

Facanha, C and Horvath, A (2006) Environmental assessment of freight transportation in the US, *International Journal of Life Cycle Analysis*, **11**, (4), pp 229–39, https://doi.org/10.1021/es070989q

Farrell, S (2016) We've hit peak home furnishings, says Ikea boss, *Guardian* (18 January)

Fathom Focus (2013) *Hull Coatings for Vessel Performance*, Fathom

Fernie, J and Grant, D B (2015) *Fashion Logistics: Insights into the fashion retail supply chain*, London: Kogan Page Ltd

Figliozzi, M (2010) Vehicle routing problem for emissions minimization, *Transportation Research Record: Journal of the Transportation Research Board*, **2197**, pp 1–7, https://doi.org/10.3141/2197-01

Figueres, C et al (2017) Three years to safeguard our climate, *Nature News*, **546** (7660), p 593, https://doi.org/10.1038/546593a

Flanders, P and Smith, R (2007) *Inhibitors to the Growth of Rail Freight*, Edinburgh: Scottish Government

Flight International (2008) As fuel costs spiral, winglets are a simple way for airlines to cut fuel consumption, *Flight International* (27 June)

Forkert, S and Eichhorn, C (2012) *Innovative Approaches to City Logistics: Inner-city night deliveries* (EU Niches Project) Brussels: Polis

Franceschetti, A et al (2013) The time-dependent pollution-routing problem, *Transportation Research Part B: Methodological*, **56**, pp 265–93, https://doi.org/10.1016/j.trb.2013.08.008

Fraunhofer IBP and IFEU (2015) *Interaktion EE-Strom, Warme und Verkehr*, Kassel: Fraunhofer-Institut fur Bauphysik

Freight Best Practice Programme (2005) *Computerised Vehicle Routeing and Scheduling (CVRS) for Efficiency Logistics*, London: Department for Transport

Freight Best Practice Programme (2006) *Fuel Management Guide*, London: Department for Transport

Freight Best Practice Programme (2007) *Focus on Double Decks*, London: Department for Transport

Freight Best Practice Programme (2010a) *Preventative Maintenance for Efficient Road Freight Operations*, London: Department for Transport

Freight Best Practice Programme (2010b) *The Malcolm Group: An award-winning multi-modal operator*, London: Department for Transport

FTA (2011) *Logistics Carbon Reduction Scheme: Recording, reporting and reducing CO_2 emissions from the logistics sector*, Tunbridge Wells: Freight Transport Association

FTA (2012) *A Natural Gas Refuelling Network for the UK*, Tunbridge Wells: Freight Transport Association

FTA (2016) *Logistics Carbon Review 2016*, Tunbridge Wells: Freight Transport Association

Galos, J et al (2015) Reducing the energy consumption of heavy goods vehicles through the application of lightweight trailers: fleet case studies. *Transportation Research Part D: Transport and Environment*, **41** (Supplement C), pp 40–49 https://doi.org/10.1016/j.trd.2015.09.010

Garnett, T (2015) The food miles debate: is shorter better? In *Green Logistics: Improving the environmental sustainability of logistics*, ed A McKinnon et al (3rd edn), London: Kogan Page Ltd

Garthwaite, J (2012) Smarter trucking saves fuel over the long haul, *Inbound Logistics* (15 January)

GEF (2017) *Funding*, Global Environmental Facility [online] https://www.thegef.org/about/funding

Gegg, P, Budd, L and Ison, S (2014) The market development of aviation biofuel: drivers and constraints, *Journal of Air Transport Management*, **39**, pp 34–40, https://doi.org/10.1016/j.jairtraman.2014.03.003

Gereffi, G et al (2001) Introduction: globalisation, value chains and development, *IDS Bulletin*, **32** (3), pp 1–8, https://doi.org/10.1111/j.1759-5436.2001.mp32003001.x

Gibbs, D et al (2014) The role of sea ports in end-to-end maritime transport chain emissions, *Energy Policy*, **64**, pp 337–48

Glaeser, K-P (2010) Performance of articulated vehicles and road trains regarding road damage and load capacity, *Proceedings of the HVTT11 Conference*, Melbourne

GLEC (Global Logistics Emissions Council) (2016) *GLEC Framework for Logistics Emissions Methodologies*, Amsterdam: Smart Freight Centre

Global Fuel Economy Initiative (2017) *Targeting Heavy Duty Vehicle Fuel Economy*, GFEI London

Goleman, D (2010) *Ecological Intelligence*, New York: Broadway Books

Gota, S (2016) *Freight Transport and Climate Change: Is freight in climate mitigation agenda*? Presented at the 95th Annual Meeting of Transportation Research Board, Washington DC

Gota, S, Huizenga, C and Peet, K (2016) *Implications of 2DS and 1.5DS for Land Transport Carbon Emissions in 2050*, Partnership on Sustainable Low Carbon Transport (SLoCaT)

Gota, S et al (2016) Nationally determined contributions offer opportunities for ambitious action on transport and climate change, *PPM /SLOCAT* [online] http://www.ppmc-transport.org/wp-content/uploads/2015/06/NDCs-Offer-Opportunities-for-Ambitious-Action-Updated-October-2016.pdf

Greene, S (2017) *Black Carbon: Methodology for the logistics sector*, Amsterdam: Smart Freight Centre

Greening, P et al (2015) *Assessment of the Potential for Demand-side Fuel Savings in the Heavy Good Vehicle (HGV) Sector*, Edinburgh: Centre for Sustainable Road Freight

Greszler, A (2009) Heavy duty vehicle fleet technologies for reducing carbon dioxide: an industry perspective, in *Reducing Climate Impacts in the Transportation Sector*, ed D Sperling and J S Cannon, Springer, (pp 101–16)

GS1 (2017) Global language of business [online] https://www.gs1uk.org/about-us/who-we-are

GT Nexus (2017) *Solution Overview: Transportation management* [online] http://mktforms.gtnexus.com/rs/gtnexus/images/GTNexus-Transportation-Management-SS.pdf

Gucwa, M and Schäfer, A (2013) The impact of scale on energy intensity in freight transportation, *Transportation Research Part D: Transport and Environment*, **23** (Supplement C), pp 41–49, https://doi.org/10.1016/j.trd.2013.03.008

Guérin, E, Mas, C and Waisman, H (2014) *Pathways to Deep Decarbonisation*, Paris: Sustainable Development Solutions Network and Institute for Sustainable Development and International Relations

Gwyther, M (2017) Have we reached peak stuff? *Management Today* (2 March)

Hao, H et al (2015) Energy consumption and GHG emissions from China's freight transport sector: scenarios through 2050, *Energy Policy*, **85**, pp 94–101

Harrabin, R (2017) Device could make washing machines lighter and greener, *BBC News* [online] http://www.bbc.co.uk/news/uk-40821915

Havenga, J H, Simpson, Z P and Bod, A D (2013) Macro-logistics trends: indications for a more sustainable economy: original research, *Journal of Transport and Supply Chain Management*, **7** (1), pp 1–7, https://doi.org/10.4102/jtscm.v7i1.108

Heath, G et al (2015) *Estimating US Methane Emissions from the Natural Gas Supply Chain: Approaches, uncertainties, current estimates and future studies*, Golden: Joint Institute for Strategic Energy Research

Heid, B et al (2017) *What's Sparking Electric Vehicle Adoption in the Truck Industry,* McKinsey & Co (McKinsey Center for Future Mobility)

Helms, H, Lambrecht, U and Hopfner, U (2003) *Energy Savings by Lightweighting: Final report*, Heidelberg: IFEU

Herman, R, Ardekani, S A and Ausubel, J H (1990) Dematerialization, *Technological Forecasting and Social Change*, **38** (4), pp 333–47, https://doi.org/10.1016/0040-1625(90)90003-E

High-Level Group on the Development of the EU Road Haulage Market (2012) *Report on the High-Level Group on the Development of the EU Road Haulage Market* [online] https://ec.europa.eu/transport/sites/transport/files/modes/road/doc/2012-06-high-level-group-report-final-report.pdf

Highways Agency (2013) *Managed Motorways: Fact sheet*, Dorking: Highways Agency

Highways England (2015) *Feasibility Study: Powering electric vehicles on England's major roads*, Guildford: Highways England

Hind, P (2014) The commercial use of biofuels in aviation, *Aviation Economics* [online] http://www.aviationeconomics.com/NewsItem.aspx?title=The-Commercial-Use-of-Biofuels-in-Aviation

Hitchcock, F L (1941) The distribution of a product from several sources to numerous localities, *Journal of Mathematics and Physics*, **20** (1–4), pp 224–30, https://doi.org/10.1002/sapm1941201224

Hjortnaes, T et al (2017) Minimizing cost of empty container repositioning in port hinterlands, while taking repair operations into account, *Journal of Transport Geography*, **58**, pp 209–19, https://doi.org/10.1016/j.jtrangeo.2016.12.015

Hoekman, B (2015) Trade and growth: end of an era, in *Global Trade Slowdown: A New Normal?* (pp 3–19), London: CEPR Press [online] http://voxeu.org/sites/default/files/file/Global%20Trade%20Slowdown_nocover.pdf

Hoffman, A (2006) *Getting Ahead of the Curve: Corporate strategies that address climate change*, Arlington, VA: Pew Center on Global Climate Change

Hogg, N and Jackson, T (2008) Digital media and dematerialization: an exploration of the potential for reduced material intensity in music delivery, *Journal of Industrial Ecology*, **13** (1), pp 127–46

Holguín-Veras, J et al (2014) The New York City off-hour delivery project: lessons for city logistics, *Procedia – Social and Behavioral Sciences*, **125** (Supplement C), pp 36–48, https://doi.org/10.1016/j.sbspro.2014.01.1454

Holguín-Veras, J et al (2016) Direct impacts of off-hour deliveries on urban freight emissions, *Transportation Research Part D: Transport and Environment*, https://doi.org/10.1016/j.trd.2016.10.013

Holmes, C D, Prather, M J and Vinken, G C (2013) The climate impact of ship NOx emissions: uncertainties due to plume chemistry, *AGU Fall Meeting Abstracts*, *11* [online] http://adsabs.harvard.edu/abs/2013AGUFM.A11C0063H

Humphrey, D G (2015) Growing yet aging – gradually, *Railway Age* (8 July)

IATA (2013) *A Blueprint for a Single European Sky*, Geneva: IATA

IATA (2014a) *Impact Assessment of WTO Bali Deal on Air Freight* (IATA Economic Briefing), Geneva: IATA

IATA (2014b) *CO_2 Emissions Measurement Methodology: Recommended practice 1678*, Geneva: IATA [online] https://www.iata.org/whatwedo/cargo/sustainability/Documents/rp-carbon-calculation.pdf

IATA (2015) *IATA Cargo Strategy*, Geneva: IATA

IATA (2017a) 2036 forecast reveals air passengers will nearly double to 7.8 billion, [online] http://www.iata.org/pressroom/pr/Pages/2017-10-24-01.aspx

IATA (2017b) *Air Freight Market Analysis: June 2017* [online] http://www.iata.org/whatwedo/Documents/economics/freight-analysis-jun-2017.pdf

ICAO (2007) *Environmental Report 2007*, Montreal: International Civil Aviation Organization

ICAO (2013) *2013 Environmental Report: Destination green*, Montreal: International Civil Aviation Organization

ICAO (2014) *ICAO Carbon Emissions Measurement Methodology Version 7*, Montreal: International Civil Aviation Organization

ICAO (2016) *On Board: A sustainable future, ICAO environmental report 2016*, Montreal: International Civil Aviation Organization

ICCT (2011) *Reducing Greenhouse Gas Emissions from Ships: Cost-effectiveness of available options*, Washington DC: International Council for Clean Transportation

ICCT (2016) Final US phase 2 heavy-duty vehicle efficiency rule sets standards that are ambitious, far-sighted and achievable (press release 16 August) International Council for Clean Transportation

IEA/UIC (International Energy Agency and International Union of Railways) (2017) *Railway Handbook 2017: Energy consumption and CO_2 emissions*, Paris

IFEU and SGKV (2002) *Comparative Analysis of Energy Consumption and CO_2 Emissions by Road Transport and Combined Transport Road/Rail*, Geneva: IRU and HGL

IGD and ECR UK (2015) *Reducing Wasted Miles*, Letchmore Heath: Institute of Grocery Distribution

IMO (2011) *International Shipping Facts and Figures: Information resources on trade, safety, security and the environment,* International Maritime Organization

IMRG/Blackaby (2016) *IMRG UK Consumer Home Delivery Review 2016*, London: IMRG

Indian National Transport Development Policy Committee (2014) *India Transport Report: Moving India to 2032*, New Delhi and London: Routledge

ING (2017) *3D Printing: A threat to global trade* [online] https://think.ing.com/reports/3d-printing-a-threat-to-global-trade/

Institute of Mechanical Engineers (2017) *Increasing Capacity: Putting Britain's railways back on track*, London: Institute of Mechanical Engineers

Inter-American Development Bank (2015) *Freight Transport and Logistics Statistics Yearbook*, Washington DC: IADB [online] https://publications.iadb.org/handle/11319/6885

International Energy Agency (2007) Fuel Efficiency of HDV's Standards and other Policy Instruments: Towards a plan of action. *Summary and proceedings of a workshop on 21-22 June*, Paris: IEA

International Energy Agency (2009) *Transport, Energy and CO_2: Moving towards sustainability*, Paris: IEA [online] https://www.iea.org/publications/freepublications/publication/transport2009.pdf

International Energy Agency (2013) *Technology Roadmap: Wind energy*, Paris: OECD Publishing

International Energy Agency (2014) *Energy Technology Perspectives 2014*, Paris: OECD Publishing

International Energy Agency (2016) *Energy Technology Perspectives 2016*, Paris: OECD Publishing

International Energy Agency (2017a) *The Future of Trucks: Implications for energy and the environment*, Paris: IEA

International Energy Agency (2017b) *World Energy Statistics 2017*, Paris: OECD Publishing

International Standards Organization (2013) *Greenhouse Gases – the Carbon Footprint of Products: Requirements and guidelines for quantification and communication* (ISO 14067) Geneva: ISO

International Transport Forum (2010) *Moving Freight with Better Trucks*, Paris: OECD Publishing

International Transport Forum (2015a) *Transport Outlook 2015*, Paris: OECD Publishing

International Transport Forum (2015b) *Urban Mobility System Upgrade*, Paris: OECD Publishing

International Transport Forum (2017a) *Managing the Transition to Driverless Road Freight Transport*, Paris: OECD Publishing

International Transport Forum (2017b) *Transport Outlook 2017*, Paris: OECD Publishing

IPCC (2014a) *Climate Change 2014: Synthesis report*, Geneva: Intergovernmental Panel on Climate Change [online] https://www.ipcc.ch/news_and_events/docs/ar5/ar5_syr_headlines_en.pdf

IPCC (2014b) *Fifth Assessment Report: Summary for policy makers*, Geneva: Intergovernmental Panel on Climate Change

Jacobs, K et al (2014) *Horizontal Collaboration in Fresh & Chilled Retail Distribution* (Test Report) Brussels: EU CO3 Project [online] http://www.co3-project.eu/wo3/wp-content/uploads/2011/12/CO3-Deliverable-D4-3-Nestlé-Pepsico-STEF-case-study-Executive-Summary.pdf

Janssen, R et al (2014) *The Impact of 3-D Printing on Supply Chain Management*, Delft: TNO.

Jardine, C N (2009) *Calculating the Carbon Dioxide Emissions of Flights*, Oxford: Environmental Change Institute, Oxford University

Johnson, H et al (2013) Will the ship energy efficiency management plan reduce CO_2 emissions? A comparison with ISO 50001 and the ISM code, *Maritime Policy & Management*, **40** (2), pp 177–90, https://doi.org/10.1080/03088839.2012.757373

Kahn Ribeiro, S et al (2007) Transport and its infrastructure, in *Climate Change 2007: Mitigation. Contribution of working group III to the IPCC fourth assessment report*, Cambridge and New York: Cambridge University Press

Kaya, Y (1990) *Impact of Carbon Dioxide Emission Control on GNP Growth: Interpretation of proposed scenarios* (Energy and Industry Sub-group) Paris: Intergovernmental Panel on Climate Change

Kellner, F (2016) Exploring the impact of traffic congestion on CO_2 emissions in freight distribution networks, *Logistics Research*, **9** (1), p 21, https://doi.org/10.1007/s12159-016-0148-5

Kewill (2013) 3D printing: supply chain game changer or new risk for LSPs? [online] http://www.kewill.com/blog/2014/01/31/3d-printing-supply-chain-game-changer-new-risk-lsps/

Knight, I et al (2008) *Longer and/or Longer and Heavier Goods Vehicles (LHVs): A study of the likely effects if permitted in the UK: final report*, Crowthorne: TRL

Kohn, C (2005) *Centralisation of Distribution Systems and its Environmental Effects*, Sweden: Linkoping University

Kohn, C and Huge-Brodin, M H (2008) Centralised distribution systems and the environment: how increased transport work can decrease the environmental impact of logistics, *International Journal of Logistics Research and Applications*, **11** (3), pp 229–245, https://doi.org/10.1080/13675560701628919

KPMG (2016) *Omnichannel Retail Survey 2016*, London: KPMG

Kristensen, H O H and Hagemeister, C (2011) Environmental performance evaluation of ro-ro passenger ferry transportation, *Selected Proceedings from the Annual Transport Conference at Aalborg University*

Kveiborg, O and Fosgerau, M (2007) Decomposing the decoupling of Danish road freight traffic growth and economic growth, *Transport Policy*, **14** (1), pp 39–48, https://doi.org/10.1016/j.tranpol.2006.07.002

Lai, Y-C, Barkan, C P L and Önal, H (2008) Optimizing the aerodynamic efficiency of intermodal freight trains, *Transportation Research Part E: Logistics and Transportation Review*, **44** (5), pp 820–34, https://doi.org/10.1016/j.tre.2007.05.011

Lammert, M P et al (2014) *Effect of Platooning on Fuel Consumption of Class 8 Vehicles over a Range of Speeds, Following Distances and Mass*, presented at the Commercial Vehicle Engineering Congress, Rosemount, Illinois

Laney, J (2017) *Biomethane in Commercial Vehicles*, presented at the 4th International Workshop on Sustainable Road Freight, Cambridge

LEK Consulting (2017) The uberization of freight – perhaps but it will be a long haul [online] https://www.lek.com/sites/default/files/insights/pdf-attachments/1908_Uberization_for_Freight.pdf

Leonardi, J et al (2014) Increase urban freight efficiency with delivery and servicing plan, *Research in Transportation Business & Management*, **12**, pp 73–79, https://doi.org/10.1016/j.rtbm.2014.10.001

Leonardi, J, Cullinane, S and Edwards, J (2015) Alternative fuels and freight vehicles: status, costs and benefits, and growth, in *Green Logistics: Improving the environmental sustainability of logistics* (3rd edn) ed A McKinnon et al, London: Kogan Page Ltd, pp 278–92

Lettenmeier, M, Liedtke, C and Rohn, H (2014) Eight tons of material footprint: suggestion for a resource cap for household consumption in Finland, *Resources*, **3**(3), pp 488–515, https://doi.org/10.3390/resources3030488

Liimatainen, H et al (2014) Decarbonizing road freight in the future: detailed scenarios of the carbon emissions of Finnish road freight transport in 2030 using a Delphi method approach, *Technological Forecasting and Social Change*, **81** (Supplement C), pp 177–91, https://doi.org/10.1016/j.techfore.2013.03.001

Lindstad, H et al (2015) *GHG Emission Reduction Potential of EU-related Maritime Transport and on its Impacts* (TNO Report R11601) Delft: TNO

Lindstad, H, Asbjørnslett, B E and Jullumstrø, E (2013) Assessment of profit, cost and emissions by varying speed as a function of sea conditions and freight market, *Transportation Research Part D: Transport and Environment*, **19** (Supplement C), pp 5–12, https://doi.org/10.1016/j.trd.2012.11.001

Lindstad, H, Sandaas, I and Steen, S (2014) Assessment of profit, cost, and emissions for slender bulk vessel designs, *Transportation Research Part D: Transport and Environment*, **29**, pp 32–39, https://doi.org/10.1016/j.trd.2014.04.001

Liu, W et al (2014) Minimizing the carbon footprint for the time-dependent heterogeneous-fleet vehicle routing problem with alternative paths, *Sustainability*, **6** (7), pp 4658–84, https://doi.org/10.3390/su6074658

Lloyd's Register (2012) *LNG-Fuelled Deep-Sea Shipping*, London: Lloyd's Register

Logistics Manager (2008) Collaboration brings savings for Kellog's and Kimberly-Clark, *Logistics Manager* (13 October)

Lovell, A, Saw, R and Stimson, J (2005) Product value-density: managing diversity through supply chain segmentation, *The International Journal of Logistics Management*, **16** (1), pp 142–58, https://doi.org/10.1108/09574090510617394

LowCVP (2015) *Low-Emission Van Guide*, London: Low Carbon Vehicle Partnership

Lütken, S et al, (2011) *Low Carbon Development Strategies: A primer on framing nationally appropriate mitigation actions (NAMAs) in developing countries*, Roskilde: UNEP Riso Centre

Lynas, M (2007a) Can Shopping Save the Planet? *Guardian*, 17 September

Lynas, M (2007b) *Six Degrees: Our future on a hotter planet*, London: Harper Perennial

Maersk (2014) *Emissions,* Copenhagen: 2014 [online] https://www.maersk.com/explore/fleet/triple-e/the-hard-facts/emissions

Maister, D H (1976) Centralisation of inventories and the 'Square Root Law', *International Journal of Physical Distribution*, **6** (3), pp 124–34, https://doi.org/10.1108/eb014366

Maloni, M, Paul, J A and Gligor, D M (2013) Slow steaming impacts on ocean carriers and shippers, *Maritime Economics & Logistics*, **15** (2), pp 151–71, https://doi.org/10.1057/mel.2013.2

Mander, S (2017) Slow steaming and a new dawn for wind propulsion: a multi-level analysis of two low carbon shipping transitions, *Marine Policy*, **75**, pp 210–16, https://doi.org/10.1016/j.marpol.2016.03.018

McCann, P (1998) *The Economics of Industrial Location: A logistics-costs approach*, Springer Science & Business Media

McGill, R, Remley, W and Winther, K (2013) *Alternative Fuels for Marine Applications*, Paris: IEA Advanced Motor Fuels Organization

McKinnon, A C (1996) The empty running and return loading of road goods vehicles, *Transport Logistics*, **1**(1), pp 1–19, https://doi.org/10.1163/156857096300150518

McKinnon, A C (1998) The abolition of quantitative controls on road freight transport: the end of an era? *Transport Logistics*, **1** (3), pp 211–24

McKinnon, A C (1989a) The growth of road freight in the UK, *International Journal of Physical Distribution & Materials Management*, **19** (4), pp 3–13, https://doi.org/10.1108/EUM0000000000312

McKinnon, A C (1989b) *Physical Distribution Systems*, London: Routledge

McKinnon, A C (2003) Influencing company logistics management, in *Managing the Fundamental Drivers of Transport Demand*, Paris: European Conference of Ministers of Transport/OECD (pp 63–71)

McKinnon, A C (2005) The economic and environmental benefits of increasing maximum truck weight: the British experience, *Transportation Research Part D: Transport and Environment*, **10** (1), pp 77–95, https://doi.org/10.1016/j.trd.2004.09.006

McKinnon, A C (2007a) *CO_2 Emissions from Freight Transport in the UK*, London: Department for Transport

McKinnon, A C (2007b) Decoupling of road freight transport and economic growth trends in the UK: an exploratory analysis, *Transport Reviews*, **27** (1), pp 37–64, https://doi.org/10.1080/01441640600825952

McKinnon, A C (2007c) Increasing fuel prices and market distortion in a domestic road haulage market: the case of the UK [online] https://www.openstarts.units.it/handle/10077/5937

McKinnon, A C (2009) The present and future land requirements of logistical activities, *Land Use Policy*, **26**, S293–S301, https://doi.org/10.1016/j.landusepol.2009.08.014

McKinnon, A C (2010a) *European Freight Transport Statistics: Limitations, misconceptions and aspirations*, Brussels: ACEA

McKinnon, A C (2010b) *Britain Without Double-deck Lorries: An assessment of the effects on traffic levels, road haulage costs, fuel consumption and CO_2 emissions*, Logistics Research Centre, Heriot-Watt University

McKinnon, A C (2010c) Product-level carbon auditing of supply chains: environmental imperative or wasteful distraction? *International Journal of Physical Distribution & Logistics Management*, **40** (1/2), pp 42–60, https://doi.org/10.1108/09600031011018037

McKinnon, A C (2012a) Measuring carbon emissions from supply chains: getting the level right, in *Outlook on the Logistics and Supply Chain Industry 2012*, Geneva: World Economic Forum (pp 20–21)

McKinnon, A C (2012b) *An Assessment of the Contribution of Shippers to the Decarbonisation of Deep-Sea Container Supply Chains*, presented at the 2012 Low Carbon Shipping Conference, University of Newcastle-upon-Tyne, 12 September 2012 [online] http://www.alanmckinnon.co.uk/story_layout.html?IDX=580&s=y

McKinnon, A C (2014a) Optimising the movement of freight by road, in *Global Logistics: New directions in supply chain management*, ed D Waters and S Risler (7th edn), London, Kogan Page Ltd (pp 282–99)

McKinnon, A C (2014b) *Options for Reducing Logistics-Related Emissions from Global Value Chains,* Florence: European University Institute [online] http://papers.ssrn.com/sol3/papers.cfm?abstract_id=2422406

McKinnon, A C (2014c) The possible influence of the shipper on carbon emissions from deep-sea container supply chains: an empirical analysis, *Maritime Economics & Logistics*, **16** (1), pp 1–19, https://doi.org/10.1057/mel.2013.25

McKinnon, A C (2015) Opportunities for improving vehicle utilisation, in *Green Logistics: Improving the environmental sustainability of logistics*, ed A McKinnon et al (3rd edn), London: Kogan Page Ltd

McKinnon, A C (2016a) Freight transport deceleration: its possible contribution to the decarbonisation of logistics, *Transport Reviews*, **36**, (4), pp 418–36, https://doi.org/10.1080/01441647.2015.1137992

McKinnon, A C (2016b) The possible impact of 3D printing and drones on last-mile logistics: an exploratory study, *Built Environment*, **42** (4), pp 617–29, https://doi.org/10.2148/benv.42.4.617

McKinnon, A C (2016c) Freight transport in a low carbon world, *Transportation Research News*, **306**, November–December 2016, pp 8–15

McKinnon, A C et al (2009) Traffic congestion, reliability and logistical performance: a multi-sectoral assessment, *International Journal of Logistics Research and Applications*, **12** (5), pp 331–45, https://doi.org/10.1080/13675560903181519

McKinnon, A C et al (2017) *Logistics Competences, Skills and Training: A global overview*, Washington DC: World Bank

McKinnon, A C, Allen, J and Woodburn, A (2015) Development of green vehicles, aircraft and ships, in *Green Logistics: Improving the environmental sustainability of logistics*, ed A McKinnon et al (3rd edn) London: Kogan Page Ltd

McKinnon, A C and Ge, Y (2004) Use of a synchronised vehicle audit to determine opportunities for improving transport efficiency in a supply chain, *International Journal of Logistics Research and Applications*, **7** (3), pp 219–38, https://doi.org/10.1080/13675560412331298473

McKinnon, A C and Ge, Y (2006) The potential for reducing empty running by trucks: a retrospective analysis, *International Journal of Physical Distribution & Logistics Management*, **36** (5), pp 391–410, https://doi.org/10.1108/09600030610676268

McKinnon, A C and Piecyk, M (2009) Logistics 2050: moving goods by road in a very low carbon world, in *Supply Chain Management in a Volatile World*, ed E Sweeney, Dublin: Blackrock

McKinnon, A C and Piecyk, M (2010) *Measuring and Managing Carbon Emissions from European Chemical Transport*, Brussels: European Chemical Industry Council (CEFIC)

McKinnon, A C and Piecyk, M I (2012) Setting targets for reducing carbon emissions from logistics: current practice and guiding principles, *Carbon Management*, **3** (6), pp 629–39, https://doi.org/10.4155/cmt.12.62

McKinnon, A C and Woodburn, A (1994) The consolidation of retail deliveries: its effect on CO_2 emissions, *Transport Policy*, **1** (2), pp 125–36, https://doi.org/10.1016/0967-070X(94)90021-3

McKinnon, A C and Woodburn, A (1996) Logistical restructuring and road freight traffic growth, *Transportation*, **23** (2), pp 141–61, https://doi.org/10.1007/BF00170033

McKinsey & Co (2009) *Pathways to a Low Carbon Economy: Version 2 of the global greenhouse gas abatement cost curve*, McKinsey

McKinsey & Co (2010) *Building India: Transforming the nation's logistics infrastructure*, New Delhi: McKinsey

Meinshausen, M et al (2009) Greenhouse-gas emission targets for limiting global warming to 2°C, *Nature*, **458** (7242), p 1158, https://doi.org/10.1038/nature08017

Merk, O (2014) *Shipping Emissions in Ports*, International Transport Forum Discussion Paper 2014/20, Paris: OECD Publishing

Mervis, J (2014) The information highway gets physical, *Science*, **344** (6188), pp 1104–07, https://doi.org/10.1126/science.344.6188.1104

Meszler, D, Lutsey, N and Delgado, O (2015) *Cost Effectiveness of Advanced Efficiency Technologies for Long-haul Tractor-trailers in the 2020–2030 Time Frame*, Washington DC: International Council for Clean Transportation

Met Office (2018) Five year forecast indicates further warming [online] https://www.metoffice.gov.uk/news/releases/2018/decadal-forecast-2018

Milliot, J (2016) As e-book sales decline digital fatigue grows, *Publishers Weekly* (17 June) [online] https://www.publishersweekly.com/pw/by-topic/digital/retailing/article/70696-as-e-book-sales-decline-digital-fatigue-grows.html

Ministry of Infrastructure and Environment (2016) *European Truck Platooning Challenge 2016*, Hague: Dutch Ministry of Infrastructure and Environment

Mofor, L, Nuttall, P and Newell, A (2015) *Renewable Energy Options for Shipping: Technology brief*, International Renewable Energy Association

Monbiot, G (2017) The car has a chokehold on Britain. It's time to free ourselves, *Guardian* (1 August)

Monios, J (2015) Integrating intermodal transport with logistics: a case study of the UK retail sector, *Transportation Planning and Technology*, **38**, (3), pp 347–74

Montreuil, B (2011) Toward a Physical Internet: meeting the global logistics sustainability grand challenge, *Logistics Research*, **3** (2–3), pp 71–87, https://doi.org/10.1007/s12159-011-0045-x

Mooney, D C and Dennis, B (2016) The world is about to install 700 million air conditioners. Here's what that means for the climate, *Washington Post* (31 May)

Morash, E A, Dröge, C and Vickery, S (1996) Boundary spanning interfaces between logistics, production, marketing and new product development, *International Journal of Physical Distribution & Logistics Management*, **26** (8), pp 43–62, https://doi.org/10.1108/09600039610128267

Morgan, J (2014) Putting heavy-duty truck aerodynamics to the test, *Fleet Equipment* (14 January) [online] http://www.fleetequipmentmag.com/putting-heavy-duty-truck-aerodynamics-test

Morganti, E and Browne, M (2018) Technical and operational obstacles to the adoption of electric vans in France and the UK: an operator perspective, *Transport Policy*, **63**, pp 90–97, https://doi.org/10.1016/j.tranpol.2017.12.010

Moskowitz, R (2008) *Consumer Energy Alliance: Trucking industry update*, American Trucking Associations [online] http://consumerenergyalliance.org/cms/wp-content/uploads/2008/12/trucking-industry-update_ata_11dec08.pdf

Mottschall, M (2016) *Decarbonisation of Road Transport*, presented at the Presentation of Oko Institute (Institute of Applied Energy) Berlin, webinar presentation (8 June)

Muncrief, R and Sharpe, B (2015) *Overview of the Heavy-Duty Vehicle Market and CO_2 Emissions in the European Union* (Working Paper 2015-6) Berlin: International Council for Clean Transportation

Muylaert, K and Stoffers, L (2014) *Driving Sustainability through Horizontal Supply Chain Collaboration: An example from P&G and Tupperware*, presented at the EU CO3 Project presentation, Brussels

MyGrid (2017) *Electricity Data: Last 12 months* [online] http://www.mygridgb.co.uk/last-12-months/

NACFE (2011) *Executive Report: Speed limiters*, North American Council for Freight Efficiency

NACFE (2014) *Confidence Report: Idle reduction solutions*, North American Council for Freight Efficiency

NACFE (2015a) *Confidence Report: Maintenance*, North American Council for Freight Efficiency [online] www.truckingefficiency.org

NACFE (2015b) *Trucking Efficiency Confidence Report: Downspeeding*, North American Council for Freight Efficiency [online] http://www.truckingefficiency.org/sites/truckingefficiency.org/files/reports/RMICWR__Trucking_Downspeeding_ExecSummary_10292015.pdf

NACFE (2016) *Annual Fleet Fuel Study 2016*, North American Council for Freight Efficiency

Nahlik, M J et al (2016) Goods movement life cycle assessment for greenhouse gas reduction goals, *Journal of Industrial Ecology*, **20** (2), pp 317–28, https://doi.org/10.1111/jiec.12277

NASA (2017) Arctic sea ice, *NASA Earth Observatory* [online] https://earthobservatory.nasa.gov/Features/SeaIce/page3.php

National Bureau of Statistics of China (2016) *China Statistical Yearbook 2016*, Beijing: National Bureau of Statistics of China

National Infrastructure Commission (2017) *Congestion, Capacity and Carbon: Priorities for national infrastructure*, London: National Infrastructure Commission

NEA (2011) *Longer and Heavier Vehicles in Practice: Economic, logistical and social effects*, The Hague: Directorate General for Public Works and Water Management (Rijkswaterstaat)

NESCCAF et al (2009) *Reducing-Heavy Duty Long Haul Combination Truck Fuel Consumption and CO_2 Emissions*, Boston/Washington

Neslan, A (2016) Delay to curbs on toxic shipping emissions 'would cause 200,000 extra premature deaths', *Guardian* (7 October)

Netherlands Economic Institute (1997) *Relationship between Demand for Freight Transport and Industrial Effects (REDEFINE)* (EU project deliverable) Rotterdam: NEI

Network Rail (2009) *Network RUS: Electrification*, London: Network Rail

Network Rail (2013) *Long-Term Planning Process: Freight market study*, London: Network Rail

Network Rail (2017) *Freight Network Study: Long-term planning process*, London: Network Rail

Nicolaides, D, Cebon, D and Miles, J (2017) Prospects for electrification of road freight, *IEEE Systems Journal*, **Pp** (99), pp 1–12, https://doi.org/10.1109/JSYST.2017.2691408

NOAA (2017) *Ballast Water: A pathway for aquatic invasive species*, National Oceanic and Atmospheric Administration

Notteboom, T and Cariou, P (2013) Slow steaming in container liner shipping: is there any impact on fuel surcharge practices? *The International Journal of Logistics Management*, **24** (1), pp 73–86, https://doi.org/10.1108/IJLM-05-2013-0055

NTM (2017) Fuels and power supply data, *Network for Transport Measurement* [online] https://www.transportmeasures.org/en/methods-and-data/fuels-and-power-supply-data//

OBB et al (2014) *Energy Efficient Rolling Stock: Is dual hybrid the future?* Presented at the UIC Energy Efficiency Days, 16–29 June, Antwerp [online] http://www.energy-efficiency-days.org/spip.php?article19

Odhams, A M C et al (2011) Active steering of a tractor–semi-trailer, *Proceedings of the Institution of Mechanical Engineers, Part D: Journal of Automobile Engineering*, **225** (7), pp 847–69, https://doi.org/10.1177/0954407010395680

Oeser, G and Romano, P (2016) An empirical examination of the assumptions of the Square Root Law for inventory centralisation and decentralisation, *International Journal of Production Research*, **54** (8), pp 2298–319, https://doi.org/10.1080/00207543.2015.1071895

Office of National Statistics (2017a) Retail Sales: Great Britain November 2017 [online] https://www.ons.gov.uk/businessindustryandtrade/retailindustry/bulletins/retailsales/november2017

Office of National Statistics (2017b) *UK Index of Services Statistical Bulletins*, London: Office of National Statistics

Office of Rail and Road (2017) *Rail Infrastructure, Asset and Environmental 2016–17 Annual Statistical Release*, London: Office of Rail and Road

Olmer, N et al (2017) *Greenhouse Gas Emissions from Global Shipping: 2013–2015*, Washington DC: International Council for Clean Transportation

Orcutt, M (2016) Airplanes are getting lighter thanks to 3D printed parts, *MIT Technology Review* (18 April)

Oxford Business Group (2011) *The Report: Morocco 2011* [online] http://bit.ly/2EdlOnR

Palmer, A (2007) *The Development of an Integrated Routeing and Carbon Dioxide Model for Goods Vehicles* (PhD Thesis), Bedford: Cranfield University

Palmer, A, Dadhich, P and Greening, P (2016) *Reconfiguring Logistics Networks (STARFISH II) Final Report*, Edinburgh: Centre for Sustainable Road Freight

Palmer, A and McKinnon, A (2011) An analysis of the opportunities for improving transport efficiency through multi-lateral collaboration in FMCG supply chains, in *Proceedings of the Logistics Research Network Annual Conference*, University of Southampton

Palmer, A and Piecyk, M (2010) Time, cost and CO_2 effects of rescheduling freight deliveries, in *Proceedings of the Annual Conference of the Logistics Research Network*, ed A Whiteing

Palmer, A et al (2012) *Characteristics of Collaborative Business Models*, Brussels: EU CO3 Project [online] http://www.elupeg.com/downloads/2395%5E

Pan, S, Ballot, E and Fontane, F (2013) The reduction of greenhouse gas emissions from freight transport by pooling supply chains, *International Journal of Production Economics*, **143** (1), pp 86–94, https://doi.org/10.1016/j.ijpe.2010.10.023

PCF Project Germany (2009) *Product Carbon Footprinting: The right way to promote low-carbon products and consumption habits*, Berlin: Thema 1 GmbH

Peeters, P (2017) *Climate and Aviation: The fundamental issues*, presented at the Presentation to EU High-Level Panel on European Decarbonisation Pathways Initiative, Brussels (2 October)

Peters, G P et al (2011) Growth in emission transfers via international trade from 1990 to 2008, *Proceedings of the National Academy of Sciences*, **108** (21), pp 8903–08, https://doi.org/10.1073/pnas.1006388108

PhysOrg (2016) Lightweight materials provide opportunities for the next generation of railway vehicles, *PhysOrg* (16 March) [online] https://phys.org/news/2016-03-lightweight-materials-opportunities-railway-vehicles.html

Piecyk, M (2015) Carbon auditing of companies, supply chains and products, in *Green Logistics: Improving the environmental sustainability of logistics*, ed A McKinnon et al (3rd edn), London: Kogan Page Ltd (pp 55–79)

Piecyk, M I and McKinnon, A C (2010) Forecasting the carbon footprint of road freight transport in 2020, *International Journal of Production Economics*, **128** (1), pp 31–42, https://doi.org/10.1016/j.ijpe.2009.08.027

Piecyk, M I and McKinnon, A C (2013) Application of the Delphi Method to the forecasting of long-term trends in road freight, logistics and related CO_2 emissions, *International Journal of Transport Economics*, **40** (2)

Pinard, I (2010) *Guidelines for Vehicle Overload Control in Eastern and Southern Africa*, International Bank for Reconstruction and Development/World Bank

Planet Ark (2017) Steel recycling factsheet, *Planet Ark Environmental Foundation* [online] http://recyclingweek.planetark.org/documents/doc-186-steel-factsheet.pdf

Potter, A, Mason, R and Lalwani, C (2007) Analysis of factory gate pricing in the UK grocery supply chain, *International Journal of Retail & Distribution Management*, **35** (10), pp 821–34, https://doi.org/10.1108/09590550710820694

Punte, S and Bollee, F (2017) *Smart Freight Leadership: A journey to a more efficient and environmentally sustainable global freight system*, Amsterdam: Smart Freight Centre

PwC (2013) *Going Beyond Reshoring to Right-Shoring* [online] http://www.inprojects.ru/assets/files/going-beyond-reshoring.pdf

PwC (2014) *The Future of 3D Printing: Moving beyond prototyping to finished products* [online] http://www.pwc.com/en_US/us/technology-forecast/2014/3d-printing/features/assets/pwc-3d-printing-full-series.pdf

PwC (2016) *Global Forest, Paper and Packaging Industry Survey 2016 Edition: Survey of 2015 Results*, Price Waterhouse Coopers LLP

Pye, S et al (2015) *Pathways to Deep Decarbonization in the UK*, Paris: Sustainable Development Solutions Network/ Institute for Sustainable Development and International Relations/UCL Energy Institute

Quanlin, Q (2017) Fully electric cargo ship launched in Guangzhou, *China Daily*

Rabobank International (2012) *The Incredible Bulk: The rise of the global bulk wine trade* (No. Rabobank Industry Note #296), Utrecht

Rai, D et al (2011) Assessment of CO_2 emissions reduction in a distribution warehouse, *Energy*, **36** (4), pp 2271–77, https://doi.org/10.1016/j.energy.2010.05.006

Rail Delivery Group (2014) *Keeping the Lights on and the Traffic Moving* [online] https://www.raildeliverygroup.com/files/Publications/archive/2014-05_keeping_the_lights_on.pdf

Rail Technology Magazine (2008) How can the railway industry become more efficient as sustainability and CO_2 standards rise? *Rail Technology Magazine*, (16 July)

Railway Technology (2013) Ricardo and E showcase fuel saving potential of Class 70 locomotives in UK, *Railway Technology* (24 January).

Rantasila, K and Ojala, L (2012) *Measurement of National-level Logistics Costs and Performance* (Working Paper No. 2012–4) International Transport Forum Discussion Paper, https://doi.org/10.1787/5k8zvv79pzkk-en

Raunek (2013) How air lubrication system for ships works? *Green Shipping* [online] https://www.marineinsight.com/green-shipping/how-air-lubrication-system-for-ships-work/

Reidy, S (2017) Eco-responsible rail freight, *Arviem* [online] http://arviem.com/eco-responsible-rail-freight/

Ricardo (2016) *The Role of Natural Gas and Biomethane in the Transport Sector* (Report for Transport and Environment (T&E)), Brussels

Ricardo et al (2014) *Heavy Vehicle Platoons on UK Roads*, London: Department of Transport.

Ricardo-AEA (2012) *Opportunities to Overcome the Barriers to Uptake of Low-Emission Technologies for Each Commercial Vehicle Duty Cycle*, London: Low Carbon Vehicle Partnership

Ricardo-AEA (2014) *Update of the Handbook on External Costs of Transport*, Brussels: Report for DG Move European Commission

Ricardo-AEA (2015) *Light Weighting as a Means of Improving Heavy Duty Vehicles' Energy Efficiency and Overall CO_2 Emissions*, Brussels: EU DG Clima

Ries, J M, Grosse, E H and Fichtinger, J (2017) Environmental impact of warehousing: a scenario analysis for the United States, *International Journal of Production Research*, **55** (21), pp 6485–99, https://doi.org/10.1080/00207543.2016.1211342

RightShip (2013) *Calculating and Comparing CO_2 Emissions from the Global Maritime Fleet*, RightShip

Risk Solutions (2017) *The GB Longer Semi-Trailer Trial: 2016 annual summary report*, London: Department for Transport

Rivoli, P (2005) *The Travels of T-shirt in the Global Economy*, Hoboken, NJ: John Wiley and Sons

Rodrigue, J-P, Comtois, C and Slack, B (2017) *The Geography of Transport Systems* (4th edn), New York: Routledge

Rodriguez, F et al (2017) *Market Penetration of Fuel-efficiency Technologies for Heavy-Duty Vehicles in the Europea Union, the United States and China*, Washington DC: International Council for Clean Transportation

Roland Berger Consultants (2016) *Automated Trucks: The next big disruptor in the automotive industry*, Chicago and Munich: Roland Berger

Rolko, K and Friedrich, H (2017) Locations of logistics service providers in Germany: the basis for a new freight transport generation model, *Transportation Research Procedia*, **25** (Supplement C), pp 1061–74, https://doi.org/10.1016/j.trpro.2017.05.479

RoSPA (2010) *Increasing the Uptake of Eco-driving Training for Drivers of Large Goods Vehicles and Passenger Carrying Vehicles*, London: Royal Society for the Prevention of Accidents

Royal Academy of Engineering (2013) *Future Ship Powering Options: Exploring alternative methods of ship propulsion*, London: Royal Academy of Engineering

Rüdiger, D, Schön, A and Dobers, K (2016) Managing greenhouse gas emissions from warehousing and transshipment with environmental performance indicators, *Transportation Research Procedia*, **14** (Supplement C), pp 886–95, https://doi.org/10.1016/j.trpro.2016.05.083

Rumpke, C (2010) *Natural Gas and Biomethane in the Future Fuel Mix*, Berlin: DENA, German Energy Agency

Saenz, M J (2016) The Physical Internet: logistics re-imagined, *Supply Chain Management Review* (23 March) [online] http://www.scmr.com/article/the_physical_internet_logistics_reimagined

Saleh, K (2017) Global online retail spending: statistics and trends, *Invespcro* [online] https://www.invespcro.com/blog/global-online-retail-spending-statistics-and-trends/

Sallez, Y et al (2016) On the activeness of intelligent Physical Internet containers, *Computers in Industry*, **81**, pp 96–104, https://doi.org/10.1016/j.compind.2015.12.006

Samuelsson, A and Tilanus, B (1997) A framework efficiency model for goods transportation, with an application to regional less-than-truckload distribution *Transport Logistics*, **1** (2), pp 139–51, https://doi.org/10.1163/156857097300151660

Sánchez-Díaz, I, Georén, P and Brolinson, M (2017) Shifting urban freight deliveries to the off-peak hours: a review of theory and practice, *Transport Reviews*, **37** (4), pp 521–43, https://doi.org/10.1080/01441647.2016.1254691

Sanchez-Rodrigues, V S, Harris, I and Mason, R (2015) Horizontal logistics collaboration for enhanced supply chain performance: an international retail perspective, *Supply Chain Management: An International Journal*, **20** (6), pp 631–47, https://doi.org/10.1108/SCM-06-2015-0218

Sanchez-Rodrigues, V, Potter, A and Naim, M M (2010) The impact of logistics uncertainty on sustainable transport operations, *International Journal of Physical Distribution & Logistics Management*, **40** (1/2), pp 61–83, https://doi.org/10.1108/09600031011018046

Sanders, U et al (2015) Think outside your boxes: solving the global repositioning puzzle, *Bcg Perspectives* [online] http://img-stg.bcg.com/BCG-Think-Outside-Your-Boxes-Nov-2015_tcm9-88599.pdf

Santén, V (2016) *Towards Environmentally Sustainable Freight Transport: Shippers' logistics actions to improveload factor performance* (Doctoral Thesis), Chalmers University of Technology [online] http://publications.lib.chalmers.se/publication/243257-towards-environmentally-sustainable-freight-transport-shippers-logistics-actions-to-improve-load-fac

Saunders, C and Barber, A (2007) *Comparative Energy and Greenhouse Gas Emissions of New Zealand's and the UK's Dairy Industry* (No. Research Report 297), Lincoln: Lincoln University

Schellnhuber, H J (2009) Tipping elements in the Earth System, *Proceedings of the National Academy of Sciences*, **106** (49), pp 20561–63, https://doi.org/10.1073/pnas.0911106106

Schipper, L and Marie, C (1999) *Transport and CO_2 Emissions*, Washington DC: World Bank

Schneider, F, Kallis, G and Martinez-Alier, J (2010) Crisis or opportunity? Economic degrowth for social equity and ecological sustainability: introduction to this special issue, *Journal of Cleaner Production*, **18** (6), pp 511–18, https://doi.org/10.1016/j.jclepro.2010.01.014

Schroten, A, Warringa, G and Bles, M (2012) *Marginal Abatement Cost Curves for Heavy-Duty Vehicles*, Delft: CE Delft

Schuckert, M (2016) *Daimler's Advances in Fuel Efficiency and Zero Emission Activities*, Presentation to the joint IEA/EU JRC workshop on 'The Future Role of Trucks for Energy and Environment', Brussels, 8 November [online] https://www.iea.org/workshops/the-future-role-of-trucks-for-energy-and-environment.html

SCI Verkehr (2012) Electric loco market primed for growth, *International Railway Journal* (29 February)

Scripps Observatory (2018) The Keeling Curve [online] https://scripps.ucsd.edu/programs/keelingcurve/

Seabury Consulting (2017) *The Future of Sea Freight and its Impact on the Air Cargo Industry*, presented at the CNS Partnership Conference, Orlando

Searle, S (2017) *How Rapeseed and Soy Diesel Drive Palm Oil Expansion.* Washington DC: International Council for Clean Transportation

Shabani, B and Andrews, J (2015) Hydrogen and fuel cells, in *Energy Sustainability Through Green Energy,* Springer, New Delhi (pp 453–491), https://doi.org/10.1007/978-81-322-2337-5_17

Shah, V P, Debella, D C and Ries, R J (2008) Life cycle assessment of residential heating and cooling systems in four regions in the United States, *Energy and Buildings*, 40 (4), pp 503–13, https://doi.org/10.1016/j.enbuild.2007.04.004

Sharman, J (2017) Construction waste and sustainability, *Sustainability, Health and Safety* (25 January)

Sharpe, B and Muncrief, R (2015) *Literature Review: Real-world fuel consumption of heavy-duty vehicles in the United States, China and the European Union* (White Paper) Berlin: International Council for Clean Transportation

Shumaker, C and Serfass, J (2017) *Hydrogen and Fuel Cell On-Road Freight Workshop Report*, Long Beach: California Hydrogen Business Council

Simon, D and Mason, R (2003) Lean and green: doing more with less, *ECR Journal*, 3 (1), pp 84–91

Sims, R et al (2014) *Transport* (*Climate Change 2014: Mitigation of climate change*, Contribution of Working Group III to the Fifth Assessment Report of the Intergovernmental Panel on Climate Change), Cambridge and New York: Cambridge University Press

Sirkin, H, Rose, J and Zinser, M (2012) *The US Manufacturing Renaissance: How shifting global economics are creating an American comeback*, Boston Consulting Group/The Wharton School

Sisario, B and Russell, K (2016) In shift to streaming, music industry has lost billions, *New York Times* (24 May)

Skanska AB (2011) *Project Carbon Footprinting*, Skanska [online] https://group.skanska.com/4a22a4/globalassets/sustainability/green/our-journey-to-deep-green/how-we-define-green/projectcarbonfootprinting.pdf

Sloman, L, Hopkinson, L and Taylor, I (2017) *The Impact of Road Projects in England*, London: Council for the Protection of Rural England

Smart Freight Centre (2017) Smart Freight Solutions Map [online] http://www.smartfreightcentre.org/smart-freight-solutions-map

SmartWay (2015) *SmartWay Tips: Trends, indicators and partner statistics*, US Environmental Protection Agency [online] https://www.epa.gov/sites/production/files/2016-05/documents/smartway_tips_trends_indicators_partner_stats.pdf

Smith, A et al (2005) *The Validity of Food Miles as an Indicator of Sustainable Development: Final report*, London: Department of the Environment, Food and Rural Affairs

Smith, T et al (2014) *Reduction of GHG Emissions from Ships: Third IMO GHG study 2014 – final report*, London: International Maritime Organisation

Smith, T et al (2016) *CO_2 Emissions from International Shipping: Possible reduction targets and associated pathways*, UMAS: London

Smithers, R (2010) Tesco's pledge to carbon label all products set to take centuries, *Guardian*, (13 October)

SMMT (2017) *Vehicle Registration Data*, Society of Motor Manufacturers and Traders [online] www.smmt.co.uk

Smokers, R et al (2017) *Decarbonising Commercial Road Transport*, Delft: TNO

Sripad, S and Viswanathan, V (2017) Performance metrics required of next generation batteries to make a practical electric semi-truck, *ACS Energy Letters*, **2**, pp 1669–73

Standing Advisory Committee on Trunk Road Assessment (1994) *Trunk Roads and the Generation of Traffic*, London: Department for Transport

SteadieSeifi, M et al (2014) Multimodal freight transportation planning: a literature review, *European Journal of Operational Research*, **233** (1), pp 1–15, https://doi.org/10.1016/j.ejor.2013.06.055

Steer Davies Gleave Ltd (2013) *A Review of Megatrucks*, Brussels: Transport and Tourism Committee of the European Parliament

Steer Davies Gleave Ltd (2015) *Freight on Road: Why EU shippers prefer truck to train*, Brussels: Report for the Transport and Tourism Committee of the European Parliament

Stern, N (2015) *Why Are We Waiting: The logic, urgency and promise of tackling climate change*, Cambridge, MA: MIT Press

Sternberg, H and Norrman, A (2017) The Physical Internet: review, analysis and future research agenda, *International Journal of Physical Distribution & Logistics Management*, **47** (8), pp 736–62, https://doi.org/10.1108/IJPDLM-12-2016-0353

Stettler, M E J et al (2016) Greenhouse gas and noxious emissions from dual fuel diesel and natural gas heavy-goods vehicles, *Environmental Science & Technology*, **50** (4), pp 2018–26, https://doi.org/10.1021/acs.est.5b04240

Stodolski, F (2002) *Railroad and Locomotive Technology Roadmap*, Chicago: Center for Transportation Technology, Argonne National Laboratory

Strauss, B H, Kulp, S and Levermann, A (2015) *Mapping Choices: Carbon, climate and rising seas – our global legacy*, Princeton, NJ: Climate Central

Subbarao, S and Lloyd, B (2011) Can the Clean Development Mechanism (CDM) deliver? *Energy Policy*, **39** (3), pp 1600–11, https://doi.org/10.1016/j.enpol.2010.12.036

Tapio, P (2005) Towards a theory of decoupling: degrees of decoupling in the EU and the case of road traffic in Finland between 1970 and 2001, *Transport Policy*, **12** (2), pp 137–51, https://doi.org/10.1016/j.tranpol.2005.01.001

Taptich, M N and Horvath, A (2015) Freight on a low-carbon diet: accessibility, freightsheds, and commodities, *Environmental Science & Technology*, **49** (19), pp 11321–28, https://doi.org/10.1021/acs.est.5b01697

Tavasszy, L (2017) *Decarbonization and Synchromodality*, presented at the ALICE Collaborative Innovation Day, 31 March, Brussels

Tavasszy, L, Behdani, B and Konings, R (2015) Intermodality and synchromodality, *SSRN* [online] https://ssrn.com/abstract=2592888

Tavasszy, L and van Meijeren, J (2011) *Modal Shift Target for Freight Traffic above 300 km: An assessment*, Brussels: ACEA

TCI, and IIM (2012) *Operational Efficiency of National Highways for Freight Transportation in India*, New Delhi: Transport Corporation of India and Indian Institute of Management

T&E (2017) *Roadmap to Climate-Friendly Land Freight and Buses in Europe*, Brussels: European Federation for Transport and Environment

The White House (2014) *Improving the Fuel Efficiency of American Trucks*, Washington DC

Thompson, L (2010) *A Vision for the Railways in 2050* (Forum Paper 4), Paris: International Transport Forum

Tian, Y et al (2014) Analysis of greenhouse gas emissions of freight transport sector in China, *Journal of Transport Geography*, **40** (Supplement C), pp 43–52 https://doi.org/10.1016/j.jtrangeo.2014.05.003

Tong, F, Jaramillo, P and Azevedo, I M L (2015) Comparison of life cycle greenhouse gases from natural gas pathways for medium and heavy-duty vehicles, *Environmental Science & Technology*, **49** (12), pp 7123–33, https://doi.org/10.1021/es5052759

TransDEK (2015) *Double Deck Trailers: Custom designed and built to maximise cubic capacity*, TransDEK Ltd

Transport and Mobility Leuven (2017) *Commercial Vehicle of the Future: A roadmap towards fully sustainable truck operation*, Geneva: International Road Transport Union (IRU)

Transport and Mobility Leuven et al (2008) *Effects of Adapting the Rules on Weights and Dimensions of Heavy Commercial Vehicles as Established with Directive 96/53/EC*, Brussels: European Commission

Transport for London (2008) *London Construction Consolidation Centre*, London

Transport for London (2014) *Five Reasons to Go for FORS*, Transport for London

Transport Research Support (2009) *Freight Transport for Development Toolkit: Air freight*, Washington DC: World Bank/DFID

UIC (International Union of Railways) (2016) *Technologies and Potential Developments for Energy Efficiency and CO_2 Emissions in Rail Systems*, Paris: UIC

UIC (International Union of Railways) (2017) *Railway Statistics: 2016 Synopsis*, Paris: UIC

UK Warehousing Association (2010) *Save Energy, Cut Costs: Energy efficient warehouse operation*, UKWA, London

UNCTAD (2017) *Review of Maritime Transport 2017*, Geneva: United Nations

UNEP (2016) *Emissions Gap Report*, Nairobi: United Nations Environment Programme

UNFCCC (2017) *Global Warming Potentials*, United Nations Framework Convention on Climate Change [online] http://unfccc.int/ghg_data/items/3825.php

UN Statistics Division (2017) *Sustainable Development Goals Indicators: Freight transport*, Washington DC [online] http://data.un.org/Search.aspx?q=Freight transport

Upham, P, Dendler, L and Bleda, M (2011) Carbon labelling of grocery products: public perceptions and potential emissions reductions, *Journal of Cleaner Production*, **19** (4), pp 348–55, https://doi.org/10.1016/j.jclepro.2010.05.014

UPS (2016a) *Committed to More: 2015 corporate sustainability report*, Atlanta: UPS

UPS (2016b) ORION: the algorithm proving that left isn't right, *UPS Compass* (October) [online] https://compass.ups.com/ups-fleet-telematics-system/

US Bureau of Transportation Statistics (2017) *Freight Facts & Figures 2017*, Washington DC

van Essen, H and Martino, A (2011) *Potential of Modal Shift to Rail Transport*, presented at the CE Delft and TRT presentation [online] http://www.cedelft.eu/assets/upload/file/Presentaties/2011/Presentation_PotentialOfModalShift24June2011_4255.pdf

van Lier, T and Macharis, C (2014) Assessing the environmental impact of inland waterway transport using a life-cycle assessment approach: the case of Flanders, *Research in Transportation Business & Management*, **12**, pp 29–40, https://doi.org/10.1016/j.rtbm.2014.08.003

van Loon, P et al (2014) The growth of online retailing: a review of its carbon impacts, *Carbon Management*, **5** (3), pp 285–92, https://doi.org/10.1080/17583004.2014.982395

van Loon, P et al (2015) A comparative analysis of carbon emissions from online retailing of fast moving consumer goods, *Journal of Cleaner Production*, **106**, pp 478–86, https://doi.org/10.1016/j.jclepro.2014.06.060

van Marle, G (2012) Carriers could cut significant cost by pooling, says grey box pioneer, *The LoadStar* (15 November).

van Rooijen, T and Quak, H (2010) Local impacts of a new urban consolidation centre – the case of Binnenstadservice.nl, *Procedia – Social and Behavioral Sciences*, **2** (3), pp 5967–79, https://doi.org/10.1016/j.sbspro.2010.04.011

Vanek, F M and Morlok, E K (2000) Improving the energy efficiency of freight in the United States through commodity-based analysis: justification and implementation, *Transportation Research Part D: Transport and Environment*, **5** (1), pp 11–29, https://doi.org/10.1016/S1361-9209(99)00021-8

Vernon, D and Meier, A (2012) Identification and quantification of principal–agent problems affecting energy efficiency investments and use decisions in the trucking industry, *Energy Policy*, **49**, pp 266–73, https://doi.org/10.1016/j.enpol.2012.06.016

Verweij, K (2011) *Synchromodal Transport: Thinking in hybrid cooperative networks* (Vol. EVO Logistics Yearbook 2011 edition), Zoetermeer: Evo

Victor, P A (2012) Growth, degrowth and climate change: a scenario analysis, *Ecological Economics*, **84** (Supplement C), pp 206–12, https://doi.org/10.1016/j.ecolecon.2011.04.013

Wahyudi, W et al (2013) Impact of axle load overloading on freight vehicles towards the increasing of greenhouse gas emissions by oxides of carbon, *Proc. Eastern Asia Society for Transportation Studies*, **9**

Walker, G and King, D (2008) *The Hot Topic*, London: Bloomsbury

Waller, M A and Fawcett, S E (2014) Click here to print a maker movement supply chain: how invention and entrepreneurship will disrupt supply chain design, *Journal of Business Logistics*, **35** (2), pp 99–102, https://doi.org/10.1111/jbl.12045

Walmart (2017) *Global Responsibility Report*, Walmart Corporate, Bentonville

Wang, H and Lutsey, N (2013) *Long Term Potential for Increased Shipping Efficiency Through the Adoption of Industry-leading Practice*, Washington DC: International Council for Clean Transportation

Wang, Y and Pettit, S (eds) (2016) *E-Logistics: Managing your supply chains for competitive advantage*, London: Kogan Page Ltd

WBCSD/WRI (2004) *The Greenhouse Gas Protocol: A corporate accounting and reporting standard*, Geneva and Washington DC: World Business Council for Sustainable Development /World Resources Institute

WBCSD/WRI (2011) *Corporate Value Chain (Scope 3) Accounting and Reporting Standard*, Geneva: World Business Council for Sustainable Development/World Resources Institute

Weetman, C (2016) *Circular Economy Handbook for Business and Supply Chains*, London: Kogan Page Ltd

Wei, T et al (2016) Quantitative estimation of the climatic effects of carbon transferred by international trade, *Scientific Reports*, **6**, p 28046, https://doi.org/10.1038/srep28046

White, M et al (2017) *Freight Industry Collaboration Study*, Crowthorne: TRL

Winnes, H, Styhre, L and Fridell, E (2015) Reducing GHG emissions from ships in port areas, *Research in Transportation Business and Management*, **17**, pp 73–82

Wiedmann, T O et al (2015) The material footprint of nations, *Proceedings of the National Academy of Sciences*, **112** (20), pp 6271–76, https://doi.org/10.1073/pnas.1220362110

Wiesmann, B et al (2016) Drivers and barriers to reshoring: a literature review on offshoring in reverse, *European Business Review*, **29** (1), pp 15–42, https://doi.org/10.1108/EBR-03-2016-0050

Wilmsmeier, G and Spengler, T (2016) *Energy Consumption and Container Terminal Efficiency*, Bulletin 350 no. 6, Santiago: UN ECLAC

WMO (2017a) Climate breaks multiple records in 2016, with global impacts, *World Meteorological Office* [online] https://public.wmo.int/en/media/press-release/climate-breaks-multiple-records-2016-global-impacts

WMO (2017b) *WMO Statement on the State of the Global Climate in 2016*, Geneva: World Meteorological Office

Woodburn, A (2007) Appropriate indicators of rail freight activity and market share: a review of UK practice and recommendations for change, *Transport Policy*, **14** (1), pp 59–69, https://doi.org/10.1016/j.tranpol.2006.09.002

Woodburn, A (2011) An investigation of container train service provision and load factors in Great Britain, *European Journal of Transport and Infrastructure Research*, **11** (2), pp 147–65

Woodburn, A (2017) An analysis of rail freight operational efficiency and mode share in the British port-hinterland container market, *Transportation Research Part D: Transport and Environment*, **51** (Supplement C), pp 190–202, https://doi.org/10.1016/j.trd.2017.01.002

Woolford, R and McKinnon, A C (2011) The role of the shipper in decarbonising maritime supply chains, in *Current Issues in Shipping, Ports and Logistics*, ed T Notteboom, Antwerp: UPA University Press (pp 11–24)

World Bank (2014) *Policy Options for Liberalising Philippine Maritime Cabotage Restrictions*, Washington DC: World Bank Group

World Bank (2017a) *Commodity Markets Outlook*, Washington DC: World Bank Group

World Bank (2017b) *World Development Indicators*, World Bank Group [online] https://data.worldbank.org/data-catalog/world-development-indicators

World Bank (2017c) *World Development Indicators: Climate variability, exposure to impact, and resilience*, World Bank Group [online] http://wdi.worldbank.org/table/3.11

World Bank, Ecofys and Vivid Economics (2017) *State and Trends of Carbon Pricing 2017*, Washington DC: World Bank Group

World Economic Forum (2010) *Consignment Level Carbon Reporting: Background to the guidelines*, Geneva

World Economic Forum/Accenture, (2009) *Supply Chain Decarbonization: Role of transport and logistics in reducing supply chain carbon emissions*, Geneva: World Economic Forum

WRAP (2011) *Bottling Wine in a Changing Climate*, Banbury: Waste and Resources Action Programme

WRAP (2012) *Reducing your Construction Waste*, Banbury: Waste and Resources Action Programme

WTO (2016) G20 trade restrictions reach highest monthly level since the crisis, *World Trade Organization* [online] https://www.wto.org/english/news_e/news16_e/trdev_21jun16_e.htm

WTO (2017) *Trade Facilitation*, World Trade Organization [online] https://www.wto.org/english/tratop_e/tradfa_e/tradfa_e.htm

Zalasiewicz, J et al (2017) Scale and diversity of the physical technosphere: a geological perspective, *The Anthropocene Review*, **4** (1), pp 9–22, https://doi.org/10.1177/2053019616677743

Zhang, M and Pel, A J (2016) Synchromodal hinterland freight transport: model study for the port of Rotterdam, *Journal of Transport Geography*, **52** (Supplement C), pp 1–10, https://doi.org/10.1016/j.jtrangeo.2016.02.007

INDEX

CPSIA information can be obtained
at www.ICGtesting.com
Printed in the USA
LVHW07s0400050718
582683LV00027B/405/P